GEOTECHNOLOGY

*An Introductory Text for
Students and Engineers*

Related Titles of Interest

GRIFFITH and KING:
Applied geophysics for geophysicists and geologists (2nd Edition)

CONDIE:
Plate tectonics

SIMPSON:
Geological maps

MAURER:
Novel drilling techniques

RAUDKIVI:
Loose boundary hydraulics (2nd edition)

GEOTECHNOLOGY
An Introductory Text for Students and Engineers

by

A. ROBERTS

*Mackay School of Mines, The University of Nevada,
Reno, Nevada 89507, U.S.A.*

PERGAMON PRESS

OXFORD · NEW YORK · TORONTO · SYDNEY · PARIS · FRANKFURT

U.K.	Pergamon Press Ltd., Headington Hill Hall, Oxford OX3 0BW, England
U.S.A.	Pergamon Press Inc., Maxwell House, Fairview Park, Elmsford, New York 10523, U.S.A.
CANADA	Pergamon of Canada Ltd., 75 The East Mall, Toronto, Ontario, Canada
AUSTRALIA	Pergamon Press (Aust.) Pty. Ltd., 19a Boundary Street, Rushcutters Bay, N.S.W. 2011, Australia
FRANCE	Pergamon Press SARL, 24 Rue des Ecoles, 75240 Paris, Cedex 05, France
WEST GERMANY	Pergamon Press GmbH, 6242 Kronberg-Taunus, Pferdstrasse 1, Frankfurt-am-Main, West Germany

First edition 1977

Library of Congress Cataloging in Publication Data

Roberts, Albert F 1911-
Geotechnology.

1. Engineering geology. I. Title.
TA705.R56 1976   624'.151   76-45440
ISBN 0-08-019602-0 Hard Cover
ISBN 0-08-021594-7 Flexi Cover

Printed in Great Britain by A. Wheaton & Co., Exeter

Contents

Contents

3. TIME-DEPENDENT BEHAVIOR OF EARTH MATERIALS

4. FAILURE CRITERIA FOR SOILS AND ROCKS

5. THE ENGINEERING PROPERTIES OF SOILS

6. FLUIDS IN SOILS AND ROCKS

Contents

8. THE STRENGTH OF ROCK MATERIALS

Contents

10. DETERMINATION OF THE STATE OF STRESS IN ROCK MASSES

Contents

Contents

xiv

Preface

Geotechnology is applied geology in the context of engineering. It deals with engineering in and on earth materials, and it is a prime concern of the geophysicist, the geological, the mining, and the civil engineer. The author's interest in the subject stems from his training and professional experience as a mining engineer, as a university lecturer both in engineering geology and in mining engineering, and as the director of a university postgraduate school which, over the years in which it was active, made a significant contribution to the emergence of rock mechanics as a science in its own right. As founder of the *International Journal of Rock Mechanics and Mining Sciences*, and its Editor-in-Chief for the past decade, it has been the author's privilege to have been in close contact with research workers, institutes, and industrial concerns interested in geotechnology, all over the world. During this period, two matters have become increasingly in evidence. First, a growing awareness, on the part of the community-at-large, of the effects of engineering and technology on the human environment, and second, a widening antipathy between many of those who are pursuing the development of geotechnology, particularly rock mechanics, as an academic field of study and research, and the engineers who are concerned with its practical application.

As to the first matter, there are signs that emphasis in the activities of the environmentalists is changing from talk to action, as it must do if the environment is to be protected or improved. But wise action is usually a matter of compromise, based on knowledge and understanding of the issues involved. This is what the education and training of the engineer must fit him to do, and if "environmental geology" is ever to be anything more than a name, then the geologist must also be an engineer.

As to the second matter, it is true that the great bulk of published material in rock mechanics consists of scientific papers written by experts for other experts in the subject. It is comparatively seldom that a paper with obvious practical application is published, although many of the scientific papers have a practical application and it is unfortunate that the investigator seldom takes it upon himself to show what that application might be. It is also necessary to dispel the impression, sometimes given, that the geotechnical expert spends half his time before the event telling the world what might happen, and another half of his time describing what happened and how, after the event, leaving the site engineer to shoulder the responsibility of attempting to deal with the event itself.

Those who are involved with the teaching of geotechnology at college level sometimes may have, in one class, students whose background covers only part of the broad spectrum of interests concerned. At one extreme there will be geologists with little or no knowledge of the basic engineering sciences, such as materials science or applied mechanics. At the other extreme there will be civil engineers, knowledgeable in structural design but to whom geology has hitherto been a closed book. Another group may be geophysicists, specialized only in a particular aspect of applied geology and physics, and in the middle may be geological and mining engineers with a broader, albeit a somewhat superficial background, over a wider range. There was a time when universities and senior institutes of learning were permitted to be organized in such a way that these various groups could be taught separately, and sometimes even individually, in tutorial fashion, but those days, alas!, are almost gone. Now even our universities must be geared to mass production.

Preface

Faced with such a situation the lecturer is posed a problem in how best to bring his diversified audience to a common level of participation in his subject and to maintain their interest while doing so. It is the purpose of the author, in this text, to pick up the various threads in geophysics, materials science, soil mechanics, rock mechanics, applied geology, civil and mining engineering, and attempt to weave them into a coherent and interrelated fabric of common interest.

The work is essentially an introductory text and it is suggested that it can form the basis of one year's or two semesters' classwork for first-year students at university or engineering college. Much of the content deals with fundamental matters, and with systems of instrumentation and measurement. It is intended that this will be followed by a second volume, dealing with applications to specific problems in geotechnology, which will form the basis for second-year student work. The two volumes will also be useful to practising engineers who work with earth materials, but whose college studies did not include geotechnology and who may hence have a need for a simplified review treatment of the subject, in relation to their everyday professional duties. If the gap between theory and practice in rock mechanics, for example, is to be bridged then the bridge must be built from both sides, the research specialist and the site engineer working together with a common aim.

The treatment throughout is descriptive and non-mathematical, and makes no assumptions as to the reader's previous background of knowledge. It may therefore be read and understood by the layman. At the same time the subject matter is pointed towards the numerous sources of reference used in compiling the text, and details of these are given at the end of each chapter. If, while he is reading this text, the serious student is also concurrently acquiring a basic knowledge of engineering science and geology, he will be able to pick up these references, together with the many more specialized texts that already exist in various aspects of geotechnology. In this way he will be able, should he so desire, to develop his interest in whatever branch of the subject he may choose to take up, in more advanced study.

CHAPTER 1

Engineering in Earth Materials

The Scope of Geotechnology

Geotechnology is a field of study and engineering in which the applied geologist, the civil engineer, and the mining engineer have a common interest. It is only during the past 35 years or so that the subject has been pursued systematically with the aim of developing it as a science in its own right. Traditionally mining has always been regarded as an art rather than a science, and while applied geology in the context of geophysics and geochemistry has made substantial advances in analytical treatment, engineering geology has remained very largely descriptive in its approach. Soil mechanics and foundation engineering emerged as a specialist field of expertise in the late 1930s, as a branch of civil engineering, to be followed more recently by rock mechanics. The three allied fields, Engineering Geology (now more appropriately termed Geological Engineering), Mining Engineering, and those aspects of Civil Engineering concerning earthworks and tunneling, comprise what may now be termed Geotechnology: engineering in earth materials.

Soil Mechanics

Soil mechanics is the science of unconsolidated material. It deals with the behavior of soils consisting of discrete particles, with intervening voids or pore spaces, and with the interaction of the particles and the pore fluids. It is the task of soil mechanics to predict the effect on the soil of a given system of forces such as might be produced, for example, by the construction of a heap of material standing at its own angle of repose, or controlled by the erection of retaining walls designed to restrict the space occupied by the heap.

Soil mechanics is also applied in foundation engineering to provide a basis for rational design of the foundations of engineering structures. The increasing size of bridges, dams, tower blocks, etc., requiring deep and extensive foundations, brings civil and structural engineers into more frequent contact with some earth materials which, in their undisturbed state, are rocks rather than soils, and with other materials that are unquestionably rocks, in any environment. Confidence in the engineering properties of material that conventional methods of site exploration have indicated apparently to be solid, hard, rock has sometimes been rudely shattered by the occurrence of failures of catastrophic proportions. These include, during recent years, the Malpasset Dam (where the cause of failure was found to lie in a material weakness of a key section of the rock foundations) and also the collapse of large rock and earth masses (such as occurred during the Alaskan and Peruvian earthquakes) or which may occur as the result of some change in the ground water characteristics (which is thought to have been partly responsible for the Vaiont catastrophe). One result of these and other occurrences of a like nature, has been that engineers previously concerned with soils are now also giving their attention to rocks, and rock mechanics, or the study of the engineering properties of rocks and rock masses, is now included with soil mechanics as the fundamental basis of foundation engineering.

Rock Mechanics

Rock mechanics is the theoretical and applied science of the mechanical behavior of rocks, as materials and in the mass. Interest in the science comes from many different directions, from Geophysics in relation to seismic phenomena, from Structural Geology in relation to earth structures and geotectonics, from Civil Engineering as a logical extension of soil mechanics and in connection with tunneling and earthworks, from Applied Geology in the context of geological engineering, and from Mining Engineering in relation to the control of ground movement, the control and support of excavations, and in relation to techniques of rock and earth excavation by mining equipment. To all these special interests must be added the general growth of public concern at the despoliation of amenities by open-pit mining and the disposal of surface waste materials, the periodic occurrence of dam failure, problems of slope stability in large rock cuts and open-pits, the distress and damage caused by natural disasters such as landslides and earthquakes, and the obvious need for more knowledge as to their causes, prediction and limitation. Other factors include the increasing scope and scale of underground civil and military engineering works. The advent of large diameter rock-boring machines has brought a new dimension to tunneling operations in rock, in which rock mechanics must take a leading role, both in relation to excavation and in support. In the 30 years that have elapsed since the advent of atomic energy an entirely new technology of nuclear engineering construction has evolved. In so doing it has promoted the use of conventional explosives, on a scale never before contemplated, to rock excavation and geological engineering. Another new technology, dealing with excavation and construction in frozen earth materials, is emerging as the result of the economic development of Siberia, Alaska, and Northern Canada.

There is much going on in geotechnology today that should stimulate the interest and imagination of high school and university students. Many of these young people are frequently heard to express concern at the state of our society as they see it, with their desire to be in some way, in the movement to promote change directed towards more idealized conditions. Amongst these groups, none are more vociferous than those who wish to limit the pollution and destruction of amenities that, all too often, can be seen to have resulted from uncontrolled or careless industrialization and urban development. But while it is right that the social scientists and the humanists should raise these questions, and prod the politicians into legislative action where this is necessary, in the end it is the economist and the engineer who, between them, will have to produce practicable solutions to the many problems that exist. It is the aim of this book to introduce the reader to some of the engineering problems involved when dealing with the soils and rocks that are part of mankind's natural environment.

The Classification of Earth Materials

At the onset we must define what it is that we are to discuss. By earth materials is meant soils and rocks, and we may look at existing classification systems to see how they are listed and described. One such listing can be made on the basis of the mode of formation. It describes the parent rock as being either igneous, sedimentary, or metamorphic in origin, and the derived soil as the products of weathering, disintegration, and decomposition of the parent rock, either residual at the parent site, or transported from that site by the processes of erosion: wind, water, ice, and gravity.

Such a classification tells us nothing about the constituents of the materials concerned.

For this we must look at the mineralogical and petrological characteristics, on the basis of which another classified subdivision may be made. This separates the igneous rocks into groups, either acid or basic, depending upon their relative content of silica, the aluminium silicates, and the ferromagnesian silicates. The same form of grouping has the sedimentary rocks as being either sands, clays, or limestones, while the metamorphic rocks fall into two broad classes: (a) quartzite and marble, both crystalline, massive, and non-foliated, and (b) the foliated slates, phyllites, and schists. (The classification as given here is not complete, but the reader will now have the broad outline of the system, and is referred to any introductory text on geology for a detailed description.)

When we look at the soil products of weathering and erosion that are derived from our new subdivision of parent rocks we see that while the acid constituent quartz breaks down to form sands, and while the felspars and the ferromagnesian silicates break down to form clay minerals, other factors come into play to determine the character of the residual soil type. These factors are climate, topography, and time.

For example, an arid climate will allow decomposition products such as the carbonates of calcium and magnesium to accumulate and so produce an alkaline soil, whereas the breakdown products of the same parent material in a wet and humid climate would be characterized by a concentration of iron and aluminium, the soluble calcium and magnesium products being leached out, resulting in an acid soil. The same parent material in a hot climate would produce the red laterite soil that is characteristic of tropical regions, but this would not happen in a cold climate, where the siliceous breakdown products would accumulate, together with the clay silicate minerals, to produce sticky soils of a predominantly blue-grey color. A cold climate, too, would slow down the rate of decay of organic materials, and promote the accumulation of mosses and peat.

Our classification has now extended beyond that possible on the basis of mode of occurrence, to include mineral composition and some chemical and physical properties. From an engineering viewpoint, however, we need also to know something about those characteristics of the soil that will affect its engineering properties. Basically, what we must know is: how will the soil behave when it is loaded by the weight of a structure erected upon it?, or how will it respond to the gravitational loads imposed by its own mass, if, for example, we cut a trench through it or build a soil embankment with it? Will the structure stand? Will the sides of the trench and the slopes of the embankment remain stable?, or will they collapse? Many factors must be explored before a reasonably assured answer may be given to such questions. We must know something of how the soil compacts under the weight of an applied load, how porous it is, and how permeable it is to water flow when compacted, and, since the water content of a soil has a critical bearing upon its strength, what is its shear strength and compressibility when compacted.

The Classification of Soils for Engineering Purposes

The engineering strength properties of an earth material are not always constant. It is possible for a material which under a given set of circumstances may be solid, to be changed, in a very short space of time, into a fluid mass, simply because during that short time interval one or more of the basic controlling factors has changed in order of magnitude. The factor that is most often charged with the responsibility for such transformations is the pore fluid pressure within the material, but there are also other influences. It could be, for example the presence of a particular type of mineral material, combined with the effect of dynamic shock, such as could result from earthquake, or in mining technology, a rock burst, which is the sudden fracture of a large rock mass, or it could be the result of blasting.

Consistency Limits

The critically controlling influence of pore fluid pressure can be included in a soils classification in terms of the consistency limits, in which four states of consistency are recognized: (1) liquid, (2) plastic, (3) semi-solid, and (4) solid.

Liquid limit

Moisture in an unconsolidated earth material, in a limited amount, promotes intergranular cohesion, but at the same time it builds up the pore pressure and, in a greater amount, it may turn the whole mass into a fluid. Such a fluidized mass has no shear strength, that is, it cannot resist deformation. If, however, the water content is reduced and the material is dried out. a point will be reached when the granular mass begins to exert resistance against the change of shape produced by an applied force, and if that force is removed the mass is seen to have suffered permanent deformation. It is then acting as a plastic solid. The moisture content at which the mass ceases behaving as a liquid and starts to behave as a plastic is termed its liquid limit (LL).

Plastic limit

With further drying the mass becomes a harder material, and its shear strength increases, ultimately until the material displays brittle failure characteristics in response to increase of imposed load. The moisture content at the point where the material ceases to display plasticity and begins to behave as a brittle solid is the plastic limit (PL) and the range of water contents over which the material is in the plastic state is called its plasticity index.

The Unified Soil Classification

The Unified Soil Classification classifies soils on the basis of the factors texture, and liquid limits (see Table 1.1). The system is comprised of fifteen soil groups, each identified by a two-letter symbol. Soils are classified in terms of particle size, coarse-grained soils being sands and gravels, while fine-grained soils are silts and clays. Gravel is defined as having a particulate grain size ranging from 76.2 mm to 4.76 mm, and sand from 4.76 mm to No.200 sieve size, while clay and silt have a component grain size less than 200 sieve, which is about the minimum individual grain size recognizable by the unaided human eye.

The first letter in each two-letter symbol indicates into which of these soil types a given material belongs. The second letter indicates the general gradation and plasticity of the soil, clay being more plastic than silt. Thus, W represents clean, well-graded, materials with a regular grain size, while P represents clean, but poorly graded, materials. M represents fine materials of a silty character and C represents clay fines. The three types of fine-grained soils — inorganic silts, inorganic clays, and inorganic silts/clays — are further subdivided in terms of their liquid limits. L classifies soils with LL less than 50, having low to medium compressibility and plasticity. H indicates materials with LL greater than 50 and comprises soils of high compressibility and plasticity.

The great merit of the Unified Soil Classification is its flexibility and its adaptability to cover materials of intermediate characteristics between the classified groups, for which purpose

TABLE 1.1. *The Unified Soil Classification System*

Major divisions			Group Symbols	Typical names
Coarse-grained soils (More than half the material above No. 200 sieve size)	Gravels (More than half of coarse fraction above No. 4 sieve size)	Clean gravels	GW	Well-graded gravels, gravel-sand mixtures, little or no fines
			GP	Poorly graded gravels, gravel-sand mixtures, little or no fines
		Gravels with fines	GM	Silty gravels, poorly graded gravel–sand–silt mixtures
			GC	Clayey gravels, poorly graded gravel–sand–clay mixtures
	Sands (More than half of coarse fraction less than No. 4 sieve size	Clean sands	SW	Well-graded sands, gravelly sands, little or no fines
			SP	Poorly graded sands, gravelly sands, little or no fines
		Sands with fines	SM	Silty sands, sand–silt mixtures
			SC	Clayey sands sand–clay mixtures
Fine-grained soils (More than half the material less than No. 200 sieve size)	Silts and clays (Liquid limit less than 50)		ML	Inorganic silts and very fine sands, rock flour, silty or clayey fine sands with slight plasticity
			CL	Inorganic clays of low to medium plasticity, gravelly clays, sandy clays, silty clays, lean clays
			OL	Organic silts and organic silt-clays of low plasticity
	Silts and clays (Liquid limit greater than 50)		MH	Inorganic silts, micaceous or diatomaceous fine sandy or silty soils, elastic silts
			CH	Inorganic clays of high plasticity, fat clays
			OH	Organic clays of medium to high plasticity
	Highly organic soils		PT	Peat and other highly organic soils

Boundary classifications: Soils possessing characteristics of two groups are designated by combinations of group symbols.

For example, GW–GC: well-graded gravel-sand mixture with clay binder.

dual symbols may be used. When used in reports to describe materials encountered in the field, say in a borehole or a trench section, it is more meaningful than is descriptive nomenclature alone, since it conveys information about some of the essential factors that are important from an engineering viewpoint. It enables the engineer to make a preliminary assessment of the suitability of the soil for the engineering project concerned, before a detailed site investigation is commissioned.

TABLE 1.2. *Engineering Uses for Various Soils, Graded in Order of Relative Suitability (Wagner)*

Soil group	Important properties — Permeability when compacted	Shear strength when compacted and saturated	Compressibility when compacted and saturated	Workability as a construction material	Rolled earth dams — Homogeneous embankment	Core	Shell	Canal sections — Erosion resistance	Compacted earth lining	Foundations — Seepage important	Seepage not important	Roadways / Fills — Frost heave not possible	Frost heave possible	Surfacing
GW	Pervious	Excellent	Negligible	Excellent	—	—	1	1	—	—	1	1	1	3
GP	Very pervious	Good	Negligible	Good	—	—	2	2	—	—	3	3	3	—
GM	Semi-perv. to impervious	Good	Negligible	Good	2	4	—	4	4	1	4	4	9	5
GC	Impervious	Good to fair	Very low	Good	1	1	—	3	1	2	6	5	5	1
SW	Pervious	Excellent	Negligible	Excellent	—	—	3A	6	—	—	2	2	2	4
SP	Pervious	Good	Very low	Fair	—	—	4A	7A	—	—	5	6	4	—
SM	Semi-perv. to impervious	Good	Low	Fair	4	5	8A	8A	5B	3	7	8	10	6
SC	Impervious	Good to fair	Low	Good	3	2	—	5	2	4	8	7	6	2
ML	Semi-perv. to impervious	Fair	Medium	Fair	6	6	—	—	6B	6	9	10	11	—
GL	Impervious	Fair	Medium	Good to fair	5	3	—	9	3	5	10	9	7	7
OL	Semi-perv. to impervious	Poor	Medium	Fair	8	8	—	—	7B	7	11	11	12	—
MH	Semi-perv. to impervious	Fair to poor	High	Poor	9	9	—	—	—	8	12	12	13	—
CH	Impervious	Poor	High	Poor	7C	7	—	10	8C	9C	13C	13C	8C	—
OH	Impervious	Poor	High	Poor	10	10	—	—	—	10	14	14	14	—
PT	—	—	—	—	—	—	—	—	—	—	—	—	—	—

Notes: A. If gravelly.   B. Erosion critical.   C. Volume change critical.

Engineering Classification of Rocks and Rock Masses

As yet there is no classification system applicable to rocks comparable to the Unified Classification for Soils. There are many types of rock, and many variants of each type. Each variant is a different material. One of the primary tasks in geotechnology is therefore to classify rocks and to define them, not only in terms of their essential physical and geological characteristics, but also in terms of their mechanical behavior in response to various forms of applied force.

A classification of the rock material could be based on mineralogical and petrological structure, including the rock fabric on scales ranging through the molecular, microscopic, and macroscopic sizes. The existence and nature of any discontinuities are critically important, both in the rock material and in the rock mass, and they must also be explored, again ranging in scale from inter- and intra-crystalline and granular boundaries, through stratification and sedimentation features, to large-scale physical and tectonic phenomena.

When we have defined and described the working material, then we must proceed to the next step, which is to define its engineering properties. This requires an exploration of the deformation characteristics of the material under load, sometimes up to the point of fracture. The purpose of this investigation is that it should be possible for the engineer to estimate the strength of the rock and its ability to withstand the loads that it will be required to sustain in the particular circumstances of the engineering problem concerned. This is not easy, in fact there is as yet no general agreement as to what is a valid criterion of failure in a rock, and it is most unlikely that a universally applicable criterion will ever be found. The response of an earth material to dynamic loads, for example, is very different from that which it displays in response to static loads. The environment of a rock mass, which includes the degree and the manner in which the rock is constrained, and the pore fluid pressures, are major factors of critical importance.

A realistic and functional engineering classification of rocks cannot be made without reference to the engineering objective concerned. Some rock properties will be more important than others, but their relative importance will not always be the same, when considered in relation to different objectives. This fact was recognized in early classification systems which were intended only for a restricted range of application.

Terzaghi's rock classification

Terzaghi proposed a classification of rocks as an aid to assessing the support requirements when tunneling in rock, as follows:

(1) Intact rock. Contains no joints.
(2) Stratified rock. With little strength between the beds.
(3) Moderately jointed rock. A jointed rock mass, strongly interbedded and cemented, so that a vertical wall will stand without support.
(4) Blocky and seamy rock. A jointed rock mass, the joints being weakly cemented and weakly interlocked, so that a vertical wall will not stand without support.
(5) Crushed rock. Rock fractured and pulverised but suffering no chemical change.
(6) Squeezing rock. Containing minerals with low swelling characteristics when wet.
(7) Swelling rock. Containing minerals with high swelling characteristics when wet.

USBM classification

A classification system adopted by the U.S. Bureau of Mines for underground excavations

7

in rock was:

    (1) Competent rock - in which an excavation requires no added support.
        (a) Massive-elastic. (Homogeneous and isotropic.)
        (b) Bedded-elastic. (As (a) but laminated, with the bed thickness less than the span
            of the openings, and little cohesion between the beds.)
        (c) Massive-plastic. (Rock that will creep or flow under low stress.)
    (2) Incompetent rock — in which an excavation cannot be maintained without added support.

These classifications were not intended for general application. For this a much wider range of rock properties must be included in the system. More recently Coates has proposed a general classification in which, in conjunction with the geological name which conventionally identifies the rock, five characteristics are used to categorize it from an engineering aspect. These characteristics are:

    (1) Uniaxial Compressive Strength of the Rock Substance.

(a) Weak	less than 10,000 psi	(68.95 MN/m^2)
(b) Strong	10,000 - 20,000 psi	$(68.95 - 137.9 \text{ MN/m}^2)$
(c) Very strong	above 20,000 psi	(137.9 MN/m^2)

    (2) Pre-failure Deformation of the Rock Substance.
        (a) Elastic.
        (b) Viscous. (If at a stress of 50% of the uniaxial compressive strength the strain
            rate is greater than 2 microstrains per hour.)
    (3) Failure Characteristics of the Rock Substance.
        (a) Brittle.
        (b) Plastic. (If more than 25% of the total strain before failure is permanent.)
    (4) Gross Homogeneity.
        (a) Massive.
        (b) Layered. (Generally including sedimentary and schistose, as well as other, layered
            effects which produce parallel lines of weakness.)
    (5) Continuity of the Rock Substance.
        (a) Solid. (Joint spacing more than 2m.)
        (b) Blocky. (Joint spacing from 1m to 2 m.)
        (c) Broken. (In fragments that will pass through a 75 mm sieve.)

Burton extends Coates' criteria (4) and (5), classifying (4) as

    (a) homogeneous
    (b) heterogeneous
    (c) heterogeneous-welded (layered rock but strongly cemented, so that the layers
        are not separated by planes of weakness)

and classifying (5) more specifically, to include discontinuities such as bedding planes in sedimentary rock and foliation planes in metamorphic rocks, as planes of weakness that do not differ in any significant way from joints. On this basis, criterion (5) (continuity of the rock substance) becomes:

    (a) Intact. (No planes of weakness.)
    (b) Tabular. (One group of weakness planes.)
    (c) Columnar. (Two groups of weakness planes.)
    (d) Blocky. (Three groups of weakness planes.)

(e) Fissured, or Seamy. (Planes of weakness irregularly disposed, generally associated with faulting.)

(f) Crushed. (In fragments smaller than 76.2 mm.)

Stapledon reports on a proposed Australian standard rock classification based on unconfined compressive strength. This coincides with Coates' classification at the "strong" level, but includes a "medium-strong" class, roughly corresponding to the range of strength displayed by concretes (see Table 1.3).

TABLE 1.3. *Classification of Rock Materials, Based on Unconfined Compressive Strength (Stapledon)*

Range of unconfined compressive strength*				Range of strength of some common rock materials
Term	Symbol	psi	kg/cm^2	
Very weak §	VW	<1000	<70	Granite, Basalt, Gneiss, Schist, Quartzite, Marble, Sandstone, Limestone, Siltstone, Slate, Concrete
Weak	W	1000–3000	70–200	
Medium strong	MS	3000–10,000	200–700	
Strong	S	10,000–20,000	700–1400	
Very strong	VS	$>20,000$	>1400	

*Tested to Australian Standard. For rocks showing planar anisotropy, the long axis of the samples normal to the fabric planes.

§Some overlap in strength with very strong cohesive soils, e.g. hard desiccated clays. Distinguish by soaking in water, after which soils can be remolded.

Deere and Miller's classification system

Compressive strength also forms the basis of Deere and Miller's engineering classification of intact rock (see Table 1.4(i)). In this classification system the several strength categories, classed A to E, follow a geometric progression, the dividing line between the categories A and B being chosen at 32,000 psi (220.63 MN/m^2) which is about the upper limit of strength for rocks. A few rocks, such as the quartzites, diabase, and the dense basalts, fall into the A class, while the B class, with strengths ranging from 16,000 to 32,000 psi (110.32-220.63 MN/m^2) includes most of the igneous rocks, the stronger metamorphics, most limestones and dolomites, and the denser sediments. Category C, 8000–16,000 psi (55.16-110.32 MN/m^2) includes the weaker metamorphics and the more porous sediments, while the D and E classes comprise the low strength, porous, friable, and weathered, rocks.

A second criterion in Deere and Miller's classification is the *Modulus Ratio*, which is defined as the ratio Modulus of Elasticity/Uniaxial Compressive Strength for the rock concerned. Rocks are classified as having high, average, or low (H, M, or L) modulus ratio, the boundaries of the average zone laying below 500/1 and above 200/1 (see Table 1.4(ii)). A typical classification plot by Deere and Miller, for various igneous intrusive rocks, is shown in Figure 1.1.

Rock-quality designation

Deere and Miller suggest an index of rock quality based on the percentage core recovery when drilling into the rock with 57.15 mm or larger diameter diamond core drills. Assuming that a consistent standard of applied drilling technique can be maintained, the percentage

TABLE 1.4(i). *Engineering Classification of Intact Rock on Basis of Strength (Deere and Miller)*

Class	Description	Uniaxial compressive strength psi
A	Very high	Over 32,000
B	High	16,000–32,000
C	Medium	8,000–16,000
D	Low	4,000 – 8,000
E	Very low	Less than 4,000

TABLE 1.4(ii). *Engineering Classification of Intact Rock on Basis of Modulus Ratio (Deere and Miller)*

Class	Description	Modulus ratio *
H	High	Over 500
M	Average	200–500
L	Low	Less than 200

* Modulus ratio = E/σ where E = tangent modulus at 50% ultimate strength; σ = uniaxial compressive strength
Rocks are classified by both strength and modulus ratio, e.g. AM, BL, BH, CM, etc.

of solid core recovered will be dependent on the inherent strength of the rock substance and on the nature and frequency of any discontinuities, or planes of weakness, within the material structure. An index, termed the Rock Quality Designation (RQD), is obtained by examining the core recovered from a borehole, discarding sections of core less than 10 cm long, and expressing the remainder as a percentage of the total length drilled.

C-factor

Hansagi describes a core-sample method of rock classification which is based, not only on the total length of the intact fragments and the average length of total core recovery, but also on the number of cylindrical pieces obtained, the lower limit of which is linked with the core diameter. A classification index is obtained which Hansagi calls the "fissuration factor" or C-factor, calculated from:

$$C = 1/2S(pH + k/n)$$

where S = unit length of the borehole (dependent on core diameter and rock strength),
p = number of cylindrical samples of length S that can be obtained from the recovered cores,
H = height of the core sample used to determine the compressive strength,
k = total length of the core fragments recovered, each of cylinder length greater than the core diameter,
n = the number of these core fragments.

TABLE 1.5. *The Range of Problems in Geotechnology*

Petroleum engineering		Oil and gas reservoir engineering
		Gas and oil storage underground
		Rock drilling
		Rock cutting
Mining engineering		Mine design
		Strata control
		Support of excavations
		Caving and subsidence
		Rock bursts
		Rock and gas outbursts
		Blasting
Geophysics	Geological Engineering	Ground vibrations
		Earthquakes
		Earthquake engineering
		Earth and rock slope stability
		Soil mechanics
		Foundations
Civil engineering		Tunneling and earthworks
		Dams and waterways
		Water supply
		Hydro-geology
		Geotectonics
Geology		Structural geology
		Petrofabrics

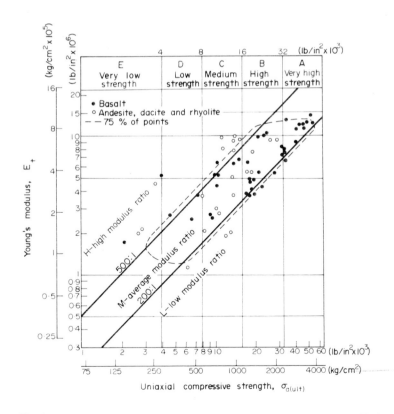

Fig. 1.1. Engineering classification for igneous intrusive rocks (Deere and Miller).

Hansagi's C-factor has evolved from a study of the strength and character of the iron ores of Kiruna, Sweden. It is applied to observe and record the strength properties of the strata penetrated in exploration boreholes. The C-factor at Kiruna generally ranges around 0.2 to 0.4. In heavily fissured areas it can drop to around 0.05 to 0.1, and rises to 0.5. in slightly fissured rock. The index is being extended to the general classification of the strata to be penetrated by the new underground railroad system at Stockholm, adapted in a form suitable for computer processing. Some recent work correlates the C factor with RQD as detailed in Table 1.6.

TABLE 1.6. *Correlation of C-factor with RQD (Hansagi)*

Strata characteristics	C−Factor	RQD %
Very poor	0.00 – 0.15	0 – 25
Poor	0.15 – 0.30	25 – 50
Fair	0.30 – 0.45	50 – 75
Good	0.45 – 0.65	75 – 90
Excellent	0.65 – 1.00	90 – 100

Ege's stability index

Ege has proposed a method of assessing rock quality from drill cores by developing an equation, including a number of factors, to yield what he terms the Stability Index.

Index number = 0.1 times core loss (length drilled less total core recovery) + number of fractures per foot length + 0.1 times broken core (core fragments less than 7.5 cm long) + weathering (graded 1 to 4 from unweathered to severely weathered) + hardness (graded 1 to 4 from very hard to incompetent).

The index number was then related to a rock grade classification from 10 (good rock, index < 8) to 1 (incompetent rock, index > 18). Faulted rock was described as ϕ.

Franklin point load index

Franklin, Broch, and Wilson combine two separate indices, namely compressive strength and the spacing of discontinuities, to yield a combined "rock-quality index". Seven classes of rock, in strength ranging from "very weak" to "very strong", are plotted horizontally along the top of a square classification diagram, on which, vertically on one side, is plotted the frequency of fractures or jointing, in six classes ranging from "very low" to "extremely high", and on the other side is plotted the frequency of bedding plane or lamination spacing, also in six classes of magnitude. On the base of the diagram is plotted a "strength anisotropy index", I_s, defined as the ratio between maximum and minimum values of the Franklin Point Load Index (see Chapter 7), measured in different directions through the material.

For general application a classification index can be obtained by zoning the square as shown on Figure 1.2(a). Again, if rock excavation is the object in view the subdivision could show the capability limits for various modes of excavation, or for the application of various types of machine. As yet there is no accumulated store of data observed from practical tests of the system that could provide a firm basis for subdividing such a diagram, but the method has interesting possibilities for the future.

The Franklin point load index, with two alternative systems of subdivision and nomenclature, together with the correlation between uniaxial and point load strengths, and typical

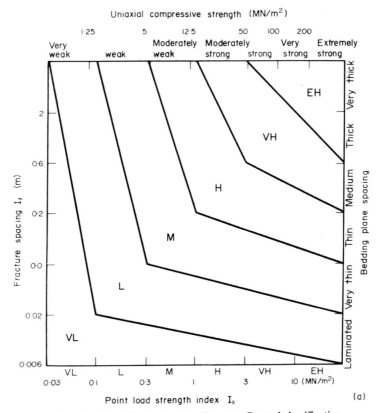

Fig. 1.2(a). Rock quality designation diagrams. General classification.

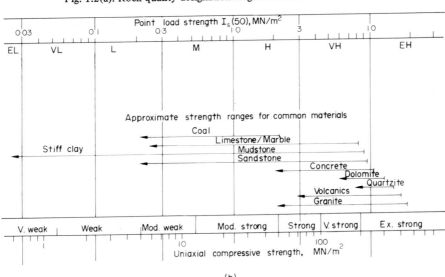

(b)

Fig. 1.2(b). Strength classification. Two alternative systems of subdivision and nomenclature are shown, together with the correlation between uniaxial and point load strengths and typical ranges of values for common rocks (Wilson, Brocht, and Franklin).

13

ranges of values for common rocks are shown in Figs. 1.2(a) and 1.2(b).

Systematic Description of Geological Factors

In reports and site investigations it is useful to have a standard nomenclature for geological details of strata encountered in boreholes, trenches, and test pits. A London working party on logging of rock cores for engineering purposes proposed the following sequence:

(i)   weathered state;
(ii)   structure;
(iii)   colour;
(iv)   grain size   [iv a] subordinate particle size,
                      [iv b] texture,
                      [iv c] alteration state
(v)   rock material strength
(vi a) mineral type;
(vi)   NAME.

Since it was considered that the system details are more important than the rock name, the details are recorded first. For example, (i) fresh, (ii) foliated, (iii) dark grey, (iv) coarse, (v) very strong, (vi a) hornblende, (vi) GRANITE.

A classification for the weathered state was suggested as:

(i)   Fresh [F], no visible evidence of weathering.
(ii)   Faintly weathered [FW], weathering limited to major discontinuities.
(iii)   Penetratively weathered [PW], weathering developed on all discontinuities but only slight weathering of the rock material.
(iv)   Moderately weathered [MW], weathering extends to the rock material but some fresh rock is present.
(v)   Highly weathered [HW], weathering extends throughout the rock mass and little fresh rock remains, but the structure and the texture are still intact.
(vi)   Completely weathered [CW], rock is wholly decomposed and in a friable, disintegrated, condition but the rock structure is still preserved.

Appraisal of rock mass structure

In sedimentary rocks the dominant discontinuity is likely to be bedding, and this can be recognized and defined in borehole cores. A proposed classification is shown in Table 1.8.

TABLE 1.7. *Strata Classification for Support Purposes (Hansagi)*
(Roadway width 5 m. Height 3.6 m)

Compressive strength (Kp/cm^2)	Support requirements
0 – 50	Immediate gunite or shotcrete, followed by rock bolting and reinforced concrete lining
50 – 150	Rock bolting and gunite or shotcrete
150 – 250	Rock bolting with wire mesh lining
250 – 350	A few rock bolts
350 and up	No artificial support needed

TABLE 1.8. *Classification of Bedding*

Description	Bedding plane spacing		
Very thick	more than 2 m		
Thick	0.6	–	2 m
Medium	0.2	–	0.6 m
Thin	60 mm	–	0.2 m
Very thin	20 mm	–	60 mm
Laminated	6 mm	–	20 mm
Thinly laminated	less than 6 mm		

In rocks other than sediments the discontinuities are likely to take a more irregular form, possibly including several systems of cross joints. Identification of the discontinuous characteristics is then not normally possible from borehole core alone, and can be made only from rock exposures, using the nomenclature suggested by Coates and by Burton.

The Range of Problems in Geotechnology

Since it is most improbable that any one classification system will suffice to embrace all the various aspects of geotechnology (see Table 1.5) a practical course of action is to limit our attention to a restricted part of the field, appropriate to the particular circumstances that we may be concerned with at the time. For example, in connection with mining engineering, not only must we explore the stress-strain characteristics of rocks, but we must do this in relation to the method of attack, by impact or cutting if one is studying techniques of breaking ground, or the response of rocks to long-term strata pressures, constraints, and reinforcement techniques, if strata control and the support of excavations are the matters to be considered.

The significance of rock mechanics to mining technology is shown in Fig. 1.3. It is sometimes necessary for the engineer in the field to take stock, when he is considering fundamental matters, and to relate these to the practical end in view. Rock mechanics is an absorbing subject in itself, and an abundantly documented materials science. Its practical applications to mining and civil engineering technology are not always very apparent, so that the engineer may be forgiven if, when he is searching through the literature for information that may help him in a practical problem, he is heard to complain that his view of the forest is obscured by a veri-

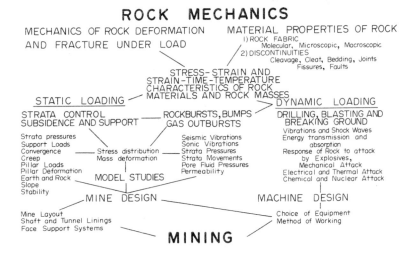

Fig.1.3. The application of rock mechanics in mining technology

G—B

table thicket of trees! The study of rock petrofabrics and the mechanics of rock deformation and fracture may seem to be academic matters, but they are fundamental to processes of strata control, the support of excavations, and to the construction of excavations. Through better understanding of the properties of the rocks and soils that are the geotechnical engineer's working materials he may confidently expect to improve his technology in terms of underground layout, the control and support of excavations, the design and application of excavating machinery, and the promotion of economic, safe, and efficient systems of working. The interrelationships of the various aspects of rock mechanics with applied mining technology are thus shown on the diagram. Similar interrelationship diagrams could be drawn for the civil engineer and the geological engineer.

Selected References for Further Reading

BROCH, E. and FRANKLIN, J. A. The point load strength test. *Int. J. Rock Mech. Min. Sci.*, vol. 9, pp. 669–698 (1972).

BURTON, A. N. Classification of rocks. *Int. J. Rock Mech. Min. Sci.*, vol. 2. p. 105 (1965).

CASAGRANDE, Arthur. Classification and identification of soils. *Trans. Am. Soc. Civil Engrs.*, vol. 113, pp. 901–991 (1948).

COATES, D. F. Classification of rocks for rock mechanics. *Int. J. Rock Mech. Min. Sci.*, vol. 1, pp. 421–429 (1964).

COATES, D. F. and PARSONS, R. D. Experimental criteria for classification of rock substances. *Int. J. Rock Mech. Min. Sci.*, vol. 3, pp. 181–189 (1966).

DEERE, D. U. and MILLER, R. F. *Engineering Classification and Index Properties for Intact Rock*, Tech. Rept. No. AFWL–TR–65–116. Air Force Weapons Lab., Kirtland AFB, New Mexico (1966).

EGE, J. R. Stability index for underground structures in granitic rock in Nevada test site. *Mem. Geol. Soc. Am.*, no. 110, pp. 185–198 (1968).

FRANKLIN, J. A., BROCH, E. and WALTON, G. Logging the mechanical character of rock. *Trans. Inst. Min. Met. London*, vol. 80A, pp. 1–8 (1971).

HANSAGI, I. Numerical determination of mechanical properties of rock and of rock masses. *Int. J. Rock Mech. Min. Sci.*, vol. 2, pp. 219–223 (1965).

HANSAGI, I. A method of determining the degree of fissuration of rock. *Int. J. Rock Mech. Min. Sci., & Geomech. Abst.*, vol. 11, pp. 379–388 (1974).

HOLTZ, Wesley, G. *Soil as an Engineering Material*, U. S. Dept. of the Interior, Bureau of Reclamation, Water Resources Technical Publication, Report No. 17 (1969).

STAPLEDON, D. H. Discussion of D. F. Coates' rock classification. *Int. J. Rock Mech. Min. Sci.*, vol 5, pp. 371–373 (1968).

TERZAGHI, K. Introduction to tunnel geology. In *Rock Tunnelling*, R. Proctor and T. White. Youngstown Printing Co., Youngstown, Ohio, 1946.

Tests for Liquid Limit of Soils. ASTM D423–66.

Tests for Plastic Limit and Plasticity Index of Soils. ASTM D424–59.

Tests for Shrinkage Factor of Soils, ASTM D427–61.

WAGNER, A. A. The use of the Unified Soil Classification System. *Proc. 4th Int. Conf. on Soil Mechs. and Foundation Engineering*, vol. 1, pp. 125–134 (1957).

Unified Soil Classification System. U. S. Army Engineers Waterways Experiment Station, Technical Memorandum No. 3–357 (1953).

Working party report on "The Logging of Rock Cores for Engineering Purposes". *Q. Jl Engng. Geol.* vol 3, pp. 1–24 (1970).

CHAPTER 2

The Behavior of Earth Materials under Static Load

Some Fundamental Concepts

In geotechnology we are concerned with the behavior of earth materials under load. The load may originate in various ways — for example, it can be the result of gravitational forces acting upon the mass of the earth material itself and upon the overlying strata. Such loads are important in matters concerning the support of excavations and the control of landslides. Other loads may be generated by tectonic forces acting upon the Earth's crust, as evidenced by geological faulting and folding, or by inherent forces within the mineralogical and petro-fabric structure of a rock material, as evidenced by crystal twinning, cleavage and foliation. Yet another source of load comes from the pressure generated by fluids acting within the pores and interstices of the earth material. All these forces are primary, that is, they exist before any engineering or excavation is done, and they are withstood by the inherent strength properties of the soils and rocks concerned.

The creation of an excavation, or the erection of a structure on a soil or rock foundation, in an earth material at an original state of equilibrium, disturbs the original force field and a new, redistributed, force field is produced. In the case of an excavation, the earth or rock walls tend to move and they try to fill the excavation from all sides, while a structural foundation imposes added loads upon the soils and rocks on which it stands. Those soils and rocks then tend to yield, from regions of high pressure to regions of lower pressure. If a condition of equilibrium and stability is to be maintained, the inherent strength properties of the earth materials must also withstand the added loads, and resist these generated tendencies to yield. It is this secondary or generated force field that must be controlled in underground support systems, in the control of caving and ground subsidence, in the stabilization of earth and rock slopes, and in foundation engineering.

Another form of secondary force field is that generated by mining tools such as rock cutters and picks, drills, explosives, and blasting materials, which are applied to break rock in the process of excavation.

The response of an earth material to these forces is described in terms of *load*, *stress*, and *strain*. The load at a point within a rock mass, being the resultant of all the forces acting upon the material at that point, is sometimes termed the *strata pressure*.

Stress

The resultant force induces a state of stress in the earth material. Stress may be defined as the force per unit area. The concept involves consideration of a plane cutting through an element of the material concerned, on which the force P acts (Fig. 2.1). Since in nature the application of a force necessarily involves the generation of an equal and opposite reaction, the force can be either compression or tension, depending upon whether it is directed towards, or away from, the plane considered.

17

The force P produces a uniform stress $\sigma = P/A$ acting on a normal cross-section of area A, and the concept of stress acting at a point is the limiting value of $\delta P/\delta A$ as A approaches zero at the point concerned.

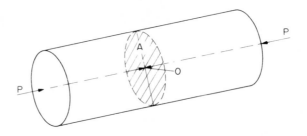

Stress on plane A = σ = P/A

Stress at point O = $\delta P/\delta A$ at limit when A → zero

Fig. 2.1. Stress at a point

Normal and shear stress components

If in Fig. 2.1, we imagine the cross-section to be inclined at an angle θ to the normal, as shown in Fig. 2.2, the area of this cross-section is $A/\cos\theta$.

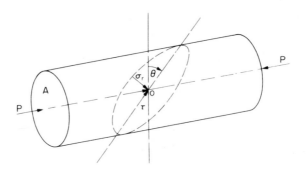

Normal stress σ_n and shear stress τ, components of stress at point 0.

Fig. 2.2. Normal and shear components of stress

The force P can be resolved into two components, P_n acting normal to the cross section of area $A/\cos\theta$, and P_s acting parallel with the surface of the section, where $P_n = P\cos\theta$ and $P_s = P\sin\theta$.

The area of the inclined section, to which P_n is normal, is $A/\cos\theta$.

Therefore the normal stress $\quad P_n = P/A.\cos^2\theta$.
And the surface shear stress $\quad P_s = P/A.\sin\theta\ \cos\theta$.
Hence, at the limit, at the point 0,
normal stress $\qquad\qquad\qquad \sigma_n = \cos^2\theta$
and shear stress $\qquad\qquad\qquad \tau = \sin\theta\ \cos\theta$.

Principal stresses

Considering all possible values of θ in Fig. 2.2, corresponding to various angles at which the cross-section is taken, maximum and minimum values of σ_n occur when $\theta = 0°$ and $90°$, for which angles τ = zero. The stresses σ when $\theta = 0°$ and $90°$ are termed principal stresses and the planes inclined at $0°$ and $90°$ are principal planes. On these principal planes $\zeta = 0$, i.e. no shear stress exists.

Maximum values of τ occur when $\theta = 45°$, i.e. along planes inclined at $45°$ to the principal planes.

The principal stresses in two dimensions are usually denoted by σ_1 and σ_2 where $\sigma_1 > \sigma_2$. For convenience in notation compressive stresses are usually regarded as positive, and tension as negative.

Stress in three dimensions

The state of stress in a material is essentially a three-dimensional problem. In Fig. 2.1 and 2.2 the magnitude of the resultant stress P and the direction in which it acts are known. If that were not so, the resultant stress magnitude and direction could be deduced if three mutually perpendicular normal stresses (principal stresses) and the corresponding principal planes were known. Conversely, the resultant stress at a point can be resolved into three mutually perpendicular principal stresses σ_1, σ_2, and σ_3, acting at that point, where $\sigma_1 > \sigma_2 > \sigma_3$.

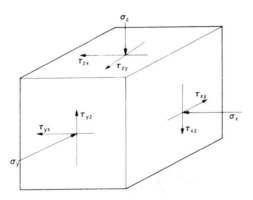

Fig. 2.3. Components of stress on an elemental cube of material
(The stresses on those sides of the cube not shown are equal and opposite to those shown on this view.)

In the general case the resultant state of stress may be represented as shown in Fig. 2.3, where the three mutually perpendicular axes x, y, and z, do not coincide with the principal stress directions. Hence the faces of the elemental cube are not parallel to the principal planes of stress and so they experience finite values of shear stress. The resultant stress at the point of intersection of the axes x, y, and z can then be represented by those on the elemental cube at the limit, when the side length of the cube approaches zero. It can be seen that this resultant is represented by nine stress components

$$\sigma_x \quad \tau_{xy} \quad \tau_{xz}$$
$$\sigma_y \quad \tau_{yx} \quad \tau_{yz}$$
$$\sigma_z \quad \tau_{zx} \quad \tau_{zy}$$

involving six quantities, since (because the cube is in equilibrium and therefore there is no turning moment on any of its faces)

$$\tau_{xy} = \tau_{yz},$$
$$\tau_{xz} = \tau_{zx},$$

and

$$\tau_{yz} = \tau_{zy}.$$

Strain

The effect of stress on an earth material is to tend to produce deformation, that is, to produce change in length, change in volume, or a change in shape. This deformation is measured in terms of strain, which is a non-dimensional quantity, represented by the ratios:

$$\text{Longitudinal strain } (\frac{\text{change of linear dimension}}{\text{original length}}) = \delta l / l$$

$$\text{Volumetric strain } (\frac{\text{change of volume}}{\text{original volume}}) = \delta v / v$$

$$\text{Shear strain } (\frac{\text{angular displacement}}{\text{length}}) = \frac{d}{l} = \tan \varphi.$$

Infinitesimal strain

Classical theory, concerning the behavior of structural materials under load, assumes that the materials are elastic, and it describes the stress-strain relationships at the onset of deformation, when the strains are infinitely small. For example, the concept of shear strain in Fig. 2.4 assumes that the angle φ remains constant throughout the thickness of the material. This will only be approximately true, and for small displacements.

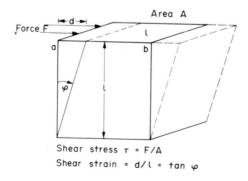

Shear stress τ = F/A
Shear strain = d/l = tan φ

Fig. 2.4. Shear strain.

Finite strain

If the strains are appreciable then the angle φ will not be constant throughout the material, so that the boundaries of the deformed rhomboid will take a curved form. In Fig. 2.5 an element in one plane of the material is pictured to have a circular shape when unstrained, but in the strained condition the circle becomes an ellipse.

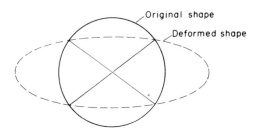

Fig. 2.5. The strain ellipse.

The strain ellipsoid

In three dimensions the ellipse becomes an ellipsoid. The shear strain in a given direction is then measured in terms of the angle between the radius vector and the tangent to the strain ellipsoid at that point.

Pure shear and simple shear

If the strain ellipsoid results from a deformation mechanism in which both the maximum shear planes have been active, then the material is said to be in a state of pure shear (Fig. 2.6). The type of shear mechanism represented in Fig. 2.4 is known as simple shear, or simple rotational shear, in which the active shear planes on which deformation occurs are assumed to be either parallel or concentric.

Heterogeneous and homogeneous strain

The geological processes of rock deformation can be seen to have produced contorted strata from what were once regular and uniform deposits. In the course of this change, elements of the material that were originally spherical become irregular in shape, and boundaries that were once linear become curvilinear. This complicated pattern of observed strain is heterogeneous strain, and as an aid towards deduction of the processes that have contributed to what he sees, the structural geologist may sometimes consider the total heterogeneous strain to be composed of elements in which the strain is homogeneous. In such elements circles become ellipses and spheres become ellipsoids.

In the general case a strain ellipse may contain components of distortion, dilation, and rotation. The deduction of these components, in three dimensions, from the evidence presented in a rock exposure, can be very complex. Probably the most easily interpreted evidence comes from the mineral inclusions or spots in such rocks as slate. These inclusions often have a discolored oxidation zone around them which, before the rock was deformed, would be spherical around the inclusion, but now in the metamorphosed slate it is ellipsoidal. Similarly, once-spherical pebbles or oolites in a sediment may now appear as ellipsoidal constituents of a deformed rock. In these cases the end-product of deformation, the ellipsoid, can be used to provide a direct geometrical description of the resultant pattern of strain in the rock, but it can provide no information as to the relative contributions made by distortion, dilation, and rotation, in the process.

In this connection, some useful information can sometimes be gained from fossils, particularly if these are fossil shells that are known to have been originally symmetrical in shape. The symmetry of such a shell is changed by the rock deformation process, and each shell now provides a graphic picture of the strain to which it has been subjected. By collating the evidence provided by a number of such shells in a rock mass the resultant strain ellipse in the mass may be deduced.

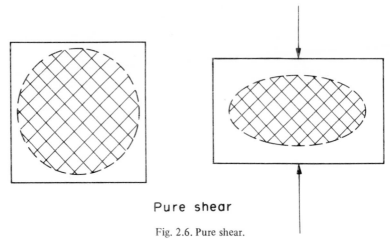

Pure shear

Fig. 2.6. Pure shear.

Plane Stress

When studying the distribution of stress and strain in a structure it is often convenient to consider a two-dimensional model, in one plane. Sometimes we are concerned only with the conditions on the surface of a solid structure such as the skin stresses on the wall of an excavation. In these cases it is assumed that the normal and shear stresses on the plane concerned are zero. This is the plane stress condition. Such an assumption would be justified if the stress situation was being pictured as one acting on the surface of a structure whose lateral dimensions, in the plane being considered, are large relative to the thickness normal to that plane. The assumption implies that there is no constraint in a direction normal to the plane of stress.

Plane Strain

If the stress situation is one acting within a solid structure, a plane section taken at some position within the structure might be chosen for study. In this case it may be more appropriate to assume conditions of plane strain, and to assume that there is no deformation normal to the plane under consideration. That is, the material is completely restrained in directions normal to the plane of measurement.

We may imagine conditions of plane strain to occur on the cross-section of a rock core cylinder, at a position remote from its ends. Another example would be the perimeter of a mine shaft or a tunnel underground. In both cases we would be justified in assuming complete restraint in a longitudinal direction, normal to the cross-section, and deformation is only possible in the plane of the section.

Springs, Dashpots, and Weights

There are two general methods of approach when studying the behavior of earth materials under load. The first is to devise methods of mechanical testing on representative samples of the soils and rocks. The tests may be conducted in the laboratory, and sometimes also in the field. In either case the results observed will be determined partly by the inherent properties of the earth material, partly by the characteristics of the testing equipment, and partly by the method followed in carrying out any particular test. The observer must be very careful in trying to establish what really are the properties of the earth material when assessing the results from such tests.

The second basic approach is to consider the behavior of imaginary materials having specific physical properties that can be precisely stated in mathematical terms. One creates, in fact, mathematical models of various types of material. The two approaches come together in that the mechanical behavior of real materials under load must be compared with the mathematical behavior of the imaginary materials. The engineer uses his mathematical models to help him forecast how real materials will behave when they are subjected to the loads imposed by the structures he designs. The degree to which his design estimates will approach a real situation in geotechnology depends upon the extent to which the behavior of earth materials approaches that of his models.

It is not uncommon for researchers in geotechnology to become so absorbed in the behavior of their models that, for them, the models assume the importance of reality. In consequence, theoretical approaches to soil and rock mechanics grow evermore complex and elaborate, while practical application of the theories lags behind. Indeed, the engineer in the field is sometimes heard to complain that the theoretical models are more difficult for him to understand than are the rocks and soils they are supposed to represent. Were it not for computers much of the theoretical structure would not exist, because the mathematical treatment would otherwise often be impracticable of solution. But the solutions yielded by computers are only to be believed in relation to the veracity of the information that is fed into them, and the engineering factors and the physical properties of rocks and soils, as they exist *in situ* are as yet obscure, to say the least.

Some of the difficulties encountered when trying to establish the engineering data required for feeding into the equations are due to the practical problems of achieving controlled conditions in the field. It is much easier to do this in the laboratory. Hence there has evolved a third line of approach in geotechnology, based on laboratory studies of physical models constructed of simulated, or sometimes "equivalent" materials. Here the behavior of the material model forms an intermediate step when attempting to correlate the mathematical model with a real situation. With these words of warning in mind, let us now look at some properties of ideal materials.

Elastic Materials

A perfectly elastic material can be imagined to behave, in response to an applied load, as does a spring, in that it responds immediately to the load. It displays an instantaneous change in dimension that is directly proportional to the stress generated by that load.

Young's modulus

In such a material the relationship stress/strain is constant. In terms of linear dimension the ratio is termed "Young's Modulus of Elasticity" (*E*),

$$E = \sigma/\epsilon$$

The greater the value of *E* for a material, the less will be the deformation produced by a given value of stress, and the stronger the material will be. In an ideal elastic material *E* will be the same both in tension and compression, but in a real material this may not necessarily be so. Real materials may also display non-linear elasticity, in which, although the stress-strain path may be the same both for unloading and loading, the path is a curvilinear one, with different values of *E* at different levels of stress. An isotropic material displays the same value of *E* in all directions, at a given value of stress, but in an anisotropic material *E* differs in different directions through the material.

Bulk modulus

An elastic material subjected to a uniform pressure from all directions would experience a hydrostatic tensile or compressive stress, and its volume would be affected by any change in the applied pressure. The extent of that change is dependent on the *Bulk Modulus* or *Compressibility* of the material.

$$\text{Bulk modulus } (K) = \text{Original volume} \left(\frac{\text{Change in pressure}}{\text{Change in volume}} \right),$$

$$K = V\left(\frac{\delta P}{\delta V}\right).$$

Modulus of rigidity

The change of shape of an elastic material subjected to a shearing force is determined by its *Modulus of Rigidity* or *Shear Modulus*.

In Fig. 2.4 consider the cube *abcd* fixed along the plane *cd* and of base area *A*. Suppose a force *F* to act in the plane *ab*, producing a shear stress $F/A = \tau$. This distorts the cube by a shear strain φ. The modulus of rigidity of the material is *G*, where

$$G = \frac{\text{shear stress}}{\text{shear strain}} = \frac{\tau}{\varphi}$$

Poisson's Ratio

Consider an element of an elastic body, say in the shape of a prism or a cylinder, and suppose this to be loaded, either in tension or compression, then the strain in the axial direction will be accompanied by strain in a transverse direction. If the applied load generates a tensile stress axially then the lateral strain is a contraction, whereas if the axial load generates an axial compression then the lateral strain is an elongation.

$$\text{The ratio } \frac{\text{lateral strain}}{\text{longitudinal strain}} = \text{Poisson's ratio } (v).$$

It is incorrect to express this change in shape in terms of the generation of lateral stress in response to and of opposite sign to the axial stress, if the material is not constrained. In such a case, if the material is uniaxially loaded the stress throughout the specimen will be either tension or compression in the axial direction. But if the material is constrained, and the element of the material is confined by other elements surrounding it, so that it is not free to deform laterally, then an axial compressive stress will generate lateral compressive stresses, the magnitude of which will depend upon Poisson's ratio for the material and the degree of constraint. The three elastic moduli, Young's modulus E, the Shear modulus G, and Poisson's ratio v, are interdependent,

$$G = \frac{E}{2(1 + v)}.$$

Rheological Properties of Earth Materials

Real earth materials do not display ideal elastic behavior, although some may approximate to it over a limited range of conditions. An elastic relationship between a stress resulting from applied load and the observed strain implies that the internal forces that hold the atomic and molecular structure of a solid material together are not overcome, so that they are able to bring the material back to its original shape and size when the applied load is removed. While in an ideal elastic material recovery is instantaneous upon relief of load, some materials may display delayed, although complete, elastic recovery.

This only happens if the stresses set up by the applied load are less than critical value, called the "*yield point*" of the material. If the applied stresses are higher than those representing the yield point then the internal elastic forces are overcome. The individual atoms and molecules of the material will then suffer permanent dislocation. Such phenomena occur within the crystal structure of the mineral constituents of crystalline rocks, but they also occur between and within the fragmented rock materials and cementing matrices of consolidated sediments, in broken rock masses, and between the granular constituents of soils. In the latter case the bond strength between the constituents is so low that yield occurs at very low values of stress, when compared with the sedimentary and crystalline rocks.

A soil is a three-phase system, composed of solid matter, water, and air. Rocks, too, are essentially multiphase materials in which internal fluid pore pressures play an important part in determining the material's response to stress. The more porous the material the more significant will be the effects of pore fluids and pore fluid pressures in lowering the respective stress levels at which occur the yield point, the onset of time-dependent deformation, and the permanent dislocation of constituent boundaries.

Rheological Models

The behaviour of materials that are not ideally elastic can be studied in terms of the behavior of rheological models. These models may be imagined as being composed of spring, dashpot, and friction elements, in various modes of combination. The spring element has elastic properties, while the dashpot is viscous and the friction element displays plastic flow (see Fig. 2.7).

The elastic element displays a constant relation between stress and strain, defined as Young's modulus E, such that

$$\sigma = E.\epsilon.$$

The viscous element displays a time-dependent relation between stress and strain, and a linear relation between stress and the rate of strain, such that

$$\sigma = d\epsilon/dt$$

25

where η is the coefficient of viscosity.

These elements may be combined to form models, the behavior of which can be described by linear equations, as shown in Fig. 2.7.

No.	Model	Characteristic	Name of investigator
I		Ideally elastic body	Hook
2		Linear liquid	Newton
3		Elastic and viscous body	Kelvin Voigt
4		Elastic liquid	Maxwell
5		Elastic and viscous body	Zener
6		Elastic liquid	Burgers

Fig. 2.7. Elements of linear rheological models.

No	Model	Characteristic	Name of investigator
I		Ideally plastic body	St. Venant
2		Plastic body	Kepes
3		Strain resistance	____
4		Velocity resistance	____
5		Plastic liquid	Bingham
6		Elastic and plastic liquid	Schwedoff
7		Elastic and plastic liquid	Prager

Fig. 2.8. Elements of non-linear rheological models.

The ideal plastic element displays no yield until the stress reaches a critical value, after which there is a constant rate of strain. This process is modeled as a dry friction sliding or St. Venant body. A more general plastic solid model is the Kepes model, in which the magnitude of the yield point is a function of strain. The Bingham, Schwedoff, and Prager bodies are other non-linear models, the first representing a plastic liquid, the second an elastic-plastic liquid, and the third an elastic-plastic liquid exhibiting delayed elastic recovery when the load is removed (Fig. 2.8).

Elastic Behavior in Real Earth Materials

When we examine the behavior of real earth materials we find that, under static loading conditions, only the hardest and dense, crystalline, non-porous igneous and metamorphic rocks approach ideal elastic behavior. The generalized stress-strain relationship for a rock is more likely to take a curvilinear form, as shown in Fig. 2.9. With increase of load from a zero condition the slope of the stress-strain curve increases as the pores and interstices of the material are closed by the applied pressure. Only after this will a rock material display approximately

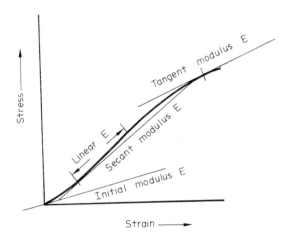

Fig. 2.9. Generalized stress-strain curve for a rock (up to the yield point).

linear elastic response and this continues until the onset of internal slip, dislocation, and microfracture within and among the rock constituents heralds the approaching yield point. The approximate linear response changes to a curvilinear relation, and the slope of the stress-strain curve diminishes after the yield point. The magnitude of the Modulus of Elasticity thus varies from a minimum at zero load to a maximum somewhere between zero load and the yield point.

We should therefore specify to what point on the stress-strain curve we are referring, when quoting the numerical value of Young's modulus for a rock material, and we should choose the most appropriate figure from a range of values to fit the circumstances involved in a particular engineering problem. For example, if we are dealing with dynamic phenomena such as occur in rock drilling or in seismic work, it is the initial dynamic modulus that we will be concerned with, but a strata control or excavation support problem will have to deal with the response of the rock over a range of stress conditions and we will be concerned with an average value over that range, in which case we want the secant modulus. Or perhaps we will be concerned with a specific value of the modulus at a particular stress level – the tangent modulus at that value of stress.

It should be pointed out here that the initial dynamic modulus will, in all probability, not be the minimum value of Young's modulus for the material, but is more likely to be near the maximum value, because the stress-strain curve under dynamic loads usually takes a convex-upwards form (see Fig. 8.16). The stress-strain relationship for an individual rock may differ from that shown on the generalized curve, depending upon its special characteristics. Hendron quotes the curves shown in Fig. 2.10, in which the type I curve applies to basalts, quartzite, diabase, dolomites and strong massive limestones, all of which approximate to elastic behavior. The type II curve applies to softer limestones, siltstone, and tuff. This curve displays a decreasing E value from an initial maximum, due to high strength plastic yield, and decreases in strength as failure is approached. The type III curve begins with an initial minimum E-value, which increases as the material consolidates slightly under load, and the initial plastic phase is followed by the major elastic phase. This would apply to fine, massive, sandstones and some granites. The type IV curve shows a transition to plastic deformation as failure is approached

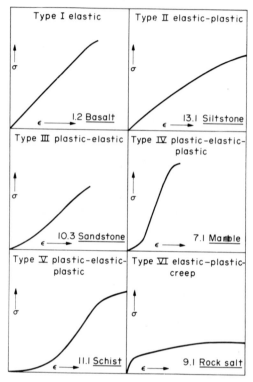

Fig. 2.10. Typical stress-strain curves for rock in uniaxial compression (Hendron, after Miller)

typical of fine metamorphics, marble and gneiss. Type V includes the porous metamorphics and sediments, and type VI the saline evaporite rocks.

Rock hysteresis

On relief of load the strain-stress relationship of a linearly elastic rock will follow that displayed on loading. Any departure from linearity and ideal elasticity shows up as a "hysteresis loop" on load cycling. If the elastic recovery is not instantaneous some residual strain will remain on relief of load. That part of the residual elastic strain recovered in the course of time is *"delayed elastic recovery"*. Should there be any plastic deformation on increase of load this will not be recovered on unloading and the width of the hysteresis loop will be increased, as a result. The effect of load cycling with progressively increasing increments of load is shown in Fig. 2.11(a) for a granite and in Fig. 2.11(b) for a porous sediment.

Stress-strain relationships in soils

Stress-strain relationships in unconsolidated earth materials in the form of assemblages of fractured rock and soil, are important in many aspects of geotechnology. The justification for applying elastic design theory to earth materials in general is that after several load-unloading cycles, either during large-scale *in-situ* tests or during laboratory tests, an identifiable stress-

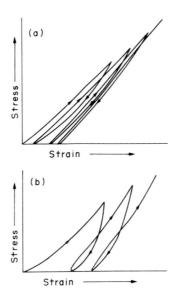

Fig. 2.11. Stress-strain curves during load cycling:
(a) on a fine-grained igneous rock, (b) on a porous sedimentary rock.

strain relation usually becomes apparent and this, although seldom linear over the whole range of observation, can often be approximated to a linear relation over a limited range of load. Hence even a non-linear earth material such as a soil can be approached by elastic design theory if the "*deformation modulus*" for the particular stress distribution can be identified.

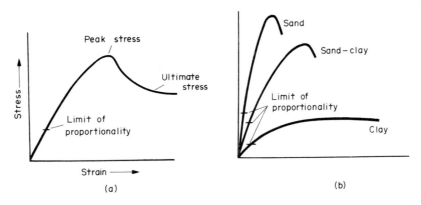

Fig. 2.12. (a) General stress-strain relationship for a soil. (b) Typical stress-strain relationships for soils on initial loading in compression.

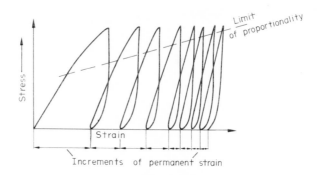

Fig. 2.13. Stress-strain curves during load cycling on a sandy-clay soil

Typical stress-strain relationships for a single application of compressive load to various soils are shown in Fig. 2.12. Soils containing clay often display linear proportionality over a limited extent from zero load, but most of the strain is not recoverable on unloading. A typical load cycling sequence is shown in Fig. 2.13, from which it can be seen that the deformation modulus and the limit of proportionality increase with each progressive cycle, while the increment of permanent strain per cycle diminishes. In the early sequences the increment of permanent strain is large, but it becomes progressively less per cycle as the material is made more dense by the closure of interstices and pores under load. As load-cycling progresses further, either the hysteresis loop will ultimately close (and the material will then behave elastically over that range of stress), or the increment of residual strain per cycle will attain a constant value (and the material will be behaving plastically).

Selected References for Further Reading

FARMER, I. W. *Engineering Properties of Rocks*, Spon. London, 1968. (Chapters 2, 3 and 4.)

HENDRON, A. J. Mechanical properties of rock. In *Rock Mechanics in Engineering Practice*, Stagg and Zienkiewicz (Eds.), Wiley, London, 1968.

HOLISTER, G. S. *Experimental Stress Analysis*, University Press, Cambridge, 1967. (Chapter 1, Stresses, strains, and stress–strain relationships.)

KIDIBINSKI, A. Rheological models of Upper Silesian Carboniferous rocks. *Int. J. Rock Mech. Min. Sci.*, vol. 3, pp. 279–306 (1966).

OBERT, L. and DUVALL, W. K. *Rock Mechanics and the Design of Structures in Rock*, Wiley, New York, 1967. (Chapters 1 and 2.)

RAGAN, Donal M. *Structural Geology – An Introduction to Geometrical Techniques*, Wiley, New York, 1968. (Chapter 5, Strain in rocks.)

CHAPTER 3

Time - Dependent Behavior of Earth Materials

Creep in Rocks

Time-dependent behavior in earth materials is commonly referred to as creep. In fact, creep is dependent on other factors also besides time such as temperature, pore pressure, and the ambient stress level. The effect is usually pictured as a strain–time curve representing the deformation of the material at constant stress. Idealized creep curves for a rock take the form shown in Figs. 3.1(a) and 3.1(b). After the initial component of instantaneous elastic

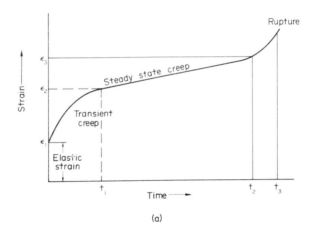

(a)

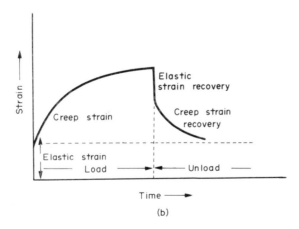

(b)

Fig. 3.1(a). Idealized creep curve for a rock material,
(b) Time-dependent creep and recovery in a rock material.

response to load the strain–time relation becomes curvilinear, with decreasing slope or strain rate over the length of time t_1. This is termed the primary or transient creep. If the stress is removed during this period the elastic component of strain is recovered immediately, and the creep strain is recovered by a time-dependent process, at a decreasing strain rate.

The transient creep phase is followed by a secondary stage of creep, at constant strain rate. This steady-state creep occurs over the time t_1 to t_2, and the deformation during this time is permanent. It is not recovered if the applied stress is removed. If the stress level is sufficiently high the secondary creep phase may be followed by a tertiary phase, displaying an accelerating creep rate as the rock loses strength at the approach to rupture, over the time t_2 to t_3. From a practical engineering viewpoint, when considering the stability of earthworks such as tunnels, underground support structures, and rock and earth slopes, the tertiary creep stage is of little importance. Our interest then lies mainly with primary and secondary creep.

If the deformation process has reached the tertiary stage, the situation, from an engineering point of view, has gone beyond control. The rock mass is approaching failure at an accelerating rate, and usually the engineer can do little else but watch things happen. Unless there is some radical change in one or more of the operative factors, complete collapse is inevitable. Sometimes such a change does occur, as for example in a rock mass that is moving under the influence of gravity. During a period of heavy rainfall an earth or rock slope may accelerate in movement and threaten to slide, but a change in the drainage characteristics or a climatic change which lowers the fluid pore pressure, or otherwise dries out the mass, might allow at least a period of temporary stability to be resumed. The acceleration period would thus only be of limited extent. It would be followed by a period of deceleration lasting until the secondary creep phase was resumed.

In general one may expect the stronger, fine-grained igneous and metamorphic rock to creep very little at the magnitudes of stress and the ambient temperatures that will prevail during conventional engineering operations. Sedimentary rocks such as sandstones and limestones may display some creep when exposed in shafts and tunnels, or in rock cuts, and the engineer will need to take this into account when constructing the support linings of shafts and tunnels, and when considering the stability of his rock cuts and slopes. Creep is liable to be very considerable indeed, even at low stress levels, in rocks such as the saline evaporites and the more porous sediments, particularly when water or other fluids are present in the rock interstices (see Fig. 3.2).

Creep Laws in Rock Materials

The total strain at time t is then given by Griggs' relationship

$$\epsilon_t = A + B.\log t + Ct$$

where A represents the instantaneous elastic strain,

$B.\log\ t$ represents the primary, or transient, creep,

Ct is the secondary or steady-state creep.

$$\frac{d\epsilon}{dt} = \frac{\sigma}{\eta}$$

where η is the coefficient of viscosity. ϵ_t is the total fractional deformation equal to $(L_t-L_0)L_0$ where L_0 is the gauged length at zero time and L_t is the length at time t.

The constants A and B are estimated from graphical plots of the creep surve on semi-log ordinates. The slope of the line drawn through early values of ϵ gives B, and the intercept of

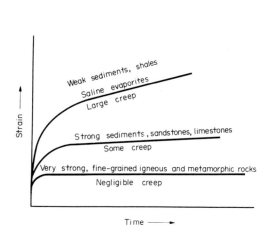

Fig. 3.2. Comparative creep characteristics for the same stress level in different types of rock.

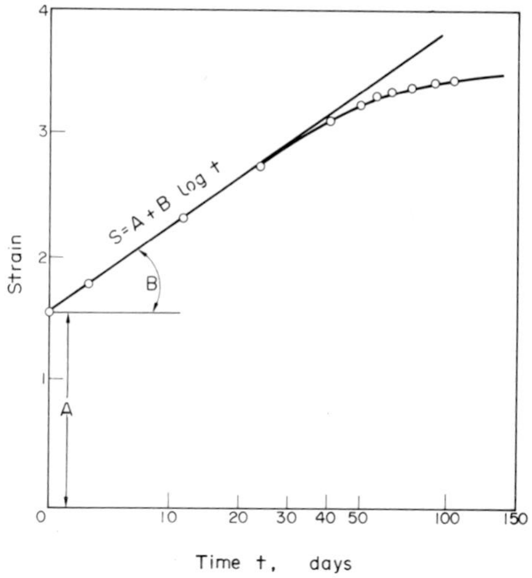

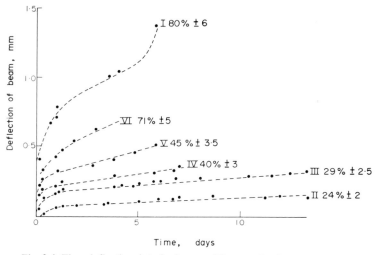

Fig. 3.3. Determination of the creep constant from the strain-time curve.

this line with the strain axis gives A, while C is the strain rate during the secondary creep stage (see Fig. 3.3). The strain rate generally decreases with decrease of ambient stress level. Typical curves observed by Price on sandstone are shown in Fig. 3.4. Price also shows the relationship between the rate of deflection and the applied load as a percentage of the instantaneous failure load, and demonstrates a linear relation between the strain rate and the applied load during secondary creep.

Fig. 3.4. Time-deflection data for beams of Pennant Sandstone (Price).

A number of investigators have observed the steady-state creep component to be relatively small, compared with transient creep, and sometimes negligible, or absent altogether. The viscosity coefficient for rocks is so high, at the temperatures that are relevant in geotechnology, that the secondary creep constant C becomes insignificant compared with transient or primary creep.

Several investigators also have observed the creep of rocks under constant stress to follow a power law, such that

$$\epsilon_t = A + Bt^n.$$

At first sight it may appear that the question as to whether the creep in a rock material follows a logarithmic law or a power law is of academic importance only, but it has some significant practical implications. Considering the mechanism of creep deformation as one of progressive dislocation and transverse sliding within and between the mineral constituents of the rock, if temperatures are low and the stress level is well below the instantaneous fracture level, dislocation and sliding will begin only at the weaker points in the material. As deformation proceeds, the resistance to further deformation will build up because progressively stronger energy barriers must be overcome. The material, in fact, becomes "work-hardened". The deformation process continues, but at a decreasing rate.

However, if the temperatures and stresses are higher, the build-up of resistance due to work hardening may be overcome, with the result that the dislocations and slip planes "climb up" or "pile up" on one another in quick succession. If this happens the work-hardening effect that helped to produce the logarithmic creep law is partially lost. Creep is then likely to take place more rapidly, and to follow a power law.

The practical implications of this are important in relation to strata control in mining, and the support of excavations in mining and civil engineering. If the conditions are such that a logarithmic law prevails, at a low strain rate, the chances of maintaining the walls of an excavation in a sound condition will be much better than would be the case when a power law holds and the strain rate is higher. In the latter case the development of dislocations may lead to the aggregation of discontinuities, general loosening of the grain structure, and ultimately to large fractures and ground movement. At the same time it must be borne in mind that some types of rock will creep extensively, even at low stress levels, during which they display steady-state creep. If the stresses remain active over a sufficiently long period of time the rock may ultimately fail, not because the ambient stress levels are high, but simply because the rock material can no longer sustain the extent of the resultant deformation. Such conditions are not uncommon in porous sediments, particularly when the energy levels to be overcome in the dislocation process are reduced by the presence of moisture or other pore fluids.

The Effect of Stress on the Strain Rate

During the transient creep phase the strain rate may follow a power law in relation to stress, such that

$$\epsilon \text{ (approx.)} = \frac{1}{t}(\frac{\sigma_1 - \sigma_3}{2G})^n$$

where G is the Modulus of Rigidity and $(\sigma_1 - \sigma_3)/2$ is the maximum shear stress.

Observed values of the exponent n in various rocks at room temperature range from around 1 to 5.

The effect of increased stress, relative to the fracture strength of the rock material, is to accelerate the strain rate. Conversely, the rate of creep of the walls in a rock excavation can be reduced in two ways: (1) by reinforcing or otherwise increasing the strength of the wall material, (2) by reducing the ambient stress in the immediate rock walls.

Practical Implications of Creep in Earth Materials

The characteristics by which the imposition of constraining forces will increase the strength of a particular earth material have a critical influence on the effectiveness of strata-reinforcement techniques such as rock bolting and tunnel lining. Layers of relatively weak ground can be bound together and secured by the elastic tension that is exerted by a rock bolt stretched between its anchor and the surface plate through which it is bolted, or by the direct compression that results from lining a tunnel with materials that are more rigid than the rock. By making the rock, or other lining reinforcement, of a high elastic modulus relative to that of the rock mass, the deformation of the reinforced ground is reduced. In this manner the imposed constraint on the rock walls increases the inherent strength of the rock, assuming that fluid pore pressures do not also increase within the rock material.

But if the stress levels are so high, relative to the strength of the rock, that significant time-dependent deformation exists, then it may not be possible to impose the necessary degree of constraint, by reinforcement or lining, that would be required to achieve permanent stability of the rock walls. In that event a rational design for the support system could only be made if the time-dependent deformation characteristics of the rock, and the state of stress within the rock walls, were known. The important parameter, which must be determined by experimental observation, is the creep constant, i.e. the slope of the time-strain relationship during the transient creep phase. Since this is a function of the stress level, the experimental tests to determine the creep constant must cover a range of stress values comparable with those that are expected to exist in the field. Field observations of the *in situ* stress level are then needed, to complete the required design data.

The rational design of strata-reinforcement and tunnel-lining structures is at an early stage of evolution. The usual method of procedure is still, very largely, intuitive, although theoretical concepts may be applied as a first approach to design. The aim of the engineer is to increase the strength of the rock walls around his excavations, by the application of direct reinforcement, and sometimes also by employing "pressure-relief" techniques, so that the rocks will support themselves over the length of time that the excavations are required to stay open. The object of direct reinforcement is primarily to increase the effective stress or effective confining pressure on the immediate wall rocks, where lateral constraint is largely removed by the excavation. This increases their yield strength and so also reduces the time-dependent deformation in them. The object of pressure relief is to deflect the imposed loads to a zone further within the rock mass, where lateral constraint is more effective. This also reduces the loads imposed upon the immediate wall rocks, from within the rock mass, and hence also reduces time-dependent deformation of the walls.

Pressure-relief technique in elastic and near-elastic rocks

The pressure-relief technique is one that has been handed down through generations of miners, wherever they are or have been involved underground at depths where support problems exist due to the existence of high strata pressures. For example, when mining a sedimentary deposit the excavations commonly take the form of tunnels of rectangular cross-section, and a network of these running at right angles to each other forms a system of "rooms" separated by supporting pillars of rock. The pillars are required to carry the weight of the overlying strata and there is an extra load imposed by the excavation of rock material from the rooms, as compared with that which existed before any mining was done. Consequently there is a redistribution of stress around the excavations, and if the excavated material is strong relative to the rock above and below it, the edges of the pillars form zones of concentrated stress to an extent which may cause the rock material in the immediate walls of the tunnels to yield and fracture.

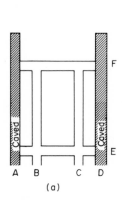

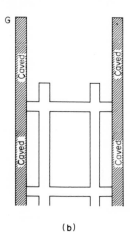

Fig. 3.5. Stress-relief and the yield-pillar technique

(a) A and D advance, B and C halted.　　　　　(b) A and D halted and caved. B and C advance.

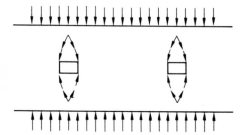

Fig. 3.5. (c) Construction of tunnels A and D in a high stress field. Stress concentrations on tunnel walls.

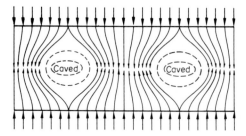

Fig. 3.5. (d) Caved ground around A and D throws stress concentrations into sides and on central pillar.

The result may take one or more of several forms.

(a) High shear stresses along and parallel with the tunnel walls may cause shear fractures along the pillar edges. The roof of the tunnels is pushed downwards, and the floor pushed upwards, relative to the pillars. From within the tunnels the general appearance is that the pillars have been pushed into the roof and floor — an occurrence which the old miners call a "thrust".

(b) If the floor strata are relatively weak, as is usually the case in coal mines — particularly if water is present — the floor strata moves plastically from underneath the pillars and into the tunnels.

The floor of the tunnels is then observed to "heave". In either case, whether it be "thrust" or "heave", the net result is that the cross-sectional area of the tunnels, or rooms, is progressively reduced. Ultimately they may close altogether, unless the broken and moving ground is excavated from within them.

The old miners were not slow to observe that if, after a tunnel had thus closed, it was recovered or reopened by making another excavation coaxially with or adjacent and parallel

36

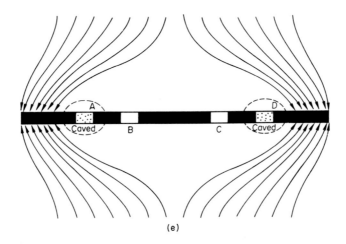

(e)

Fig. 3.5(e). Construction of tunnels B and C and yield of pillars A–B, B–C, and C–D deflect major stress concentrations further into the side zones. The central zone in which B and C must be maintained is now largely relieved of stress.

to the first, then this second excavation was usually much easier to support and maintain. In fact, very often it remained quite stable over the length of time that it was required to stay open. Arising from this there evolved the practice of constructing "pressure-relief" chambers or "caving chambers", alongside the tunnels whose stability it was desired to ensure. There are various modes of application of the technique, which vary in detail, but the general principle is common to all. It is that of deflecting the stress concentration zones away from the periphery of the tunnels and into the solid ground over, within, and beneath, the interior of the supporting pillars.

A typical example is shown in Fig. 3.5, in which four tunnels A, B, C and D are being driven parallel with one another, B and C being those whose permanent support is desired. The tunnels are connected laterally, at intervals, by "cross cuts" E, F, and G. In Fig. 3.5(a) the tunnels A and D are being driven in advance of the cross cut F and when they reach the appropriate distance to the next cross cut G they are stopped, all supports, struts, etc., within them are withdrawn, and they are allowed to collapse and close. In so doing they become "caving chambers" and the broken ground around them is partially relieved from the weight of the overlying strata. Stress concentrations are thus deflected into the solid ground on either side of tunnels A and D and towards the middle of the ground between A and D. (see Fig. 3.5(d)). However, when construction of tunnels B and C is resumed, to advance from cross cut F towards G, the ground between A and D yields slightly under the added stress concentration zones produced by excavations B and C. In so doing, the stress concentration zone around the middle of the central pillar between B and C is also deflected into the flanks on either side of A and D (Fig. 3.5(e)). The net result that is aimed at by this construction of "caving chambers" combined with "yield pillars" is to reduce the stress levels in the walls of the tunnels B and C to a point at which the rock walls will remain intact, with negligible time-dependent deformation.

A similar result is achieved, when difficulty is experienced in maintaining a narrow excavation at depth, by widening the excavation and "back filling" on either side to the finished dimensions of the tunnel. The back fill may consist of broken rock, packed sand, or poured concrete, and it has a finished strength considerably less than that of the original rock. The fill therefore yields under the weight of the overlying ground, and while continuing to support the strata immediately above and below the excavation, deflects the major strata pressures into the solid ground on either side (see Fig. 3.6).

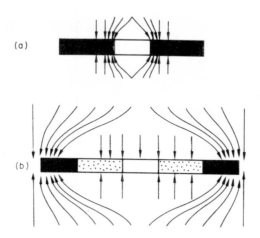

Fig. 3.6. Pressure relief by widening and backfill. (a) Narrow excavation. Stress concentrations close to sides. (b) Wide excavation backfilled. Stress concentrations thrown into solid ground. Central excavation relieved of stress.

Creep in plastic rocks

The assumption of elasticity in saline evaporite rocks, such as rock salt and potash, is only valid if the shear stresses in the material do not exceed a limiting value — the shear strength of the rock. If this limit is exceeded then plastic deformation occurs. The consequence of creep in the material is to reduce the principal stress difference $\sigma_1 - \sigma_2$, until, ultimately, the shear stress τ falls below the shear strength of the rock, at which point an elastic regime is regained. Baar maintains that strain hardening does not occur in salt rocks at depth, with the result that the deformations around underground excavations in deep potash and rock salt mines are dominated by creep, because the shear strength of the walls remains low (around 100 psi — or 689.5 kN/m^2).

Following this argument it is appearent that high stress differences cannot persist for very long around newly made excavations in salt, and a zone, which Baar terms the "stress relief creep zone", extends deep into the solid around any such opening. Within this zone the salt behaves plastically, to redistribute the stresses generated by the imposed loads, until local stresses in excess of the shear strength have been eliminated.

The initial stress relief creep may be completed within a period of hours around a new opening, but occurs over a longer period of time when a change of load takes place due to some cause, such as a change in mining conditions resulting in surface subsidence or the re-loading of earlier stress-relief creep zones, due to pillar extraction. Over a long term the creep rates of the wall rocks become constant if the loading conditions do not change, and a condition of steady-state creep maintains constant stress gradients within the walls.

The Effects of Temperature on Creep

The effect of increase in temperature is to increase the magnitude of the constants in the creep equations, and also to decrease the viscosity. However, these effects are seldom of any practical importance in engineering situations, when temperatures below 100°C are to be expected. The effects might be significant in abnormal situations, such as could exist, for example, on the occurrence of fire in a mine or tunnel. They also reach practical significance at normal working temperatures in rocks such as the saline evaporites, and they will certainly have to be taken into account when dealing with frozen earth materials.

In experiments to observe creep in model pillars of Saskatchewan potash, King observed that the increase in temperature associated with an increase in depth of mining from 3400 to 4500 ft (1036.3 – 1371.6 m) can be expected to have approximately the same effect on creep behavior as would be produced by an additional 1000 psi (6.895 MN/m^2) vertical stress on the pillar. This effect was considerably intensified when seams of clay occurred in the roof and floor strata. Under these conditions very large creep deformations can be expected to occur when mining potash by conventional methods at depths over 1400 m from the surface.

Creep of ice and frozen ground

The movement of glaciers and ice sheets demonstrates the existence of creep in ice. The dependence of strain rate upon the ambient stress in ice becomes weaker as the stress level decreases and it has been suggested that at low stresses, between 0.1 and 0.5 kg/cm^2, which are representative of the conditions in glaciers, the ice may demonstrate linear viscous behavior.

However, Mellor and Testa, on the basis of evidence provided by the results of uniaxial compressive tests on fine-grained polycrystalline ice, show that the strain rate during secondary creep is proportional to $\sigma^{1.8}$, where σ is the applied stress.

All flow processes in materials arise essentially from the thermal activation of their atomic structure and, consequently, temperature has an effect on creep. It is found experimentally that the creep strain rate bears a relationship to temperature, such that

$$\text{Strain rate } \delta\epsilon/\delta t \qquad \propto \qquad \exp\left(\frac{-Q}{\kappa t}\right)$$

where κ is a constant and Q is the activation energy.

Mellor and Testa observed an activation energy for ice, of 16.4 kcal/mol at temperatures below −10°C, and at higher temperatures the creep rate of polycrystalline ice became progressively more temperature-dependent. A single ice crystal displayed a more rapid strain rate, at all stages of creep, than did polycrystalline ice, and it is apparent that creep of ice at 0°C is influenced strongly by grain growth and crystal reorientation during the deformation process.

Creep in Soils

Creep of rock detritus and soil is frequently to be seen on hilly terrain. The motive forces in this case are almost entirely gravitational, although lateral stresses generated by expansive forces due to swelling ice and clay in the interstices and crevices of the material, and by forces due to heat and shrinkage, are sometimes important. The effects are distributed throughout the surface layers of material, to a depth through which the influences of atmospheric temperature and moisture can penetrate.

The deformation mechanism is essentially one of plastic flow, in which a slow downslope movement of the whole mass is combined with a vertical movement of fragments through the moving layer. The creep rate is not constant through the layer, but decreases with depth. Material at the base of the layer becomes compacted, while tension cracks open up at the surface. These openings form access paths and lodgement for water during periods of rainfall, so that in wet seasons the hydrostatic pressures increase within the interstices of the soil, to reduce its cohesion and increase the creep rate.

If clay minerals are present there may be considerable shrinkage in the intervening dry periods, so that desiccation cracks open up to provide further access for water in the succeeding wet season. Water soaking through the surface layers reaches the compacted sub-surface layers, which are relatively impervious, and conditions are set up that are conducive to a sheet slide, representing structural failure of the detritus material.

Selected References for Further Reading

BAAR, C. A. *The Long-term Behavior of Salt Rocks in Deep Salt and Potash Mines*, Report E 72–7, Saskatchewan Research Council (1972).

FARMER, I.W. *Engineering Properties of Rocks*, Spon, London, 1968. (Chapter 4, Rheological properties of rocks.)

GRIGGS, D. T. Creep of rocks. *J. Geol.*, vol. 47, no. 3, pp. 225–251 (1939).

KING, M. S. Creep in model pillars of Saskatchewan potash. *Int. J. Rock Mech. Min. Sci.*, vol. 10, pp. 363–371 (1973).

MELLOR, M. and TESTA, T. Effect of temperature on the creep of ice. *J. Glaciology*, vol. 8, no. 52, pp. 131–145 (1969).

MELLOR, M. and TESTA, T. Creep of ice under low stress. *J. Glaciology*, vol. 8, no. 52, pp. 147–152 (1969).

OBERT, L. and DUVALL, W. I. *Rock Mechanics and the Design of Structures in Rock*. Wiley, New York, 1967, Section 10.4, pp. 293–301, Time-dependent properties of rock.

PRICE, N. J. A study of the time-strain behavior of coal measure rocks. *Int. J. Rock Mech. Min. Sci., vol.1*, pp. 287–303 (1964).

ROBERTSON, E. C. Viscoelasticity of rocks. In *State of Stress in the Earth's Crust*, W. R. Judd (Ed.), pp. 181–234, Elsevier, New York, 1964.

CHAPTER 4

Failure Criteria for Soils and Rocks

More Fundamental Concepts

Structural Failure

The two properties that enable a soil or a rock mass to remain in equilibrium when forces are acting to disturb it, for example to resist the gravitational forces produced by its own weight, to sustain the stress concentrations produced around an excavation, or to support the weight of a structure erected upon it, are its shear strength and its tensile strength. If an earth material yields to these applied forces it can only be through the medium of processes of deformation and fracture. If the material is to fail then the cohesion of its mineral and particulate constituents must be overcome, by forces of tension or by pressure acting from within (in which case the fracture surfaces are pulled or else pushed directly apart), or by forces of shear (in which case the fracture surfaces are forced over one another at an angle and against a resistance which depends upon the internal friction characteristics of the earth material in relation to the ambient stress distribution). When we speak of the "strength" of an earth material we mean its ability to resist deformation and fracture by virtue of its properties of cohesion and internal friction. It is these properties that the conventional methods of materials testing are intended to display, under loads that generate forces of shear, tension, and compression.

It is now necessary for us to define some more fundamental concepts, added to those we have already considered.

Friction

Consider a horizontal plane, such as a bedding plane in a rock mass, supporting an elemental block, of weight N, resisting on its surface (see Fig. 4.1(a)). The weight of the block generates an equal and opposite reaction R. N and R together form a compression force normal to the plane of contact, and there is no generated tendency for the rock to move.

If a small horizontal force (not large enough to move the rock) is applied to the block, the reaction R will no longer be normal to the plane of contact. It adjusts in magnitude and direction to equal the resultant N and H. The triangle of forces represents, in magnitude and direction, the relationships between N, H, and R, and the angle θ (see Fig. 4.1(b)).

The normal force $N = R \cos \theta$, and the force H acts across the plane of contact, such that $H = R \sin \theta$.

If the shear force H is increased until the block is just about to slide, R will increase, and so will the angle θ. At the point when sliding begins, the frictional contact holding the materials stable along the plane of contact will be broken, and θ will have attained its maximum possible value, on the surface concerned.

That maximum value is φ, the *angle of friction*, and tan $\varphi = H/N$ = the *coefficient of friction*.

Geotechnology

Referring now to Chapter 2 (Fig. 2.2), where the resultant stress acting along a plane is resolved into a stress σ_n acting normal to the plane and a shear stress τ acting along the plane, if we imagine the plane to be a plane of fracture in the material, movement along that plane in response to the shearing force will be dependent upon the angle of internal friction (φ) of the material concerned, where $\tan \varphi = \tau/\sigma_n$.

Mohr's Circle of Stress

Although the state of stress in a real situation is essentially a three-dimensional problem it is often more convenient, and quite adequate for practical purposes, to simplify the conditions and to picture the problem in two dimensions, in the plane of the intermediate principal stress. The major and minor principal stresses σ_1 and σ_3, acting within a body under load, may then be represented as shown in Fig. 4.2. Consider a plane AD in the material, inclined at an angle θ to the direction of the minor principal stress σ_3.

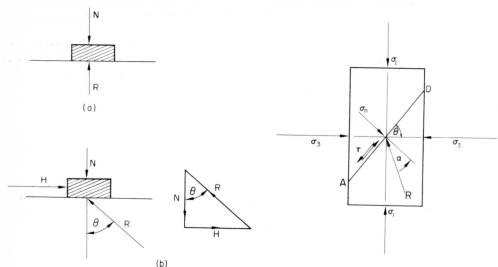

Fig. 4.1. Sliding friction.

Fig. 4.2. Two-dimensional stress pattern on a plane AD intersecting the minor principal stress direction at an angle θ.

It can be shown that the principal stresses σ_1 and σ_3 induce a stress system in that plane, and we have seen that such a system can be resolved into a shear stress acting along the plane and a normal stress acting across the plane. The magnitudes of these stresses are:

shear stress $\qquad \tau = \dfrac{\sigma_1 - \sigma_3}{2} \sin 2\theta$

and

normal stress $\qquad \sigma_n = \dfrac{\sigma_1 + \sigma_3}{2} + \dfrac{\sigma_1 - \sigma_3}{2} \cos 2\theta,$

or $\qquad\qquad\qquad \sigma_n = \sigma_3 + (\sigma_1 - \sigma_3) \cos^2 \theta.$

These relationships may be represented graphically by Mohr's circle of stress (see Fig. 4.3).

The following convention is adopted. It is assumed that the direction of the major principal stress is parallel with OY, that is, OX lies in the major principal plane. All principal stresses are then plotted along OX, as are all normal stresses. All shear stresses are plotted along OY.

Procedure

Lay down the axes OX and OY, set off OA and OB along the OX axis, representing σ_3 and σ_1, respectively, to scale.

Construct a circle on the diameter AB. The coordinates of all points on the circumference of this circle represent to scale, shear stresses and normal stresses for all planes passing through the point A, orthogonal to the plane of the diagram.

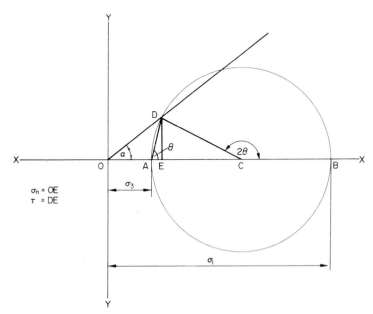

Fig. 4.3. Graphical construction of (Mohr's Circle of Stress), representing the two-dimensional stress pattern of Fig. 4.2.

For example on Fig. 4.2. the plane AD is inclined to the major principal plane at an angle θ. On Fig. 4.3. the line AD, inclined at θ to AB, cuts the perimeter of the circle at D. The coordinates of D then give the normal and shear stresses on the plane AD.

$$\sigma_n = \text{OE},$$

$$\tau = \text{DE}$$

Because   the normal stress $\sigma_n = \text{OE} = \text{OA} + \text{AE}$

$$= \sigma_3 + \text{AD} \cos \theta$$

$$= \sigma_3 + \text{AB} \cos^2 \theta$$

$$= \sigma_3 + (\sigma_1 - \sigma_3) \cos^2 \theta,$$

and   the shear stress $= \tau = \text{DE} = \text{DC} \sin (180 - 2\theta)$

$$= \text{DC} \sin 2\theta$$

$$= \frac{\sigma_1 - \sigma_3}{2} \sin 2\theta.$$

In Fig. 4.3. the triangle ODE represents the stress situation on the plane AD of Fig. 4.2. The line OD represents the resultant stress R in magnitude, to scale, while the angle DOB represents the angle of obliquity (α) that the resultant stress makes with the direction of the normal to the plane AD.

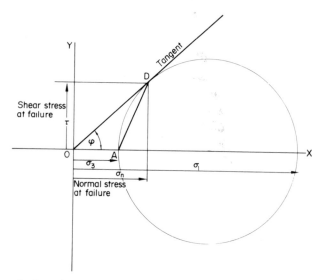

Fig. 4.4. Determination of shear stress and normal stress at failure, using Mohr's circle of stress.

Maximum Shear Strength

The maximum resistance to shear is developed when the angle of obliquity reaches a limiting value ϕ, the angle of internal friction. For this condition the line OD becomes a tangent to the stress circle, inclined at an angle ϕ to the axis OX.

Theories of Failure

Observation on earth materials in the field, in mines and tunnels and in open-pits, tells us that rocks which on some occasions may appear to be hard and brittle may, under other circumstances, display characteristics of plastic flow and creep. The structural geologist sees ample evidence of this in folded strata and in the flow structures often displayed in what now seem to be the most permanent and rigid rock formations. We know too that if an earth material, such as a rock or a soil, becomes highly loaded it may yield to the point of fracture, consequent breakdown, and ultimate collapse.

Failure of Elastic Materials

We may study the processes of deformation and fracture in the materials-testing laboratory, to determine the stress-strain and strain-time characteristics that we have already discussed. Let us now consider what happens when we load our laboratory specimens until they fail.

If the material is elastic and ductile, like a metal such as copper or mild steel, it will display a linear deformation-stress relationship up to the yield point A, shown in Fig. 4.5. It will then deform inelastically over the range AB, as the material becomes strain-hardened, but at B it attains a state of plasticity in which it ceases to resist further deformation, until it ultimately ruptures at the point C.

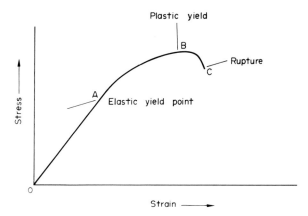

Fig. 4.5. Stress-strain relationship for an elastic-ductile material

If the material is elastic and brittle, like cast iron or glass, it will at first deform elastically under load, but it may then rupture before a discernible yield-point is reached (or very soon thereafter). Failure is liable to be instantaneous and associated with the violent release of energy from the testing mechanism. As a result of this, the specimen may be shattered completely (Fig. 4.6).

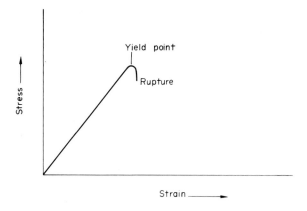

Fig. 4.6. Stress-strain relationship for an elastic-brittle material

Failure of Earth Materials

In general, earth materials such as rocks are not as ductile as the metals, and seldom so brittle as glass, but they display failure characteristics intermediate between the two extremes. The stronger rocks with high silica content and crystalline composition are usually more brittle than the porous sediments, and most rocks appear to have a discernible yield-point before they rupture. If the testing mechanism is suitably designed, to prevent the sudden release of strain energy when the rock begins to yield, the specimen may be observed to retain some resistance to deformation after the yield point (see Fig. 4.7).

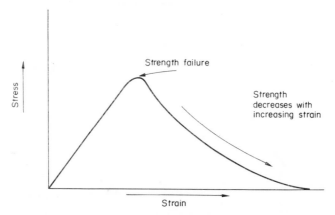

Fig. 4.7. Stress-strain relationship for a rock material loaded on a "stiff" testing machine.

We may therefore interpret the "failure" of a brittle rock material to occur at one or other of:

(a) The attainment of the yield-point.
(b) The point of rupture of the material.
(c) The point at which the material can no longer sustain the imposed load.

However, if the material is not brittle but displays a tendency to deform plastically, the point of failure might be described as:

(d) When the strain rate begins to accelerate under a constant load.

or, if both elastic and plastic deformation characteristics are displayed:

(e) When a maximum allowable percentage of residual permanent deformation occurs in the material.

Criteria of Failure

By selecting one or other of these definitions, and applying it to a specific earth material, we may specify the point at which a given material is likely to "fail" under a given distribution of stress. That is to say, we can establish a "criterion of failure" by means of which the maximum utilizable strength of the material can be identified and used for the purpose of design. The success or failure of our design will then depend upon the extent to which we are correct in our assumptions as to the mode of failure and on the degree to which we can establish the magnitudes and distribution of the stresses and strains involved.

There are many alternative criteria from which to choose, and it is important that we should select one that is appropriate to the particular design problem that we are concerned with. If we were dealing with materials such as metals, or other elastic bodies, we could apply criteria such as the maximum allowable principal stress, or maximum permissible shear stress, or maximum energy of distortion that the material could sustain. Such theories assume that the materials behave perfectly elastically with tensile strength equal to the compressive strength, and they define failure as the beginning of inelastic behavior in the material. We know that earth materials such as soils and rocks, and some constructional materials such as concretes, may behave inelastically under very low stresses and that they have very different compressive and tensile strengths. We should look towards theories of failure which do not involve assumptions that are obviously contrary to our experience. The failure criteria that are most commonly applied in geotechnology include the following:

Mohr's Theory of Failure

Mohr's theory of failure does not attempt to differentiate between failure by deformation or by actual rupture of the material. It states that the failure of a material may be represented by a functional relationship between the shear stress τ acting along the plane of failure, and the normal stress σ_n acting across that plane, such that

$$\tau = f(\sigma_n).$$

The basic assumption here is that the normal stresses, whether they be tensile or compressive contribute towards causing failure, and that shearing stresses also contribute, one being a function of the other. It is not assumed that the material is equally strong both in tension and compression, but it is inferred that in a stress field resolved into three principal stresses $\sigma_1 > \sigma_2 > \sigma_3$, the intermediate principal stress σ_2 has no influence on the failure of the material. The fundamental relationship between τ and σ_n is characteristic of the material concerned, and it must be determined by experimental tests. A graphical representation of the state of stress in the material — the Mohr circle of stress — is used, at the limit condition for failure (Fig. 4.4).

Determination of the Mohr Failure Envelope

A number of drill-core samples of the material are prepared, each with its cylindrical surface enclosed in a flexible jacket. The samples are placed, each in turn, on a triaxial testing machine so that a lateral confining pressure can be applied through the medium of a hydraulic jacket. Keeping the lateral confining pressure constant, the axial load which generates the principal stress σ_1, is increased until the specimen fails. Each test is conducted with a different value of lateral confining pressure, including the unixial ($\sigma_3 = 0$) condition. Sometimes, with rocks, the uniaxial test is also conducted with σ_1 in tension, but this requires special testing arrangements, and it is not easy to perform satisfactorily. With most earth materials, and soils, a tension test is not attempted. The results of each test are plotted as a Mohr circle, so that the complete test produces a number of Mohr circles of stress. The line formed by joining a common tangent to these circles is called the *Mohr Envelope* (Fig. 4.8).

The Mohr envelope defines the conditions for stability of the material under load. If for a given load condition the values of stress lie within the envelope, the material will be stable. But if for a load which produces a certain value of normal stress, the corresponding value of shear stress lies outside the envelope, the material will fail. Since the envelope is symmetrical about the OX axis it is customary to represent the envelope only by the half that lies above the OX axis.

The Coulomb Criterion of Failure for Soils

Referring to the Mohr graphical construction, if the angle of internal friction φ is assumed to be constant for a certain material with no uniaxial tensile strength, then the shear strength

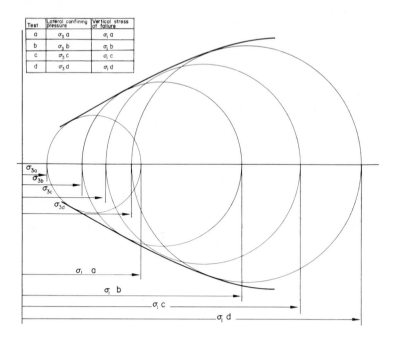

Test	Lateral confining pressure	Vertical stress at failure
a	$\sigma_3\,a$	$\sigma_1\,a$
b	$\sigma_3\,b$	$\sigma_1\,b$
c	$\sigma_3\,c$	$\sigma_1\,c$
d	$\sigma_3\,d$	$\sigma_1\,d$

Fig. 4.8. Construction of Mohr failure envelope from the results of triaxial compression tests.

can be represented by two lines passing through the origin O at angles $+\varphi$ and $-\varphi$ to the axis OX (see Fig. 4.9). These lines comprise the Mohr envelope for the material, for which circle A represents stable conditions, circle B represents incipient failure, and circle C represents

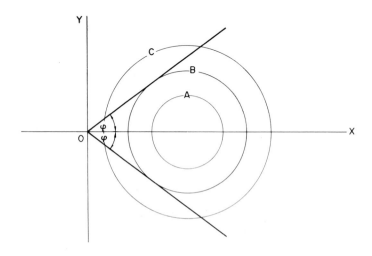

Fig. 4.9. Mohr failure envelope for an earth material with no tensile strength and no cohesion.

stress conditions beyond the limits that the strength of the material can withstand.

Cohesion

A material such as a dry unconsolidated sand will not sustain a slope angle steeper than that which can be held by the frictional contact of the constituent grains. The sides of a trench cut through such a sand will slump to this angle, and the maximum slope of a heap of the sand could not exceed the same angle. Other soil materials, such as damp sands, silts, and clays, will behave differently. The sides of a trench cut through these materials may stand unsupported, or not slump to the "angle of repose" for some considerable time, because the constituent grains are held together by some force additional to that produced by internal friction. This additional force is the *cohesion* of the material.

If a material has no cohesion, but depends entirely upon internal friction for its stability, then its Mohr envelope will intersect the OX axis at the origin O, as shown in Fig. 4.9. But a cohesive material possesses some shear strength even when the normal stress is zero. Its Mohr envelope will therefore intersect the shear stress axis at a finite value, when the normal stress is zero. This intercept on the shear stress axis is a measure of the cohesion of the material (see Fig. 4.10). The resistance to failure exhibited by such a material consists of two parts: (a) internal friction and (b) cohesion.

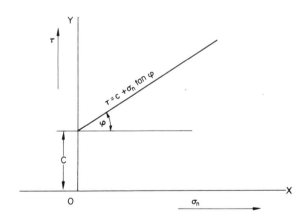

Fig. 4.10. The Coulomb criterion of failure for a soil.

A material such as a soil which is not consolidated can be assumed to possess a constant value of internal friction, no matter what is the value of the normal stress on the plane of shear. Its resistance to shear will, however, also depend on its cohesion and this too is a constant value, independent of the applied stress.

The strength envelope of a soil may thus be described by the straight-line equation

$$\tau = c + \sigma_n \tan \varphi$$

where τ =   the shear resistance at failure, i.e. the shear strength,

     c =   the cohesion,

     $\tan \varphi$ =   the coefficient of internal friction,

     σ_n =   the normal stress at failure.

This is the Coulomb criterion of failure— the basic equation in soil mechanics.

Figure 4.10 shows the characteristic Mohr envelope for a cohesive soil. The strength of a soil with no cohesion, such as a dry sand, would be described by $\tau = \sigma_n \tan \varphi$ (Fig. 4.11 (a)), while the strength of a wet clay, incapable of offering frictional resistance to deformation, and subject only to plastic flow controlled by its cohesive power, would have a Mohr envelope similar to that shown in Fig. 4.11(b).

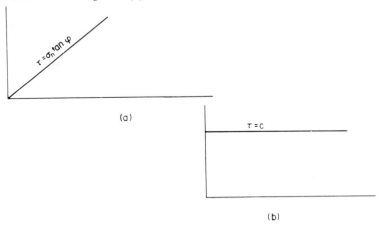

(a)

(b)

Fig. 4.11. Failure envelopes for soils. (a) Soil with no cohesion. (b) Soil with no frictional resistance to deformation.

The Coulomb-Navier Criterion of Failure for Rocks

The Coulomb criterion may also be applied to brittle solids if it is postulated that on the plane of failure the shear strength is reinforced by a frictional component of resistance to shear. Using the analogy of frictional resistance to sliding of a body resting on an inclined plane, the frictional resistance to shear failure is given by the product of the normal force acting across the plane of failure, and the coefficient of friction along the plane concerned. At the point of failure, when shear sliding is just about to begin, this frictional resistance reaches a maximum value, equal to $\mu \sigma_n$, where μ is the coefficient of internal friction ($\mu = \tan \varphi$).

The Coulomb-Navier failure criterion may then be stated:

At the point of failure the maximum shear resistance of the material (shear strength τ) equals the shear stress on the plane of failure (S_s) plus the internal frictional resistance ($\mu \sigma_n$).

Alternatively,

At the point of failure the shear stress on the plane of failure (S_s) equals the shear strength of the material (τ) minus the internal frictional resistance to shear ($\mu \sigma_n$).

The shear stress and the normal stress on a failure plane inclined at an angle θ to the minor principal stress σ_3, where the major principal stress is σ_1. are:

Normal stress $\quad \sigma_n \quad = \quad \dfrac{\sigma_1 + \sigma_3}{2} + \dfrac{\sigma_1 - \sigma_3}{2} \cos 2\theta$

and

Shear stress $\quad \tau \quad = \quad \dfrac{\sigma_1 - \sigma_3}{2} \sin 2\theta$

By substituting these values in the Coulomb-Navier equation, the criterion is expressed in a form to define the limiting stress conditions that the material can withstand under tensile and compressive loads. Conversely, if we have data concerning the shear strength, at zero confining pressure, the compressive strength, the tensile strength, and the angle of internal friction, of a particular rock material, it is possible to predict the limiting stress conditions that the material can withstand in terms of the Coulomb-Navier failure envelope.

The method is to plot circles, with radii proportional to compressive strength and tensile strength, respectively, to right and left of zero, on the normal stress axis. Plot the shear strength of the material on the shear stress axis, and insert the appropriate angle of friction by a line tangent to the compression circle. Join up the point and the line so plotted by a smooth curve, to produce the failure envelope (see Fig. 4.12).

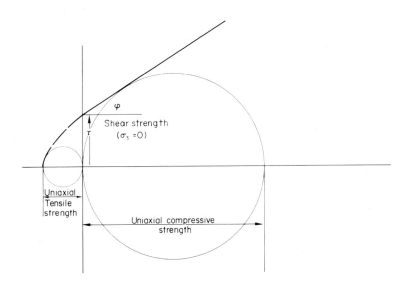

Fig. 4.12. The Coulomb-Navier failure criterion for a rock material.

This envelope may now be applied, in conjunction with the construction of Mohr circles of stress for given principal stress differences $\sigma_1 - \sigma_3$, to estimate the stress magnitudes at which shear failure will occur.

Depending upon the relative magnitudes of the various parameters represented, the envelope so obtained will be more or less curvilinear in the tensile zone, but straight in the compressive zone. This is because the Coulomb-Navier criterion assumes a constant value for the coefficient of friction, except over a limited range of stress at and near the tensile zone. Most rocks have internal friction characteristics that vary with change of loading conditions, so that the angle of friction may decrease with increase of the stress deviator ($\sigma_1 - \sigma_3$). The failure envelope is then curvilinear in the compression zone as well as in the tensile zone, and it can no longer be described by the Coulomb-Navier criterion.

The Griffith Brittle Failure Criterion

A theory to explain the failure of brittle materials was originally postulated by Griffith in 1924. Griffith worked with glass but his theories have been extended generally to other brittle materials, including rocks. In essence the theory is that fracture of brittle materials is initiated as a result of tensile failure produced by the stress concentrations that exist around the tips of micro-cracks and flaws present in the material. It is assumed that fracture extends from the boundary of an open flaw when the tensile stress on this boundary exceeds the local tensile strength of the material. It can be shown that high tensile stresses occur on the boundary of a suitably orientated elliptical opening, even under compressive stress conditions on the material as a whole.

The theory originally dealt with the mechanism of crack propagation in a uniaxial stress field, but subsequent investigators have extended Griffith's ideas to biaxial and triaxial stress conditions in rocks and also to explain the process of failure around a closed crack.

The results of all this work suggest that in an isotropic rock material, where the orientation of the cracks may be assumed to be random, fracture will occur if the uniaxial tensile strength is less than

$$\frac{-(\sigma_1 - \sigma_3)^2}{8(\sigma_1 + \sigma_3)} \quad \text{(the minus sign denotes tension)}$$

That is, the fracture criterion is

$$\frac{S_T}{\text{(Tensile strength)}} = \frac{-(\sigma_1 - \sigma_3)^2}{8(\sigma_1 + \sigma_3)}.$$

According to this criterion, when $\sigma_3 = 0$, σ_1 becomes the uniaxial compressive strength, and $\sigma_1 = 8S_T$.

Murrell has shown that the Griffith criterion of brittle fracture can be expressed as the equation of a Mohr envelope in which

$$\tau^2 - 4S_T\sigma = 4S_T{}^2.$$

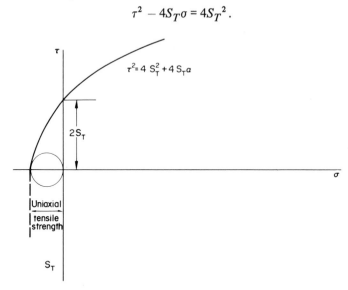

Fig. 4.13. The Griffith failure criterion.

In deriving this criterion it was assumed that the crack retains its elliptical shape until the moment of failure. When the principal stresses σ_1 and σ_3 are tensile, or where the rock is strong and not very highly stressed, this assumption may be valid. But in a weak rock, or in a highly stressed rock containing flat cracks, allowance may have to be made for closure of the cracks before the rock fails. This was done by McClintock and Walsh, by introducing the internal friction coefficient into their analysis. They then derived a modified Griffith criterion for the failure of a closed crack, in the form:

$$4S_T = [(\sigma_1 - \sigma_3)(1 + \mu^2)^{\frac{1}{2}}] - \mu(\sigma_1 + \sigma_3).$$

This criterion may be represented by a straight-line Mohr envelope with the characteristic

$$\tau = \mu\sigma + 2S_T$$

which is similar to that of the Coulomb-Navier criterion.

Selected References for Further Reading

FARMER, I. W. *Engineering Properties of Rocks* (Chapter 5, Strength and failure in rocks, pp. 55–69), Spon, London, 1968.

HOEK, E. Brittle failure of rock. In *Rock Mechanics in Engineering Practice*, Stagg and Zienkiewicz (Eds.), pp. 89–124, Wiley, London, 1968.

HOLTZ, Wesley, G. *Soil as an Engineering Material*, U. S. Dept. of the Interior, Bureau of Reclamation, Water Resources Technical Publication, Report No. 17, 1969.

MURRELL, S. A. F. A criterion for the brittle fracture of rocks and concrete under triaxial stress, and the effects of pore pressure on the criterion. *Proc. 5th Rock Mechs, Symposium, Minnesota,* C. Fairhurst (Ed.), pp. 563–577, Pergamon Press, Oxford, 1963.

SMITH, G. N. *Elements of Soil Mechanics* (Chapter 4, Shear strength of soils, pp. 73–107), Crosby Lockwood, London, 1968.

CHAPTER 5

The Engineering Properties of Soils

Pore Fluid Pressure

When describing the basic Coulomb equation for the shear failure of a soil it was assumed that the normal stress acting upon the shear plane was determined by the load imposed upon the soil from without, and that the whole of this load was carried by the particulate granular material. In reality we should also take into account the fact that part of the imposed load is carried by the pore fluids, which can be either gaseous or liquid, or sometimes both of these together. The pore fluid pressure acts to oppose the normal stress generated by the imposed load, and its magnitude depends upon the boundary conditions at the site concerned and the compressibility both of the soil structure and of the pore fluids. If the soil is being unloaded then an opposite effect takes place, so that the permeability of the soil and the boundary conditions at the site will determine the value of pore pressure at any one time.

If the boundary is impervious and the permeability of the soil is low it will take longer for the pore fluids to drain through the material than would be the case for a pervious soil within a free-draining boundary. In the latter case the pore fluids would be forced out of the material as quickly as a slowly imposed load was increased, so that significant pore pressures might not develop. On the other hand, a rapid or a dynamic build-up of load could generate high momentary pore pressures, even in a free-draining situation.

In a saturated granular material pore pressures may also exist as a result of capillary forces, holding moisture by surface tension on the soil particles. This phenomenon is more pronounced in fine-grained soils such as silts and clays, where the effect is sometimes seen as a negative pore pressure within the material.

Effective Stress

The intergranular stress which helps to determine the stability of a soil is thus the resultant of the normal stress generated by the applied load, and the pore fluid pressure.

That is, $\sigma_e = \sigma_n - u$

where σ_e is the *effective stress*,

σ_n is the applied normal stress,

u is the pore fluid pressure

Terzaghi's original concept of effective stress in saturated soils has been supplemented by Bishop for partially saturated soils as: $\sigma_e = \sigma_n [u_a - X(u_a - u_w)]$

where u_a is the pore air pressure,

u_w is the pore water pressure,

X is a parameter related to the degree of saturation.

We can now state Coulomb's basic law for the failure of a soil, in terms of the effective stress, as:

$$\tau = C_e + \sigma_e \tan \varphi_e$$

where σ_e is the effective stress and C_e and φ_e are the true cohesion and true angle of internal friction, respectively.

The Determination of Soil Strength

The Shear Box Test

To determine the shear strength of a soil by the shear box test a sample of the soil is placed in a rectangular box, the case of which is split horizontally into two halves. A vertical compressive load is applied to the top of the box and maintained at a constant value, while a horizontal force is applied to the upper half of the box. The horizontal load is increased until the contained prism of soil is caused to shear along the horizontal dividing plane of the box. The size of the test sample is usually about $150 \times 150 \times 50$mm for clays, and $300 \times 300 \times 150$mm for gravels. The test is repeated on a number of similar specimens of the soil, using different values of the vertical compression load for each test. The observed values of the shear resistance at failure are plotted graphically against the normal loads.

The test may be conducted either on undrained or on completely drained specimens, and the tests are sometimes denoted as "quick" and "slow" tests respectively. In the undrained, or quick, test the horizontal shear load is applied as soon as the vertical load has been imposed. The purpose of this is to generate a rate of applied shear that is fast enough to prevent consolidation of the soil but not so fast as to generate appreciable viscous resistance to movement. These two aims are incompatible, and a compromise is therefore arrived at by fixing on a standard rate of shear, which for a clay would probably be around 0.5mm/min. Alternatively, the rate of loading can be fixed to correspond with that anticipated in the particular field problem concerned.

Certain soils, that exhibit a limited degree of consolidation on loading, may be subjected to a "consolidated-undrained" test, in which the specimen is allowed to consolidate under the vertical load before the horizontal load is applied. In the "slow" or undrained test a saturated specimen of the soil is placed in the shear box and allowed to consolidate under the vertical load as the water drains out. The specimen is then sheared at a slow rate, so that the pore pressures remain at zero level. Such tests are easily made on sands, but they may take a considerable length of time (up to 2 days consolidation and 8 hours shearing) on clays.

Triaxial Compression Test for Soils

Highly permeable and porous materials are difficult to test satisfactorily by undrained methods in a shear box. The triaxial apparatus offers an alternative testing system. The test differs from the shear-box test in that the plane of shear failure is not predetermined. A cylindrical specimen of the soil is contained in a flexible jacket, so that it can be subjected to a lateral confining pressure, imposed by a hydraulic system surrounding the jacket in a transparent enclosure. The size of the specimen usually ranges from 38mm diameter by 76mm long for fine-grained soils to 76–102mm diameter and 150–200mm long for coarse soils. Vertical dead-weight load is applied axially to the specimen and measured by a proving ring. A dial gauge fitted on the apparatus measures the vertical deformation of the specimen.

The method of testing is to apply a constant value of confining pressure (radial stress) through the hydraulic system, and then to apply a slowly increasing vertical load until the specimen fails in shear. As with the shear-box test the procedure is repeated on various samples of the same soil, but this time using different values of the radial confining pressure for each test.

It is customary to plot the results in two ways, one recording the stress-strain characteristics (from which the modulus of deformation is determined) and the other providing a graphical representation of the failure envelope for the material.

The stress system imposed in such a triaxial test can be considered as being comprised of two components: (a) the hydraulic confining pressure σ_3 imposed in all directions and (b) the *"deviator stress"* $(\sigma_1 - \sigma_3)$ (see Fig. 5.1). If the deviator stresses are plotted against strain of the specimen for each test, a series of curves is obtained such as those shown in Fig. 5.2. When plotting these results allowance may be made for any change of volume of the specimen during the test, although for quick tests on unsaturated specimens it may be assumed that the volume of the specimen remains constant. The average cross-section at any given deformation is calculated accordingly.

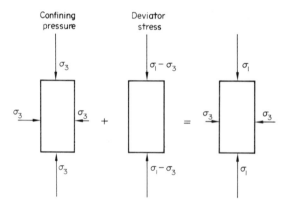

Fig. 5.1. Deviator stress and confining pressure in triaxial testing

The modulus of deformation of the soil can then be determined from the slope of the stress-deformation graph, where this is linear, or approximated for a specific stress range when the graph is curvilinear. The apparent cohesion of the soil, and its angle of frictional resistance to shear, are obtained by plotting the best common tangent to the results, as expressed in the form of a series of Mohr circles of stress, one for each test (see Fig.5.3).

As with the shear-box test, the triaxial test may be made "drained","consolidated undrained", or "undrained". In the undrained test the sample, contained in its rubber sleeve, is held between impervious end-plates at top and bottom. Vertical loading begins as soon as the radial confining pressure has been applied. Readings of deformation and load are made at frequent intervals, until the specimen fails. The rate of loading must be slow, so that viscous resistance to movement does not add appreciably to the apparent measured strength, and the test is repeated on from three to five specimens of the same material. If the pore pressures are not measured, then the c and φ parameters are obtained in terms of total stress, not effective stress.

To obtain results in terms of effective stresses either the pore pressures must be measured or, in the case of saturated sands which are highly permeable, an outlet for the pore water

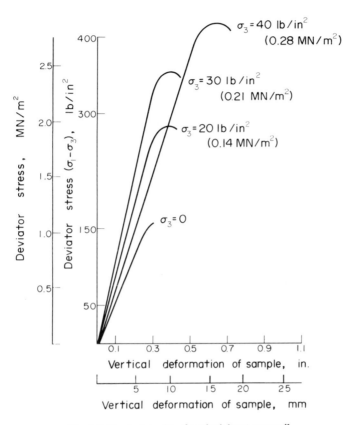

Fig. 5.2. Typical results of a triaxial test on a soil.

pressure can be provided from the specimen, connected to a burette open to the atmosphere. The pore water pressure is then maintained constant throughout the test by slightly adjusting the lateral confining pressure and observing the meniscus level in the burette while the vertical load is applied. The volume of the specimen is thus maintained constant during the test, no drainage occurs, the pore pressures remain zero, and therefore the applied vertical and lateral stresses are effective stresses.

The consolidated undrained triaxial test is made on a saturated specimen which is first allowed to drain and consolidate, keeping the lateral pressure constant. Consolidation is judged to be complete when no more water is expelled from the pore-water outlet. This outlet is then closed, and the vertical load is increased, at a strain rate of around 0.05 mm/min. until the specimen shears. This is the form of triaxial test most commonly applied to soils. Applying the deviator stress induces pore fluid pressure which must be measured to obtain effective stress by difference.

In the drained test a saturated specimen is first allowed to consolidate, under a constant value of lateral confining pressure, without applying vertical load. The deviator stress is then applied very slowly, to prevent the build-up of pore pressure and to drain out the pore fluids. Consolidation of the specimen must be complete before it is sheared, and the volume change of the specimen is measured by the volume of pore water expelled. Such tests on clay soils

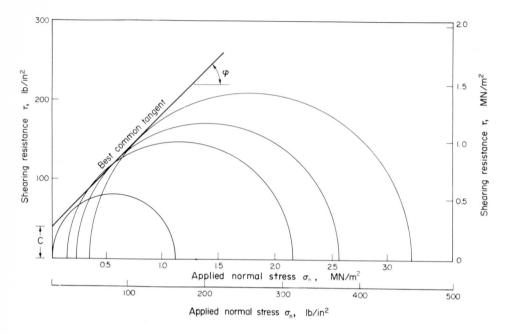

Fig. 5.3. Determination of the apparent cohesion and the frictional resistance to shear, from the results of a triaxial test on a soil.

may take up to 2 weeks to complete. Hence they are seldom conducted.

The design of earthworks in soil can be approached either from a basis of apparent, or total stresses, or from the basis of real or effective stresses. If done by using total stresses the determination of the relevant design parameters requires that very careful attention should be given to the consistent reproduction of pore-pressure conditions in the laboratory tests, if meaningful results are to be obtained. Most soils engineers prefer to design from the basis of effective stresses. The shear strength of the soil is then related to specific values of pore fluid pressure, either determined at the site, measured in the soil concerned, or estimated from previous experience in similar conditions.

The Unconfined Compressive Strength of Soils

A compression test on a sample of soil, conducted without placing lateral confining pressure on the sample, can be performed to determine the uniaxial compressive strength of the sample. The apparatus required is simple and portable, and it gives quick results. It is important that the sample be disturbed as little as possible from its native environment, so that its cohesion and pore-water content be not changed in transit. A cylindrical sample, measuring 38mm by 86mm long is cut out at the site, and the ends of the sample coned internally to give a length of 76mm, apex to apex. The sample is then mounted between conical end-plates, the purpose of which is to prevent the specimen "barreling" during the compression test. Barreling would occur if flat compression end-plates were used, because of friction end-effects at the plate contacts.

Load is applied to the specimen through the medium of a spring, controlled by a hand feed-screw. Arrangements are attached to provide for mechanically recording the longitudinal deformation of the specimen as the load is applied, and the stress in the specimen is calculated, allowing for the stiffness of the spring (which must be calibrated) and the change in area of cross-section of the specimen during the test. (It is assumed that the volume of the specimen remains constant, and that its shape remains cylindrical.)

For such a test on a saturated clay φ, the angle of internal friction is zero and the failure envelope is a straight line parallel to the normal stress axis, with the intercept c on the shear-stress axis (see Fig. 4.11(b)). The shear strength thus equals the apparent cohesion and is half the uniaxial compressive strength. For unsaturated soils φ does not generally equal zero and its value cannot be obtained from the unconfined compressive strength alone. This strength, in fact, varies widely in any one type of soil, depending upon the degree of compaction and the moisture content of the soil. A typical range of values for a sandy-clay soil is shown in Fig. 5.4.

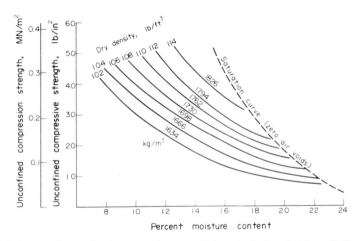

Fig.5.4. Range of unconfined compressive strength for a typical sandy-clay soil. (Imposed stress pattern constant throughout.)

Changing the pattern of stress in the soil could result in unconfined compression strength values even outside the range shown in this figure. Hence, as an engineering strength parameter for a soil the unconfined compressive strength is only likely to be useful when applied to saturated or nearly saturated clays.

The Vane Shear Test for Soils

To take out a cylindrical core-sample of a soil, without disturbing its native strength characteristics, is increasingly difficult with increasing depth of the borehole from which the sample must be taken. It would be advantageous if the shear strength could be obtained at depth on the site, on undisturbed specimens of soil. The vane shear test is aimed to achieve this. It is performed by using a rod, at the end of which two vanes are attached, at right angles to one another in a hollow cylinder. This is pushed into the soil at the bottom of the borehole. The rod is then rotated so that the vanes make a shear cut, to push the soil along the internal surface of the cylinder. The torque required to shear the soil in this way is then measured.

Consolidation of Soils

It is important for the engineer to know by how much a soil mass will compress under load, and at what rate the compression will occur. This information is required in order that he may make reliable estimates of such things as the settlement of foundations and the possible volume change of an earthwork. The compressibility of a soil mass is determined primarily by the voids and by the compressibility of the pore fluids, and only to a relatively small extent by deformation of the soil particles. If the voids contain only air the soil will compress more rapidly under load than it would if the voids contain water. In the latter event the pore fluids would require time before they could drain from the soil mass, and they would drain more quickly from a coarse permeable soil than they would from a fine dense material. This drainage-time effect determines what is known as consolidation in a soil.

A shear test on a coarse permeable material such as a sand, or a gravel, would almost certainly be made under the "drained" condition, unless drainage was deliberately prevented, because the high permeability of the material allows consolidation and drainage to take place very rapidly. But with clays the situation is different, because these materials display a wide range of consolidation characteristics. Figure 5.5. shows typical load-compression curves for normally-consolidated and over consolidated clays.

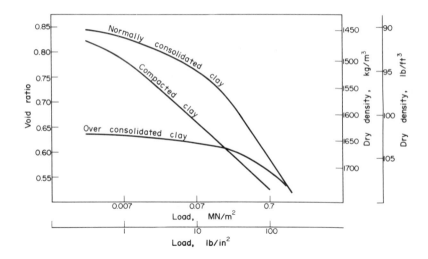

Fig. 5.5. Typical load-compression curves for a clay.

Normally consolidated clays

A normally consolidated clay is one which has never been subjected to pressures greater than that generated by the existing overburden. That is to say, it has never had to sustain more than the weight of the material now lying above it. The effective overburden pressure is the pressure generated by this weight — less the pore fluid pressure at the point concerned. A normally consolidated clay is in a state of equilibrium under the natural load conditions that have been gradually established with the passage of time during the geological sedimentation process.

Over consolidated clays

If, during the geological history of the clay, a superimposed load has existed which is greater than that now represented by the present overburden, then the soil as it is seen today will be over consolidated. The soil will have been consolidated by some geological sedimentation process, possibly supplemented by the weight of glacial ice sheets, and then partially unloaded as a result of ice melting, erosion, or possibly man-made excavation. As a result of this unloading there will have been some rebound, but this will be relatively small compared with the original consolidation. When such an over consolidated soil is again loaded it will be considerably stronger at a given pressure than it would have been if normally consolidated. Its volume change will be smaller than that of the normally consolidated clay, up to the point when the applied load equals the maximum value that had occurred during its past consolidation history.

An over consolidated clay tends to dilate when loaded to failure in an undrained shear test. Negative pore pressures are then generated, which increases the effective stress, so that the undrained strength is higher than the drained strength. This is the direct opposite to what is observed on a normally consolidated clay, in which the drained strength is higher than the undrained strength. It means that the sides of an excavation made in an over consolidated clay require careful attention, and they may need to be supported if they are to stand for any great length of time, because such a clay decreases in strength as the water drains from it.

Settlement of Foundations

The rate of settlement of a structure founded on a normally consolidated clay will be greater than that which the same structure would experience on an over consolidated clay. The consolidation characteristics of a clay foundation can be modified, and the anticipated rate of settlement can be reduced, by artificially compacting the material and so produce a characteristic intermediate between the normal and the over consolidated condition (see Fig. 5.5). To estimate the amount and rate of settlement of a foundation the engineer needs to know what are the strength characteristics of the various soil layers, their thickness and disposition, together with the soil-drainage conditions and the pore-fluid pressures. This information has to be obtained by direct observation at the site, and from laboratory tests on samples taken from the site. The consolidation-time, and voids ratio-effective pressure characteristics must also be known, and these, too, will be determined by laboratory tests.

Finally, all these properties must be assessed in relation to the stress distribution that will be imposed on the soil by the completed structure.

Compaction

Methods of compaction in the field include the application of rollers, ranging from smooth-wheeled road rollers, pneumatic tyred vehicles, rollers with sheep's-foot and club-foot projection, and track-laying vehicles, to impact hammers and vibrators. The optimum method of compaction depends upon the soil type, together with its moisture content and particle size distribution, or gradation. Smooth-wheeled rollers are generally best on cohesionless materials such as crushed rock, hardcore, gravels, and sands. Pneumatic-tyred rollers are used on fine sands of uniform gradation, and on fine-grained cohesive soils having a moisture content at or near the plastic limit. Projection-rollers, such as sheep's-foot rollers, are best suited to fine-grained cohesive soils with low moisture content.

Geotechnology

Density

In general, the greater the density of the soil the better its structural properties will be. The object of compacting a soil is to reduce the air voids and to increase the soil density. This increases its structural stability, reduces its water-absorption tendencies, and improves its resistance to settlement. The degree of compaction of a soil is measured in terms of the solid matter contained in one unit cube of material. This is the "dry density" of the soil. It ranges in value from about 2243 kg/m³ in gravels, to about 1442 kg/m³ in heavy clay. For any soil the curve of dry densities plotted against percentage water content is convex upwards. That is to say, there is an optimum moisture content at which the dry density is a maximum (see Fig. 5.6).

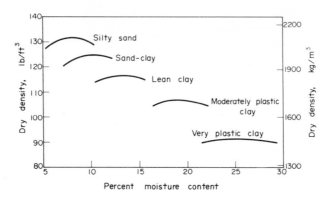

Fig. 5.6. Typical standard compaction curves for various soils.

Increasing the degree of compaction of a soil increases the dry density up to a point at which either the air voids are practically eliminated, or else the soil builds up resistance to further compaction, as shown in Fig. 5.7.

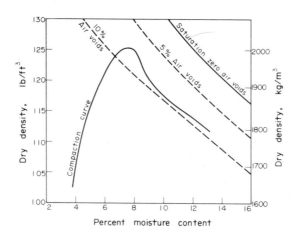

Fig. 5.7. Dry density-moisture content curve for a sandy soil compacted by machine.

Relative density of soils

Cohesionless materials such as coarse gravels and sands do not respond to compaction in the same manner as do silts and clays, so that their dry density-moisture content compaction curves do not relate so easily to their engineering strength properties. The relative density criterion may then be used to compare the actual density of such materials with the soil at its minimum and maximum density extremes. The relationship between relative density and dry density, for typical cohesionless soils, is shown in Fig. 5.8. The efficiency of compaction may then be expressed as:

$$\frac{\text{Field dry density} - \text{Loose dry density} \times 100\%}{\text{Maximum} - \text{Minimum dry densities}}$$
(as determined by standard compaction tests)

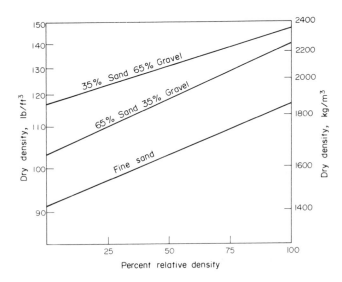

Fig. 5.8. Relationship between relative density and dry density for typical cohesionless soils.

Expansion and Shrinkage in Clay Soils

Expansion

The chemical and mineralogical constitutents of clay soils have a pronounced influence on the physical characteristics of the material. One of the most important of these properties is the potential swelling capacity of soils containing montmorillonite-type clay minerals. These substances have a strong affinity for hydration, in the process of which the crystal lattices of the minerals expand. Other clay minerals such as illite may also display expansion properties if the dry clay comes into contact with water, but to a lesser extent than montmorillonite. The third class of clay minerals — the kaolins — have fixed crystal lattices which do not expand from the dry condition if water is absorbed.

From an engineering viewpoint, montmorillonite, and to a lesser extent also illite, are troublesome in that they can produce heave and swelling in clay foundations and earthworks and lead to instability of the structures erected in contact with them. These effects may extend beyond the surface clay deposits into the underlying bedrock, if that rock contains clay-filled fissures. Percolating water enters the rock and swelling pressures may be generated in the fissures, sometimes so great in magnitude as to fracture the rock mass.

The extent of the change in volume that takes place in an expansive soil depends upon several factors, which may be categorized as being either "internal" or "external". The internal factors define the potential swellability which the clay may possess. The external factors determine whether that potential swellability will be realized, and what pattern the swelling process will follow. Among the internal factors are the initial density and water content of the soil, the type and amount of clay minerals present, and the soil structure. The external factors include the accessibility of water to the soil, the ion concentration in the water, the conditions of confinement, and the nature, intensity, and distribution of counter pressures due to external loading.

TABLE 5.1.
Expansion Potential of High Plasticity Clay Soils (Holtz)

Index properties			Probable Expansion Per cent total volume change (dry to saturated)	Degree of Expansion
Colloid content (% minus 0.001 mm)	Plasticity index	Shrinkage limit (per cent)		
28	35	11	30	Very high
20 – 31	25 – 41	7 – 12	20 – 30	High
13 – 23	15 – 28	10 – 16	10 – 20	Medium
15	18	15	10	Low

Shrinkage

Desiccation of a clay soil after it has expanded under the influence of water opens up shrinkage cracks in the clay. These form easy paths for the access and lodgement of water during the next wet season. Alternate expansion and shrinkage thus proceeds at an aggravated rate, and ultimately it may reach an extent that can seriously damage structures erected upon clay. The engineering techniques that are applied to protect structures built on expansive soils include removing the soil and replacing it by non-expansive material, to an adequate depth in the foundations. Also, piles and caissons are used to resist uplift, while protective coatings and chemical treatment may limit the ingress of water to a soil. It should be remembered that the swelling process is essentially one of electrical osmosis, so that the application of drainage protection by electrical osmosis of a clay soil is likely to be beneficial also in contributing to the reduction of potential swellability and subsequent shrinkage.

The term "osmosis" is used to describe the spontaneous movement of water, containing mineral matter in solution, from a lower to a higher concentration, when separated by a suitable membrane. The reverse process, which is sometimes employed for the desalinization of water, involves the deposition of mineral matter from solution. In electrical osmosis of a soil the process is controlled by applying an electrical potential, by means of suitable electrodes, so that the solutions contained in the soil serve as the electrolyte from which mineral matter is caused to be deposited, with the object of sealing the interstices of the soil. Electro-osmosis may also be applied to promote drainage of groundwater in a specific direction, since the water normally tends to flow towards the negative electrode.

Weathering of Clay Rocks

The processes of weathering are important in relation to their effects on the engineering properties of rocks. For example, Smart classifies weathered chalk in four grades of successive deterioration, in which the deformation modulus on first loading decreased from around 47,000 to 3700 kg/cm^2. A decrease of approximately 50% rigidity was observed during the earliest phases of weathering. Considerable attention has also been given to the effects of weathering on clay rocks, because of their significance in relation to the stability of coal-mine dumps and of clay slopes. It appears that the breakdown of such excavated materials, due to subsequent weathering, is essentially physical rather than chemical, provided that montmorillonite is not present, and it is limited to those surfaces actually exposed to the elements. The reduction to fundamental particle size is a very slow process.

Studies on the stability of mine spoil heaps, reported by Spears and Taylor, indicated that, apart from the initial and relatively rapid breakdown which occurred for a short time immediately after the material was brought to the surface and dumped, the materials in the spoil heap changed very little. Nevertheless, those changes had a very profound influence on the engineering strength properties of the materials concerned. The most highly weathered equivalents of siltstone and shale showed a decrease in triaxial compressive strength of around 90%, when compared with the unweathered parent rocks. The shear strength parameters of fissile mudstones and shales showed a fall of up to 37%, and a drop in cohesion of around 90%. Quartz-rich siltstones showed a similar fall-off in cohesion but only about 10% change in shear strength.

Selected References for Further Reading

BISHOP, A. W. and HENKEL, D. J. *The Measurement of Soil Properties in the Triaxial Test*, Arnold, London, 1962.

BREKKE, Tor L. On the measurement of relative potential swellability of hydrothermal montmorillonite clay. *Int. J. Rock Mech. Min. Sci.*, vol. 2, pp. 155–165 (1965).

HOLTZ, Wesley G. *Soil as an Engineering Material*, U. S. Bureau of Reclamation, Water Resources Technical Publication, Report No. 17 (1969).

HOLTZ, Wesley G. and GIBBS, H. J. Engineering properties of expansive clays. *Trans. Am. Soc. Civil Engineers*, vol. 21, pp. 641–663 (1956).

HOUGH, B. K. *Basic Soils Engineering* (Chapters 4, 5, and 6, pp. 83–167), Ronald Press, New York, 1957.

HOUGH, B. K. *Laboratory Shear Testing of Soils*, ASTM STP 361 (1963).

HOUGH, B. K. *Compaction of Soils*, ASTM STP 377 (1965).

SMART, P. Strength of weathered rock. *Int. J. Rock Mech. Min. Sci.*, vol. 7, pp. 371–383 (1970).

SMITH, G. N. *Elements of Soil Mechanics*, (Chapter 4, pp. 73–107), Crosby Lockwood, London, 1968.

TAYLOR, R. K. and SPEARS, D. A. The breakdown of British Coal Measures rocks. *Int. J. Rock Mech. Min. Sci.*, vol. 7, pp. 481–501 (1970).

TAYLOR, R.K. and SPEARS, D. A. The influence of weathering on the composition and engineering properties of in-situ Coal Measures rocks. *Ibid.* vol. 9, pp. 729–755 (1972).

CHAPTER 6

Fluids in Soils and Rocks

One of the most important aspects of geotechnology is an understanding of the mode of occurrence of fluids in earth materials, and their effects on the engineering characteristics of soils and rocks. The study of groundwater, with particular reference to its physics, chemistry, and geology, forms the basis of the science of hydrogeology, pursued jointly by engineers and geologists whose past interest has been mainly in regard to problems of drainage and water supply.

In recent years, large and expensive engineering projects for the construction of motorways and tunnels, high-rise buildings, dams, canals, and reservoirs, and the underground storage of water, oil, and natural gas, have extended the application of hydrogeological interest beyond that of water supply and drainage, to include the study of engineering problems relating to the mechanical strength and structural stability of earth materials.

In reviewing some of the natural and man-made disasters that have occurred periodically throughout the years — dam failures such as that at Malpasset, the Vaoint, Peruvian, and Alaskan landslides, the flow of coal-pit debris that buried the children of Aberfan, in Wales, the periodic outbursts of gas and rock that occur in some salt and coal mines throughout the world, the rockfall that destroyed the mine at Coalbrookdale, South Africa, one cannot escape the fact that all these occurrences had one factor in common. That factor is the occurrence of an abnormal change in the engineering characteristics of the earth materials, involving drainage, interstitial fluids, and pore-fluid pressures. Of course, there were other contributory factors as well, and these differed in the various occurrences, but the common factor, and sometimes very likely the critical factor, was there. Indeed, one can speculate even further and ask also whether fluid pressures in the strata at depth have more than a little to do with the origin of earthquakes. The occurrence of seismic disturbances associated with the injection of fluids at pressure into deep boreholes lends weight to such a hypothesis, in the light of evidence as to the effect of pore pressures and friction on the strength characteristics of rocks.

Groundwater

The most common fluids that are encountered in geotechnology are water and air. Hydro-carbon gases and oils are important in relation to natural gas and oil reservoir engineering, while methane and carbon dioxide are frequently associated with rock and gas outbursts in mines.

A small proportion of the water that exists in soils and rocks may have its origin in the earth's interior, to find its way to the surface through volcanic and magmatic processes. By far the greater proportion, however, originates as precipitation from the Earth's atmosphere. This percolates downwards to form sub-surface water, in which various zones can be recognized. A simple engineering classification refers to all water lying below the water table, which is the surface below which the strata are saturated, as "ground water", while the water which percolates from the surface to the water table is termed "gravitational water". Strictly speaking, however, this classification is neither adequate nor true, since all the water, including that below the water table, is under gravitational influence, some of it in a continual state of movement and flow.

A more accurate terminology refers to the zone above the water table as the *Aeration Zone*. in which three sub-zones or belts exist. These are, at the top, the *Soil Belt*, affected by plant transpiration and the evaporation of moisture to the atmosphere. At the bottom of the aeration zone, and as an upper fringe to the water table, is the *Capillary Belt* in which water is drawn up from the water table against the force of gravity, by the suction pressure generated by surface tension between the moisture and the solids in the voids of the material. The height of the capillary belt is determined by the nature of the solids and the size of the voids, and the region extending from the top of the capillary belt to the base of the soil belt is the *Intermediate Belt*. The water in the aeration zone therefore consists of two main fractions one of which is in a state of movement, either percolating downwards under the influence of gravity or being drawn upwards by evaporation, transpiration, and capillary suction, and the other being "held water" which is retained.

Similarly, below the water table there are two zones. The upper of these is the saturated zone in which "free" water moves and flows from regions of higher pressure to regions of lower pressure, This saturated zone extends underneath the water table to a considerable depth, determined by the nature of the rocks and the intensity of the strata pressures. Depending upon these factors, the strata ultimately attain a plastic condition, in which the interstices of the rocks are closed by plastic flow.

Held water

Some water is not free to move. It is "held" in the strata, and consists of

1. Water chemically combined in the crystal structure of the constituent minerals. This is relatively small in quantity. It cannot be removed from the soils and rocks by oven-drying at 110°C, but it can be removed by thermal attack at high temperatures. This fact is of great significance in relation to earthquake phenomena, the brittle-plastic transition of rocks under high confining pressures, and the reversion from plastic to brittle behavior again in some basic rocks if the high pressures are accompanied by high temperature.

2. Water adsorbed on the surface of solid particles within the soil and rock. Some, but not all, of this water can be removed by oven-drying at 110°C. If an oven-dried sample of the material is exposed to moist air while cooling it will adsorb moisture again, at a rate depending upon the humidity of the air. The amount of water that can be adsorbed in this way is determined primarily by the surface area of the solid particles within the earth material.

   Contact moisture and adsorbed water have important effects on the properties of a soil. In the case of a coarse-grained particulate material such as a sand, the surface tension which holds moisture around the points of contact of an unsaturated sand gives the material a cohesion which it does not have if the material is either completely dry or completely saturated. The damp material also has greater volume, as compared with its bulk when dry.

3. Most of the held water in unsaturated earth materials is retained by forces of surface tension in the pores and capillaries of soils and rocks, and around the points of contact of solid particles in granular material. These forces generate a suction pressure which holds the water, the suction pressure increasing as the moisture content of the material is reduced, which means that the proportion of water retained, being progressively smaller, recedes

into the smaller pores and capillaries

In the case of a fine-grained material such as a silt or a clay, the adsorbed water is held at pressures that may reach high values. Consequently this water has properties that differ very considerably from those demonstrated by water at atmospheric pressure. Its density and boiling-point are higher and its freezing point lower, so that it remains liquid in otherwise frozen ground.

Soil suction

The forces of surface tension, absorption, and adsorption, reduce the vapor pressure of the held water and generate suction pressures which may vary from zero at saturation to more than 100 lb/in² (689.5 kN/m²) in completely dry soils. Typical suction pressure-moisture content relationships for sand/clay soils are shown in Fig. 6.1. It can be seen that the moisture

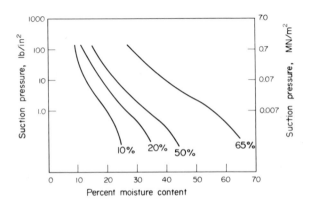

Fig. 6.1. Range of soil moisture suction pressures for typical sand-clay soils.
(Percentage clay content indicated.)

content generating a given value of suction pressure increases with increase of clay material content, and soils of different types may be in suction equilibrium at very different moisture contents. This is why, when in a trench or a borehole, different types of soil or rock are encountered at the same depth, such as lenses of sand in clay, the sand lenses are much drier than the surrounding clay.

"Held water" in rocks

Rocks differ from soils in that a soil consists of discrete particles in point-by-point contact. The whole of the pore space and capillarity is interconnected in a soil. This is not always true for rocks. Sedimentary rocks are sometimes akin to concretes in that they may be regarded as a solid skeleton traversed by a fine network of capillaries, but in crystalline rocks the voids are not all interconnected. They are more probably limited to grain-boundary cracks and flaws. Such materials have relatively low porosity.

The question as to whether the pore fluids are held in interconnected or discontinuous voids is important in its implications regarding the effects of pore pressure on effective stress, and the deformation characteristics of rocks under load.

Porosity

The porosity of a material is customarily defined as the volume of voids, expressed as a percentage of the total volume of the sample. A loose sand may have a porosity of around 40-50%, a sandstone from 10-20%, while siltstones and shales will probably rate 5-6% porosity. Igneous and metamorphic rocks generally have a low porosity, compared with sediments, but there are exceptions to this rule. For example, vesicular extrusive basalt may have 10-15%, while a compact dolomitic limestone may only have around 3-4% porosity.

The porosity that is determined by interconnected voids is sometimes termed the *Effective Porosity*. It may be measured by weighing a specimen of the material when it is dry (W_d), and then weighing the same specimen after saturating it with water (W_w).

Then the effective porosity

$$P = \frac{W_w - W_d}{V_d}$$

where V_d is the dry volume of the specimen.

Alternatively, if a volume of fluid V_f can be injected into V_d, the effective porosity = V_f/V_d.

The *Total Porosity* includes the volume of discontinuous voids, and to determine this the material must be crushed. An empirical relation by which the true porosity of sandstones, shales, clays, and silts may be estimated from measurements of the bulk density (ρ_B) of a sample is given by Davis as

$$P = \frac{2.654 - \rho_B}{2.654} \times 100$$

in which the value 2.654 represents the average grain density of the mineral assemblage in 600 samples of these materials.

In general the determination of true porosity requires that a sample be crushed to powder of a size less than that of the voids, when

$$\text{True porosity} = \frac{\rho_1 - \rho_2}{\rho_1} \times 100,$$

where ρ_1 = specific gravity of the material in powder form,
ρ_2 = specific gravity of the material in bulk.

The Flow of Groundwater

Where the soil and rock interstices are completely saturated the movement of groundwater is assumed to flow in accordance with Darcy's law of fluid flow through a porous medium. This states: the velocity of flow is directly proportional to the hydraulic gradient. By *hydraulic gradient* is meant the loss of hydraulic head per unit length of flow path. In hydrogeology the hydraulic head at a given point is sometimes referred to as *fluid potential*. This is the algebraic sum of the hydrostatic pressure potential and the gravity potential at the point concerned.

Permeability

The ability of a porous material to provide passage for a fluid will also depend upon the physical properties of the porous medium and the fluid concerned. Assuming the Darcy law applies:

$$Q = Aki$$

where Q is the quantity of fluid passing in unit time through a flow path of cross-sectional area A, i is the hydraulic gradient, k is the proportionality constant or *coefficient of permeability*.

Note that the quantity flowing is assumed to depend on the total cross-sectional area of the material, and not on its porosity or on an average area of voids. The assumed flow velocity is thus a fictitious value, sometimes referred to as *specific discharge* or *seepage velocity*. (Not to be confused with *specific yield* which is the proportion of water that can be drained from a saturated aquifer.)

k may be defined as the velocity of unit flow per unit hydraulic gradient, and may be measured in cm/sec, or m/day. In coarse gravels it may exceed 6000 m/day and in heavy clays it may be less than 0.00006 m/day. Representative values are quoted in Table 6.1.

TABLE 6.1. *Representative Values of the Permeability Coefficient k (cm/sec)*

Soils	Gravel	$1 - 10^2$
	Clean sands	$10^{-3} - 1$
	Fine sands and clay sands	$10^{-6} - 10^{-3}$
Rocks	Sandstones	$10^{-10} - 10^{-8}$
	Granite	$5 \times 10^{-11} - 2 \times 10^{-10}$
	Slate	$7 \times 10^{-11} - 2 \times 10^{-10}$
	Limestone	$7 \times 10^{-10} - 10^{-7}$

Laboratory Determination of the Coefficient of Permeability for Soils

A simple laboratory method of determing the coefficient of permeability (k) for a saturated soil makes use of the constant-head permeameter, the principle of which is illustrated in Fig. 6. 2(a). A constant head of water is maintained across the sample, such that the hydraulic gradient is h/l. The quantity of water Q, flowing through the sample in unit time, is measured, from which

$$k = Q/A \cdot l/h$$

where A is the cross-sectional area of the soil sample..

The apparatus is only suitable for a coarse-grained material. For fine-grained materials a more sensitive arrangement is needed, such as that provided by the variable-head permeameter (Fig. 6.2(b)). Here the flow of water is measured at the inflow end and after noting the height of the water column h_1 in the narrow-bore inflow tube, the flow-valve is opened at the time t_1. The time t_2 required for the water column to fall to height h_2 is then measured, from which

$$k = \frac{2.303 al \quad \log h_1/h_2}{A(t_2 - t_1)}$$

where A is the cross-sectional area of the soil sample, a is the cross-sectional area of the inflow pipe, l is the length of the flow path through the soil.

Laboratory Determination of the Permeability of Rocks

In describing the results of permeability tests made on the foundation rocks of Malpasset, Bernaix distinguishes between vacuolar rock, in which the interstices have a generally spherical

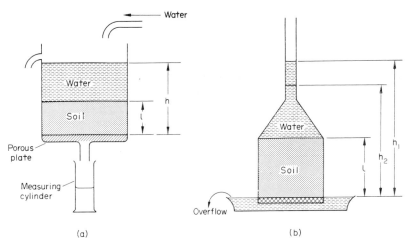

Fig. 6.2. Arrangements for laboratory tests of soil permeability. (a) constant head permeameter. (b) variable head permeameter.

shape, and "fractured" rocks in which the discontinuities take a more linear configuration. The permeability of rocks of the former type can be expected to change very little with variations in the ambient stress field, but rocks of the "fractured" type will not only have a marked anisotropic permeability but their permeability coefficient will also change markedly with change in the stress field. This is because of the variable resistance to flow presented by deformed cracks opening and closing in response to the changes in stress around the cracks.

Bernaix compared the radial percolation characteristics of the St. Vaast limestone – a vacuolar rock – with those demonstrated by the Malpasset gneiss – a "fractured" type rock. In the limestone the permeability remained constant over a wide range of percolation pressures and effective stresses, but in the gneiss there was a marked change in permeability with change in pressure. It was deduced that these changes were dependent on change of effective stress and not dependent on the hydraulic gradient.

The apparatus used in these permeability tests is shown in Fig. 6.3. The test specimens were cylindrical core-samples 60 mm diameter by 150 mm long, with a hole 12 mm diameter drilled for a length of 125 mm along the axis of the specimen, from one end. This hole was sealed at the end, over a length of 25 mm, and a tube was left through the seal to provide a passage for hydraulic fluid during the test. The test was made either by introducing the rock core-sample into a pressure-cell containing water, the central cavity being maintained at atmospheric pressure, or by injecting water at pressure into the central cavity. The two types of test enabled a study to be made of radial percolation inwards through the rock ("convergent") or outwards from the central cavity ("divergent").

Permeability tests on small samples of rock material may be conducted using the apparatus shown in Fig. 6.4. This is applicable both for liquid and gaseous fluids. The specimen, in the form of a disc 2.5 cm diameter and 1 cm long, is sealed between retaining washers. The fluid is then applied at 100 atm pressure, to one side of the disc, while a vacuum pump is applied to the other side. The volume of fluid passing through the sample is measured by the movement of a meniscus along a fine capillary tube, and the volume so collected over a given time

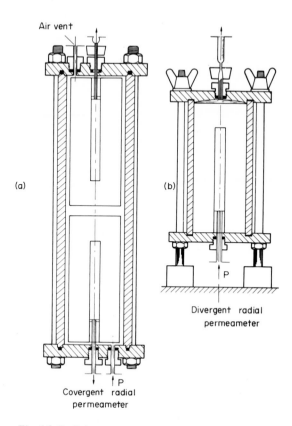

Air vent

(a)

(b)

Divergent radial
permeameter

Covergent radial
permeameter

Fig. 6.3. Radial permeameters for rock studies (Bernaix).

interval provides the necessary information from which the permeability of the rock material can be determined.

The Determination of Permeability Coefficients in the Field

Pumping-in test

Two methods may be employed to determine the coefficient of permeability of rock masses and soils in the field. The "pumping-in" test is used in drillholes that may penetrate deep into the bedrock. Its purpose is to determine the permeability of the strata encountered at various depths, as the hole is deepened. The method is to insert a perforated casing, about 2-3 m long, at the base of the hole, and then to pump water at a pressure of about 10 kg/cm² (9806 kN/m²) through the casing. In shallow drillholes, in soil, the applied pump pressure is adjusted to maintain a constant head of water above the casing in the hole, and the rate of flow is then measured.

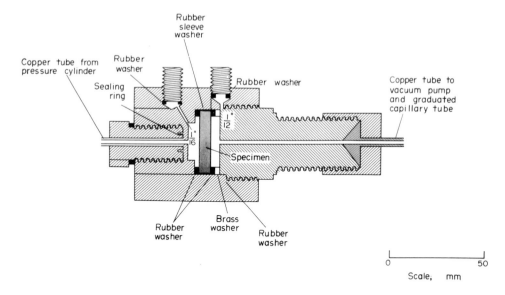

Fig. 6.4. Ohle-type permeameter (Chakrabarti and Taylor).

Lugeon test

In deep drillholes, in rock, such a procedure would, in all probability, be futile, because higher applied fluid pressures may be needed to overcome the fluid pressures and the *in situ* stresses in the rocks, before fluid can be injected. The "pumping-in" test may then become a *Lugeon test*, in which a plug seal is inserted about 5 m from the base of the hole and then water applied under pressure through a pipe extending through the plug to the base of the hole. The flow of water is measured, at various pressures. The permeability of the rock may then be assessed in terms of Lugeon units, one Lugeon being defined as a flow of 1 liter/min per m length of hole, at a standard applied pressure of 10 kg/cm^2. (1 Lugeon equals about 10^{-5} cm/sec, the precise equivalent being dependent upon the diameter of test borehole and the depth to the water table.)

However, if the *in situ* stresses are high, or the internal fluid pressures are high, the Lugeon test becomes useless as an index of the permeability coefficient, because the applied pressures that are required to force water into the interstices of the rock mass will open any fissures and interconnected discontinuities that may exist therein, and thus render the result meaningless. The Lugeon test is of value, nevertheless, as a comparative index of the progress and efficacy of sealing-off the rock, when injecting cement grout for the purpose of drainage control.

Pumping-out test

A "pumping-out" test will provide a more accurate assessment of the *in situ* permeability coefficient, but it is difficult to apply in narrow boreholes. Nor does it provide a means of determining the permeability at different depths. It gives a measure of the average permeability over the depth of a test well, to a maximum depth of about 45 m. In this test a central well, about 0.5 m diameter, is sunk to bedrock, or to an impermeable layer, and observation wells about 75 mm diameter are drilled at various radial distances from the central well. Water is then pumped from the central well, to depress the water table around the well, after which the rate of pumping is reduced to maintain the water levels in the central well and in the observation holes at constant values. From which, using the symbols depicted in Fig. 6.5, and assuming two-dimensional radial isotropic flow in a well that penetrates a confined aquifer,

$$k = \frac{Q}{2\pi b(h_2 - h_1)} \log_e r_2/r_1$$

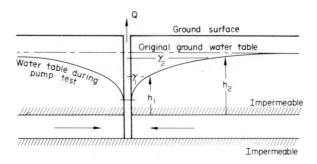

Fig. 6.5. The pumping-out test from a confined aquifer.

or, if the aquifer is unconfined, as pictured in Fig. 6.6,

$$k = \frac{Q}{\pi(h_2^2 - h_1^2)} \log_e r_2/r_1$$

where Q is the measured rate of pumping.

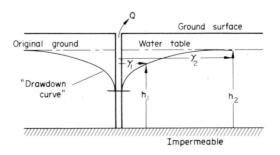

Fig. 6.6. The pumping-out test from an unconfined aquifer.

Borg describes on-site permeability measurements which were made by isolating sections of 2- and 3-in diameter drillholes, using packing, and then blowing air into the isolated portions. The required data were obtained by recording either the rate of pressure decay, or the rate of flow under steady-state conditions. Both methods were stated to give comparable results.

Fluid Potential

The resultant of the pressure head at a point in the strata and the head represented by the elevation of the point above a reference datum is termed the *fluid potential* or *piezometric head* at that point. The hydraulic gradient between two points is the ratio between the difference in fluid potential at, and the distance between, the two points concerned. The level to which water will rise in a borehole, or a well, is determined by the fluid potential. That level, in a number of wells or boreholes, defines the piezometric surface, and in most practical situations the slope of the piezometric surface defines the direction in which water moves through the strata.

Determination of fluid potential

The piezometric head may be determined by inserting a tube, perforated over the length for which information is required, into a borehole, sealing the borehole at a point above the perforations, and filling in the hole with clay. The level to which water rises in the observation tube is then noted. Such a procedure has the advantages of cheapness and simplicity but there may be practical problems in keeping the open-ended observation tube clear of debris, and there may be an appreciable time lag between the observed changes of water level in the tube and changes of fluid potential in the strata. More refined instruments are therefore commonly used, incorporating closed piezometric tubes with electrical or mechanical devices to monitor the fluid pressures in the strata at the bottom of the tube.

Flow Nets

If the permeability and the hydraulic gradient are known then the flow characteristics within a specified boundary can be estimated, assuming two-dimensional isotropic flow in homogeneous strata. This is done by means of a flow net. The flow net assumes that the fluid potential over the area can be represented by a series of contours of equal potential, or equipotential lines, such that the fluid potential between each pair of adjacent lines is constant. The equipotential lines control the direction of flow, in that the water moves in a direction across the equipotentials at right angles. Flow lines may thus be sketched in, to represent the movement of water.

The net may be constructed to represent plane horizontal flow, or plane flow over a vertical cross-section. The former could be used to study the drainage pattern towards a well, for example, or into an open-pit, in plan, while the latter would facilitate the study of flow characteristics across a dam, or underneath the foundations of an engineering structure. The characteristics of the flow net may be calculated mathematically, or determined with the aid of physical and mechanical models, or estimated on a graphical construction. However, before any of these methods can be attempted the boundary conditions must be specified. This is not always easy, especially if the rock strata in the foundations include joints, fissures, and fault discontinuities, the location and configuration of which may not be known in detail.

A comparatively simple example is commonly encountered in the form of an earth dam resting on an impervious base. This is the type of construction that is often employed to retain water, or coal slurry, or the tailings from a mine-treatment plant. The flow net for such a structure could take the form shown in Fig. 6.7. In constructing this flow net the boundary conditions are set by:

1. The upstream surface AB is a line of constant head, assuming that k is constant for the material in the dam.
2. If the base BC is impervious it must be a flow line.
3. The top flow line is at atmospheric pressure.

4. The downstream surface CD is a seepage surface. Water exudes and trickles down the dam surface over this part of the boundary.

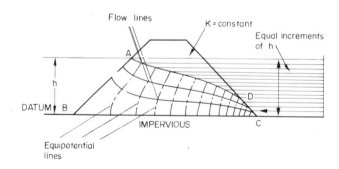

Fig. 6.7. Flow net through an earth dam.

Of these boundaries the curvature of the top flowline is unknown, and its delineation calls for some intuition, but certain rules can help. For example, (i) there must be an equal increment of height loss vertically between each point of intersection of the equipotentials with the top flow line. (Because there is no change in pressure head along this line. It is at atmospheric pressure.) (ii) The flow line must begin at right angles to the upstream face of the dam, if the dam is constructed homogeneously. (iii) The position of D is chosen by trial and error, repeating the construction until an acceptable solution is obtained. Since DC is a seepage surface, the inclination of the top flow line runs smoothly to follow the slope of the embankment at D.

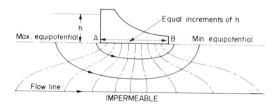

Fig. 6.8. Flow net under a gravity dam.

Another example of a simple flow net is shown in Fig. 6.8. representing the seepage of water underneath a gravity dam. In this case the hydraulic gradient is determined by the hydrostatic head due to water impounded by the dam, relative to atmospheric pressure on the downstream side. The selection of an appropriate "contour interval" for the equipotential lines may thus be made ($\triangle h$).

The boundary conditions here are:
1. The horizontal ground surface on the upstream side is an equipotential at high head, and the ground surface on the downstream side is an equipotential at low head.
2. The surface at bedrock, or the surface of the impermeable layer, is a flow line.

Having chosen the equipotential countour interval the base of the dam AB is subdivided into equal increments, representing this interval. The first flow line is then sketched in bearing in mind that the boundary conditions must be obeyed and allowing room for three or four flow paths between the base of the dam and bedrock. The first flow channel is completed by extending equipotential lines from the points marked along AB, and bearing in mind that the potential lines and the flow lines cross each other at right angles. The succeeding flow channels may then be constructed in a similar manner, to give the final flow net.

The net consists of a number of "squares" all having the ratio width/length = b/l.

In each of these "squares" the quantity of water flowing is $\triangle q$, and $\triangle q = k \triangle h. b/l$. Now, since b/l = unity, $\triangle q = k. \triangle h$ and if N_p is the number of increments of potential and N_s is the number of flow stream channels

$$\triangle h = h/N_p \quad \text{and} \quad \triangle q = q/N_s,$$

where h is the total fluid potential and q is the total quantity flowing, per unit width of the structure concerned.

Hence $$q/N_s = k.h/N_p$$

and $$q = k.h.N_s/N_p.$$

Seepage in Earth and Rock Masses

The permeability of rock masses is very largely determined by major discontinuities such as joints and fissures. The porosity of the rock material is relatively unimportant in a fissured rock mass. The coefficient of permeability, both of a rock mass and of a rock material, often varies widely with changes in the values of stress and fluid pressure. The flow characteristics are also likely to show marked anisotropy, due to the dominating effect of sedimentation and bedding planes, the orientation of microcracks and incipient failure surfaces in preferred directions, generated by the *in situ* stress system, the existence of joints, and the presence of faults and other discontinuities. Joints and fissures provide open channels for the flow of water. A fault zone often may act as a drainage surface or channel, or, on the other hand, by virtue of the gouge and clay infilling material that is compressed along the fault zone, may act as an impermeable barrier across which water will not flow.

In these circumstances idealized theoretical concepts of porosity, permeability, and flow are of limited value. Far more important becomes the need for a meticulous and comprehensive site investigation, to observe and record all the geological details, to measure the drainage, and to determine the strata and the fluid characteristics relating to the pattern of permeability and flow.

Seepage pressures

The pressures which cause the flow of water through a soil or a rock generate seepage forces. In flowing from point A at high fluid potential to point B at a lower potential there will be some loss of head due to friction, but the excess head remains as a seepage force and, in an isotropic flow pattern, the seepage force is assumed to act in the direction of flow. However, the seepage force is not dependent on the rate of flow. Seepage forces can exist in compact soils, such as clays of slight permeability and in rocks of low porosity. The effect of seepage pressures may be seen as an uplift pressure underneath the foundations or on the downstream side of a dam, or acting towards the exposed face of an excavation or cut, as a contributory factor towards causing the instability of a rock slope. The methods of protection against such effects of seepage pressures include the injection of cement grout to form a screen of reduced permeability underneath a dam, or in the walls of an excavation. This, besides reducing the direct flow of water, increases the length of the flow path round the screen and thus reduces

77

the hydraulic gradient. Drainage galleries and wells to form collecting points for water may be sited to control the direction of flow in the rocks at the site, or in the material of an earth or gravity dam.

Selected References for Further Reading

BERNAIX, J. New laboratory methods of studying the mechanical properties of rocks. *Int. J. Rock Mech. Min. Sci.*, vol. 6, pp. 43–90 (1969).

BORG, I. Y. Extent of pervasive fracturing around underground nuclear explosions. *Int. J. Rock Mech. Min. Sci.,* vol. 10, pp. 11–18 (1973).

CHAKRABARTY, A. K. and TAYLOR, R. K. The porosity and permeability of the Zawar dolomites. *Int. J. Rock Mech. Min. Sci.*, vol. 5, pp. 261–273 (1968).

DAVIS, D. H. Estimating the porosity of sedimentary rocks from bulk density. *J. Geol.*, vol. 62, pp. 102–107 (1954).

HOUGH, B. K. *Basic Soils Engineering* (Chapter 3, Soil moisture, pp. 54–82), Ronald Press, New York, 1957.

PACHER, F. The influence of fissure water on the stability of the rock abutments of arch dams. *Int. J. Rock Mech. Min. Sci.*, vol. 1, pp. 327–339 (1964).

RAMANA, Y. V. and VENKATANARAYANA, B. An air porosimeter for the porosity of rocks. *Int. J. Rock Mech. Min. Sci.*, vol. 8, pp. 29–53 (1971).

SERAFIM, J. L. Influence of interstitial water on the behaviour of rock masses. In *Rock Mechanics in Engineering Practice*, Stagg and Zienkiewicz (Eds.), pp. 55–97, Wiley, London, 1968.

SMITH, G. N. *Elements of Soil Mechanics* (Soil water, permeability and flow, pp. 20–72), Crosby London, 1968.

TODD, D. K. *Ground Water Hydrology*, Wiley, New York, 1959.

CHAPTER 7

The Laboratory Measurement of Load, Stress, and Strain in Earth Materials

When conducting tests on soils and rocks load may be applied in various ways. The application of deadweight loads, either direct or through lever systems, is simple and very convenient when the load must be maintained at a constant value, as in creep testing. When the rate of deformation of the specimen must be controlled, as for example, when determining the compressive strength, the testing machine may incorporate screw jacks or hydraulic jacks. Screw jacks are generally limited to tests of small magnitude. By far the commonest procedure today is to employ electro-hydraulic systems. These have wide variation in size, range, and flexibility, and can be applied to static or to variable loads. They may incorporate automatic systems to control rate of loading, or rate of strain, or coupled to provide a predetermined program, or they may incorporate feed-back systems and sensing devices so that the load-deformation characteristics of the specimen under examination will control either the rate of loading or the rate of strain during the test.

   In some experimental work the principle of "barodynamic loading" is used. This employs a centrifuge in which the specimen is held and revolved at variable speed. The specimen or

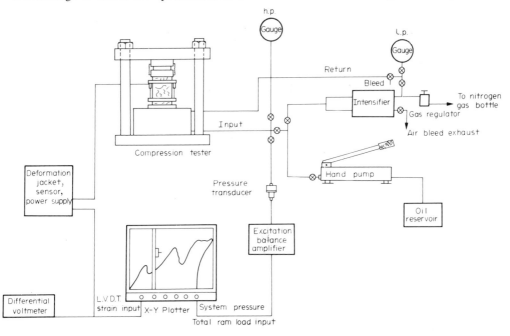

Fig. 7.1. Simple rock testing laboratory system (SBEL).

G—D

model is thus subjected to centrifugal forces which are distributed throughout its mass, just as gravitational forces are distributed throughout the mass of materials in nature.

A line diagram of a simple laboratory system for observing the stress-strain characteristics of rock specimens under load is shown in Fig. 7.1. The application of load by hand-pump is inexpensive and suitable for short-term tests under moderate loads, and for general classification work of a preliminary nature. More serious investigations, where uniformly controlled loading rates or strain rates must be maintained, require more elaborate provisions. Fluid piston pumps, electrically operated, can be programmed to controlled stress and strain. For long-term tests, air pumps, operating through itensifiers on to the oil-fluid system, offer a simple and efficient loading arrangement.

To contrast with the elementary system of Fig. 7.1. a modern automated test facility for rock mechanics investigations is shown in the block diagram (Fig. 7.2). In this system, transducers associated with the axial load **(L)**, confining pressure **(PC)**, and internal pressure **(PI)** systems, sense these controlled variables. Strain gauges **(SG)** or other strain transducers mounted on the model or specimen, sense its behavior under the applied loading conditions. Electrical signals from these transducers are connected into a signal conditioning unit, the outputs from which are monitored both by a digital data acquisition system, with digital printer output, and a bank of strip chart recorders. The investigator can thus monitor his experiment on his analog plotters while it is in progress, and also record his data on the digital print-out, for subsequent examination and analysis. Electrical links **(INTER)** between the loading control system and the digital data system make it possible for the loading system to control the operation of the data acquisition and recording system to suit the conditions of the experiment.

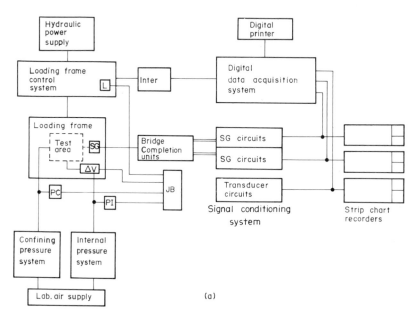

Fig. 7.2. An automated laboratory system for rock mechanics (Hardy *et al.*).
(a) Block diagram of overall testing facility.

Long-term loading, in creep tests, may also be provided with varying degrees of sophistication. A simple creep test arrangement is one which places the rock specimen between two platens, under the compressional force of calibrated springs (Fig. 7.3(a)). Another arrangement, this time employing deadweight loading through a lever system, is shown in Fig. 7.3(b). The arrangement can serve to place the rock specimen under a uniaxial tensile pull, which can be converted to a compression test by the load reverser anchored to a spring-and-jack assembly. This assembly takes up any elastic stretch in the framework and tension members and also applies the load smoothly to the rock specimen.

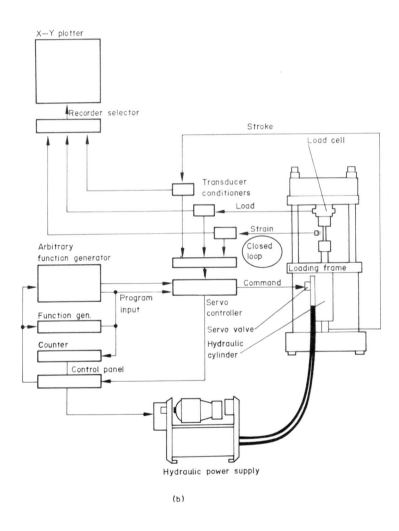

(b)

Fig. 7.2. An automated laboratory system for rock mechanics (Hardy *et al.*).
(b) Block diagram of axial loading system

In the system layout of Fig. 7.1, a sustained constant load for creep tests is obtained by means of a compressed gas bottle supply, coupled by a gas regulator and an intensifier unit to the oil-fluid system.

Fig. 7.3.(a) Spring-loaded compression creep testing machine.

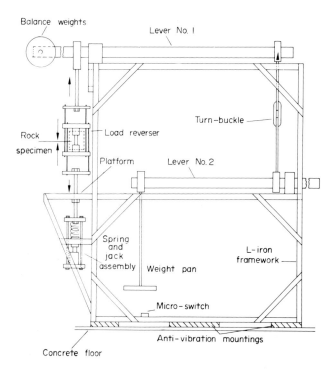

Fig. 7.3.(b) Creep test machine for rock specimens in uniaxial loading. (Murrel and Misra).

The Measurement of Load

Pressure Gauges

Various alternative devices may be used to measure the applied load. The fluid pressure in a hydraulic system is usually observed on a Bourdon type pressure gauge. The load applied by such a system, being determined by the fluid pressure and the ram area of cross-section, can be indicated directly on the face of the pressure gauge. A feature of Bourdon tubes is their tendency to creep under pressure, so that they drift from their original calibration, and their calibration characteristic is also likely to be adversely affected if the gauge is subjected to any sudden change of pressure. Where precision work is to be undertaken therefore it is necessary, in the ordinary way, to remove them from the line for recalibration at frequent intervals.

Optical gauges can remove the uncertainty associated with the Bourdon tube gauges, each one serving as an in-place check on the stability of the Bourdon tube. The optical pressure gauge is a photoelastic transducer in the form of a glass cylinder loaded by the fluid pressure through a piston on the hydraulic line. This is an extremely rigid gauge, with excellent long-term stability. A 25 mm diameter glass gauge registers the pressure in the fluid line to within 21 kN/M^2 at a working pressure of 10.34 MN/m^2.

Load Cells

Load cells are essentially structural components incorporated into the loading system. The cell deforms under the load and that deformation is measured, either directly or indirectly, and calibrated in terms of the applied load. The variation in design of load cells is considerable and new varieties, designed for specific purposes, are continually being made. Some typical examples include the fluid load cell, which consists of a hollow steel cylinder filled with mercury and loaded in the axial direction. The load is measured by virtue of the change in volume of the cell and calibrated in terms of the mercury level in a vertical capillary tube tapped into the cell. A micrometer screw also tapped into the cell serves as a means of adjusting the mercury to the index mark on the capillary tube when the cell is adjusted to zero under no load. Such load cells serve mainly as standardizing boxes, to check the calibration characteristics of laboratory testing machines. They are made to cover a range of loads from about 15t to 300t in tension and 1000t in compression.

Another hydraulic load-measurement capsule consists essentially of a hydraulic ram upon which the load is converted into hydraulic pressure, measured on a Bourdon-tube gauge. The ram operates in the cylinder which forms the body of the capsule and the design incorporates

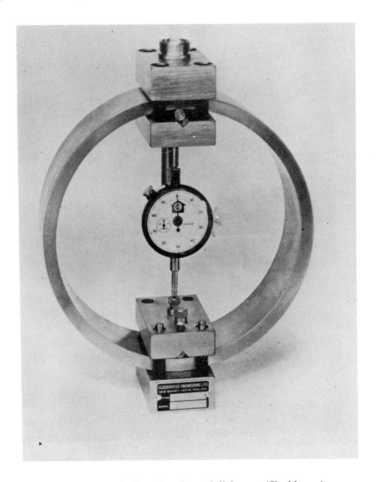

Fig. 7.4.(a) Standard proving ring and dial gauge (Clockhouse).

Fig. 7.4.(b) Proving ring with linear potentiometer (SBEL).

special features which permit the ram to function with virtually no friction over a limited range of movement, and at the same time ensure that no leakage of fluid takes place. The pressure gauge may be fitted at the end of a length of pressure-tubing, to permit readings at a remote distance from the capsule. These cells are made to measure compression loads ranging up to 1000t.

Fluctuating loads over a wide range of extreme values can be measured by load cells which incorporate piezo-electric crystals, that is, crystals whose electrical properties are pressure-sensitive. The load cell is so designed that the pressure which is to be measured is applied to the crystal, and the resulting electrical signals are monitored on a control circuit that is calibrated in terms of pressure.

Loads external to a fluid system are generally measured, if not by dead-weight, then by means of some form of elastic deformation device. Over the range up to about 2t, a helical spring can be calibrated to indicate load by measurement of its linear change in response to variation of load.

Proving Rings

In the laboratory, loads from a few kilograms up to 200 or 250t compression or about 100t tension may be measured by means of proving rings. A proving ring is a precision measuring instrument. It is generally used to check the calibration of laboratory loading arrangements, the ring being inserted on the testing machine in place of the material specimen. Sometimes a proving ring may be permanently installed to function as a dynamometer on the testing machine and placed in line with the axial load applied to the specimen.

The ring is machined from high-quality steel with integral diametrically opposed bosses, or clamped-on loading blocks, which take the imposed load. The resulting deformation of the ring under load is measured between these two bosses or blocks, as shown in Fig. 7.4(a) and 7.4(b).

The Determination of Stress

The stress on or in a specimen under load may be deduced from measurements of the applied forces and the dimensions of the specimen. Where the load is applied by fluid pressure then that pressure may be equated to the stress generated by the applied force, i.e. force per unit area. In any case, the stress within a solid specimen is a hypothetical concept, not a real one. It cannot be measured directly but must be assumed to exist as a deduced entity, by reason of the magnitudes of forces applied to or inherent within the material, and the dimensions and physical properties of the material over which those forces are distributed. For example, if we assume that a rock cylinder, in compression under uniaxial loading, behaves perfectly elastically, then the distribution of principal stresses over a longitudinal section of the cylinder would be described by the contour diagram shown in Fig. 7.5.

The assumption of perfect elasticity implies that there is a constant stress-strain relationship, which may be described in terms of linear dimension by Young's modulus, in volumetric space by the bulk modulus and Poisson's ratio, and in terms of shear deformation by the modulus of rigidity. On the basis of these assumptions it is possible to deduce the stress distribution of Fig. 7.5 from measurements of the applied axial load and the observed linear strains over the surface of the material.

There are other stress-dependent properties that may be used to deduce the state of stress in a material under load, if the characteristics of that dependence are known; for example, the velocity of sound in the material. Another property that may be used is the magnetic permeability of the material. By far the most common procedure, however, is to measure strain.

The Measurement of Strain

Strain is a non-dimensional ratio, which, in linear terms is expressed as

$$\frac{\text{extension (or compression)}}{\text{original length}}.$$

The unit is the microstrain, such that

$$\text{strain} \ (\epsilon) = \epsilon \times 10^{-6} \text{ microstrains } (\mu\epsilon).$$

Instruments used to measure strain include extensometers and strain gauges operating on mechanical, electrical, and optical principles.

Mechanical Strain Gauges and Extensometers

Dial gauges

Dial gauges are general-purpose instruments, often incorporated into other devices, such as proving rings, extensometers, and dimension indicators. The dial gauges illustrated in Fig. 7.4(a) consist essentially of a spring-loaded extensible arm, coupled to a rotary pointer by a rack and pinion device. The gauge is fixed so that the arm abuts against the surface to be observed, or spans the gauge points across which a measurement is desired. The sensitivity limit

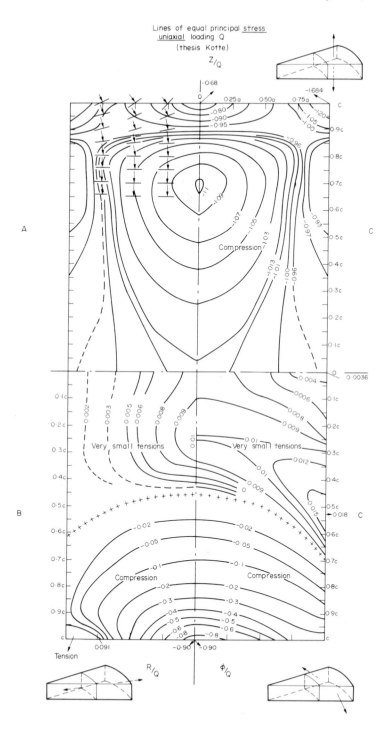

Fig. 7.5. Stress distribution within a rock cylinder under uniaxial loading (Kotte *et al.*).

depends upon the range of the indicator.

When dial gauges are affixed as component parts of other equipment, such as proving rings and extensometers, they should be regarded as an integral part of that equipment. If the gauge is removed then care must be taken to ensure that the no-load reading after replacement corresponds with that which existed at the initial calibration. Otherwise a recalibration will be necessary.

Mechanical extensometers

Various alternative designs of mechanical extensometer are available. In general they may be classified as either short gauge length (12 to 25 mm) instruments with high magnification, or longer gauge length (up to 150 or 200 mm) instruments with lower magnification. A typical representative of the former class is the Huggenberger extensometer, in which a caliper spanning the gauge points has its movement increased by a factor of up to 2000 by a spring-loaded lever-and-pointer system. The longer gauge length extensometers may embody micrometers and dial gauges attached to a measuring beam (Fig. 7.6). Another representative of this type of gauge is the Demec gauge, designed for use on concrete, but which has similar applicability to rocks, both in the laboratory and in the field. To use this instrument, small, conically ground, stainless steel pads form the gauge points and these are cemented on to the specimen or structure under

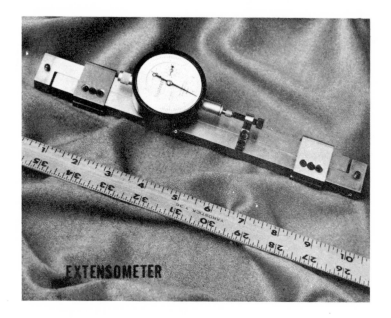

Fig. 7.6. Multi-purpose mechanical strain gauge (SBEL).

observation. The distance between the gauge points is initially established by reference to a calibration and setting-out bar. The measurement beam spans the gauge points over a length of 50, 100, or 150 mm, one end being fixed and the other pivoting on a knife-edge with its movement transferred to a dial gauge. The instrument may be read to ±3 microstrains.

Electrical strain gauges

Various electrical devices may be used to measure strain. They include instruments which measure changes in capacity, inductance, or resistance, while one very useful type in geotechnology, the vibrating-wire strain gauge, embodies both mechanical and electrical principles.

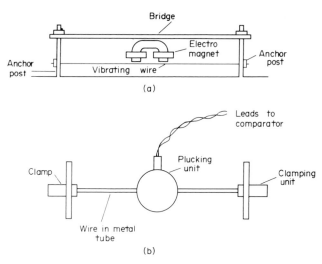

Fig. 7.7(a) The vibrating-wire strain gauge. (b) Vibrating-wire gauge for embedding into concrete. Gauge cast in concrete block in laboratory before embedding into concrete on site.

The vibrating-wire gauge, illustrated in Fig. 7.7, functions on the principle that a vibrating wire, stretched between two anchor posts forming the gauge points, has a characteristic frequency of vibration. If the tension of the wire changes, due to strain between the gauge points, then its frequency of vibration alters proportionately. Measurement of the frequency therefore serves to indicate the strain.

The frequency of vibration of the stretched wire is described by

$$f = 1/2L\sqrt{T/\rho}$$

where f = frequency,
L = gauge length,
T = tension in the wire,
ρ = density of wire material,

from which

$$\text{Strain} = \frac{\delta L}{L} = \frac{4L^2\rho.f^2}{AE}$$

where A is area of cross-section of the wire, and E is the Young's modulus of the wire material.

When in use the gauge is mounted on anchor posts attached to the surface of the specimen or, in the case of concrete, embedded within the material. The wire is set in vibration by operating a "plucking" circuit which is energized when a pulse of electrical current is passed through the coils of an electromagnet placed in proximity to the wire. The vibration of the wire sets up a variable voltage in the coil, and this is monitored on an oscilloscope screen, where it may be viewed simultaneously with another signal emitted from a reference wire. The vibration of the reference signal is measured and controlled by a variable oscillator, the frequency of which at any time is known. When the two signals are matched then the frequency of the strain gauge is the same as that which is indicated by the variable oscillator. The sensitivity of such a device is determined by the electrical control circuit. Gauge lengths of from 25 mm to 350 mm are commonly used. For a 100 mm gauge length a sensitivity of 1 microstrain over a range of measurement of 0.1 mm can be attained.

Another type of electrical strain gauge is the capacitance gauge. This consists of an electrical condenser, the plates of which are attached to the gauge points. The electrical capacity of the condenser alters when the distance between the gauge points changes. The instruments are very useful when the deformation of very weak or soft materials is to be observed, since the gauge places no restraint upon the specimen. They are also very useful for observing vibrating materials, or for use as "proximity" gauges, since no mechanical contact between the gauge points is required.

Inductance gauges may also be used as proximity gauges, when one of the gauge points is attached to an electromagnet and the other to a soft-iron armature conductor. The air gap

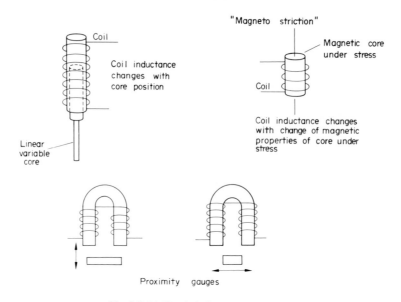

Fig. 7.8 (a) Simple inductance gauge circuit.

between the magnet and the armature alters the electrical characteristics of the control circuit in which either the variations in voltage or current may be measured. Another form of inductance gauge employs the transformer principle, in which the linear displacement of the core alters the output voltage of the transformer. In this device the core is placed in the primary coil of the transformer, whose output is gained from two secondary coils wound in opposite directions. Alternating current is fed into the primary, and when the core is at central position the resultant output voltage from the two secondaries is zero. Any subsequent displacement of the core, either to one side or to the other, will produce an output signal to + or − of the central, null, position, depending upon the direction in which the core moves (see Fig. 7.9).

The linear variable differential transformer (LVDT) output voltage may be displayed on a voltmeter, chart recorder, or oscilloscope. Since it is dependent on the voltage in the primary circuit the input voltage must be carefully controlled and maintained constant. The output voltage is calibrated as a function of linear motion by fixing it in a clamp, with the

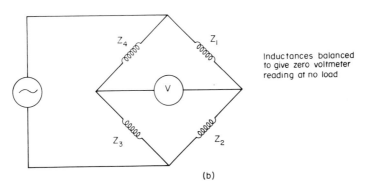

Inductances balanced to give zero voltmeter reading at no load

(b)

Fig. 7.8(b) Inductance gauge measuring bridge circuit.

end of the core-rod abutting against a precision micrometer head. Gauge lengths ranging from 12½ to 150 mm are common, and a sensitivity of ± microstrains, over a measurement range of from 1.0 to 75 mm, may be attained.

Electrical resistance strain gauges

These gauges operate on the principle that the deformation of an electrical conductor under load produces a change in the electrical resistance of the conductor proportional in magnitude to the imposed strain. If the conductor is anchored at its ends to a structure under load, then it forms an unbonded strain gauge, and may be used in a fashion similar to the vibrating-wire strain gauge. More frequently, however, the principle of resistance change related to strain is applied by means of "bonded" strain gauges which are cemented to the surface of the specimen or structure under examination. The gauge may consist of a length of wire, about 0.0254 mm diameter, wound in the form of a compact grid, and supported on a paper backing. Alternatively it may be stamped from metallic foil, supported on an epoxy resin or a bakelite base. More recently, semiconductor gauges, consisting of thin crystals of germanium or silicon, have been introduced.

The alteration of electrical resistance in a wire or foil gauge under load is described by the relationship

$$\text{Strain} = 1/K \cdot \frac{\text{Change of resistance}}{\text{Original resistance}}$$

where K is the "gauge factor".

For many of the gauges in general use the gauge factor is about 2.0.

The gauges may be mounted singly, aligned in the direction of stress for uniaxial measurements, or in various combinations of two aligned at 90°, or three aligned either at 45°; 60°, or at 120° to one another, for the determination of the magnitude and direction of the principal strains in a biaxial stress field. Such gauge combinations are known as strain gauge "rosettes".

The commonest measuring equipment that is used in conjunction with resistance strain gauges employs the Wheatstone bridge principle, in which the "active" gauge R_1, a "dummy" gauge R_2, and two known resistances R_3 and R_4 form the resistance components of the circuit

91

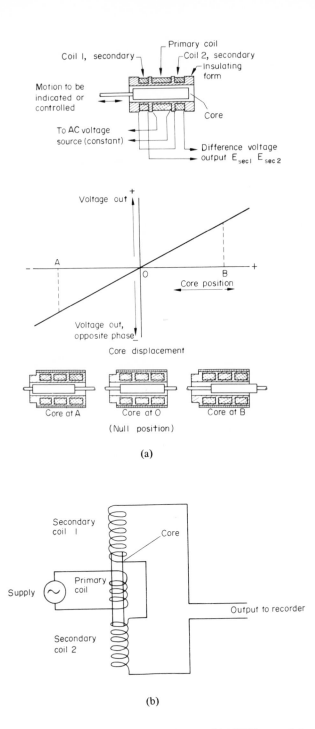

Fig. 7.9(a) Linear variable differential transformer. (b) LVDT control circuit.

(Fig. 7.10). When the bridge is balanced, so that no current flows in the galvanometer

$$\frac{R_1}{R_4} = \frac{R_2}{R_3} \text{ and } R_1 = \frac{R_2}{R_3} . R_4$$

If R_4 is a calibrated variable resistor, the bridge may be balanced with the active gauge under no-load and again under load conditions, to yield the change in resistance δR. Alternatively the bridge is balanced at no-load, and the out-of-balance current that is subsequently registered on the galvanometer, when the active gauge is strained, may be used as a measure of the change in resistance.

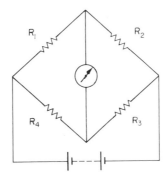

Fig. 7.10. Simple Wheatstone bridge circuit for bonded electrical resistance strain gauge operation.

The use of the "dummy" gauge opposite the active gauge in the bridge circuit eliminates the effect of change in resistance due to temperature variation on the active gauge, provided that both the active and the dummy gauges are exposed to the same temperatures at any one time, and that the dummy gauge is maintained without strain. Electrical resistance strain gauge techniques and circuitry are widely applied to an extremely varied range of problems in stress analysis and structural design generally. In geotechnology they find many useful applications in the laboratory, but they are often found to be less satisfactory for field applications.

The major problems in this regard lie in ensuring that a good bond exists between the strain gauge and the rock, and maintaining perfect insulation against the effects of atmospheric humidity and rock moisture. When they are attached directly to rock materials moisture is liable to penetrate to the underside of the gauge, via the rock, no matter how effective the external bonding, insulation, and protection of the gauge and its electrical contacts may be. In the laboratory it is possible to dry out the rock specimen before attaching the gauge, and thereafter it is wise to complete all the desired measurements in as short a time as possible. Long-term measurements are likely to be affected by deterioration in insulation properties, deterioration in mechanical and electrical contact conditions, and possible creep of the cement bond. However, improvements in strain gauge technique are constantly being made, and improved materials being found, as a result of widespread research and development. The reputable strain gauge manufacturers will supply detailed instructions as to optimum operating procedures when using their products.

One of the most successful applications of electrical resistance strain gauges is in load cells and dynamometers. Many designs are available, or may be constructed. A compression load cell may be fabricated from a solid steel cylinder, to the external surface of which the strain gauges are bonded, some axial and others in a circumferential alignment. Alternatively the body of the load cell may be a hollow steel cylinder, with the strain gauges bonded to the inner cylindrical surface. Deformation of the steel body, under load, is measured in terms of strain in the gauges, calibrated in terms of load. The elements of such a load cell are shown in Fig. 7.11,

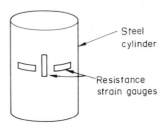

Fig.7.11. Mode of construction of a simple strain gauge load cell for measuring compressional loads.

and a typical control circuit in Fig. 7.12. In the arrangement shown, compensation for temperature change is achieved by placing the main and compensating gauges on opposite arms of the bridge, so that the net resistance measurement is a function of the load applied to the cell.

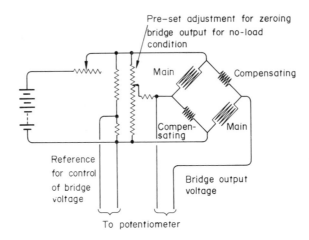

Fig. 7.12. Control circuit for a strain gauge load cell.

A fluid pressure cell embodying electrical resistance strain gauges may be constructed on the principle shown in Fig. 7.13, in which the differential pressure on either side of a flexible diaphragm is monitored by means of strain gauges bonded to the diaphragm.

Measurement of Strains on Rock Samples During Triaxial and Creep Tests

The difficulties that are sometimes associated with the application of bonded electrical resistance strain gauges to rock materials are very apparent in triaxial testing, and during long-term creep tests on rock. For routine triaxial tests the axial strain on the specimen may be measured outside the cell, being equated to the vertical movement of the loading piston. Any

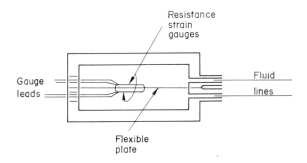

Fig. 7.13. Mode of construction of a differential fluid pressure gauge employing electrical resistance strain gauges.

lateral deformation of the rock specimen is usually measured after the test, when the specimen has been removed from the pressure-cell. A detailed study of the load-deformation characteristics of the material during the test requires that the strains on the specimen be measured from within the cell, and monitored from outside.

Monitoring by electrical means presents few problems, but the use of resistance strain gauges in the cell at high pressure is clouded by the uncertainties of insulation and the effects of the cell pressure on the cement bond. Not only is the cementing material deformed, to produce extraneous readings of apparent strain on the gauges, but the cement is also forced into the pores of the specimen, and thus affects the material properties of the rock in the vicinity of the gauges.

The problem may be solved by using mechanical caliper gauges clamped over the specimen, the clamping contact pressure being controlled by light springs. In the arrangement shown in Fig. 7.14 two such calipers are used to measure the axial compression, while a third measures diametral expansion. These measurements are made by means of linear differential transformer gauges, one mounted across the central, diametral, caliper, and the other spanning the two calipers between which the axial strain is measured.

LVDT gauges are also useful in creep testing, but an alternative here is to use optical gauges, such as the Tuckerman gauge or a Marten's type extensometer. These instruments consist of combined mechanical and optical lever systems which operate between knife-edge contacts on the specimen, or between extension arms attached to the upper and lower platens of the testing machine. The Tuckerman gauge consists of two units, the extensometer and the auto-collimator. Sometimes two extensometers are attached, one on either side of the specimen, and readings are taken from each of these in turn, using the autocollimator, which is a precision telescope. These readings are averaged to yield strain.

Photoelastic Strain Measurement

Some materials exhibit the property of double refraction, and are said to become birefringent when strained. When polarized light is passed through such materials, and observed by eye, an optical pattern is seen which completely identifies the state of strain, both in magnitude and direction, in the material. If the incremental values of strain are small enough to allow elastic analysis to be applied to the material, then those strains may be interpreted in terms of stress. The technique of photoelastic strain, and hence stress, analysis makes use of this optical

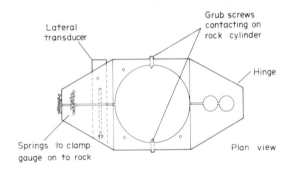

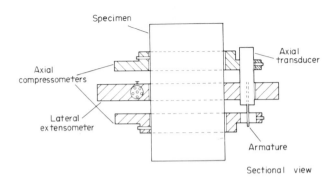

Fig. 7.14. Lateral and axial extensometers for triaxial tests on cylindrical rock specimens.

effect, in various ways.

The classical approach to photoelastic stress analysis employs plate models, fabricated out of sheets of birefringent material. These are placed in a loading frame on an optical photo-elastic bench, which enables the strain patterns to be observed in two dimensions, over the sheet model. The technique can be extended, to study conditions in three dimensions, if a complete scale model of the subject is fabricated out of stress-birefringent material and this placed under load so that it deforms in a relatively high-temperature environment. The ambient temperature around the solid model is then reduced, so that the model "freezes" while still under load. This locks the deformation permanently into the model at the lower temperature. The model is then cut into into slices which are then viewed separately on the two-dimensional photoelastic bench. The three-dimensional strain pattern and stress distribution are then built-up, to extend throughout the model, by reassembling the strain patterns viewed over the individual slices.

In more recent years the photoelastic coating technique has evolved. This makes use of layers of stress-birefringent materials that are sprayed, molded to shape, or attached as flat sheets cemented over the whole of one side, to an opaque solid model, or to the actual structure under examination. The cemented side of the birefringent material incorporates a reflective medium. Assuming that the bond is strong enough to transmit the deformation of the structure or model to the birefringent coating, then, when it is observed under reflected polarized light, the coating displays an optical pattern which identifies the strain distribution over the surface to which it is attached. Another technique is to incorporate birefringent materials into optical transducers to measure stress, strain, and load. These optical transducers make use of photo-elastic strain gauges.

Basic Principles of Photoelasticity

Light is a form of radiant energy which is visible to the human eye. A light source such as a lamp normally radiates its energy in all directions. If the light is controlled by optical means or devices that only allow the passage of radiations in specific directions, then it is said to be polarized. A beam of plane polarized light can be produced by inserting a polarizer, such as a "Polaroid" filter in the path of an ordinary light beam.

Plane polariscope

When a beam of polarized light passes through a stressed birefrigent material it is split into two component beams. These two components emerge from the birefringent material polarized in the planes of the principal stresses in that material. One of the beams is slowed down relative to the other by a time difference, or relative retardation, which is proportional to the difference in magnitude between the principal stresses at the point of observation (and hence proportional to the shear stress) and also proportional to the length of the light path through the material.

If the two emergent polarized beams are then viewed through a second plane polarizing filter, they are further controlled and resolved in the specific direction of that filter, which is known as the analyzer. The two resolved components emerge in such a way that an optical pattern is seen by the observer. If the incident light is white, then the emergent light is split into various colors by the analyzer, each color representing a specific value of relative retardation, and hence a specific value of the principal stress difference, in the birefringent material. Each color is termed an "isochromatic".

At those points in the birefringent material where one of the principal stress directions coincides with the controlling directions of the polarizer, light passes through the material without hindrance. But if the axis of polarization in the analyzer is fixed at right angles to that of the polarizer all this light will be cut off at the analyzer, and this point of the bire-fringent material will appear black. To an observer looking through the analyzer all points where the principal stresses are similarly aligned will coalesce, to form dark zones called "iso-clinics" in the field of view.

Therefore, when the polarizer and the analyzer are crossed at right angles to one another, and the birefringent material is stressed, the optical pattern seen by the observer consists of isochromatics, which are related to the magnitudes of stress in the material, and isoclinics which indicate where the directions of principal stress are parallel with the axes of the polari-scope. By rotating the polariscope about the longitudinal axis of the optical bench, with the polarizer and the analyzer in the locked "crossed" position, the isoclinics are seen to sweep over the field of view, so that the stress directions can be determined at all points over the birefringent material (see Fig. 7.15).

Crossed-circular polariscope

For some applications of photoelastic stress analysis the isoclinics are not required, either because the stress directions are rendered self-evident in the isochromatic pattern, or because they are predetermined by the conditions of loading and the shape of the birefringent material.

In such circumstances the isoclinics are a hindrance to the observer, whose interest is only in the isochromatics, and it is more convenient to remove the isoclinics entirely from the field of view. This can be done by using circularly polarized light in the polariscope.

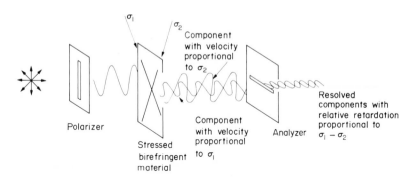

Fig. 7.15. Linear plane polariscope.

Circularly polarized light is obtained by inserting two quarter-wave plates in the polariscope, one just behind the polarizer and the other just in advance of the analyzer, relative to the incident light beam. Each quarter-wave plate is a filter that is permanently birefringent, having a relative retardation that is one-quarter of some specific wavelength. It follows that if white light is being used in the polariscope, only that portion of it which corresponds with the wavelength of the quarter-wave plates will be precisely controlled. All other wavelengths will be affected to a degree depending upon their position in the visible spectrum. On the other hand, if monochromatic light is employed on the polariscope, such as may be obtained from a sodium lamp, or from white light if a monochromatic filter is inserted in the incident light path, all the light passing through the birefringent material on the polariscope will be circularly polarized by the insertion of quarter-wave plates of the appropriate wavelength (Fig. 7.16).

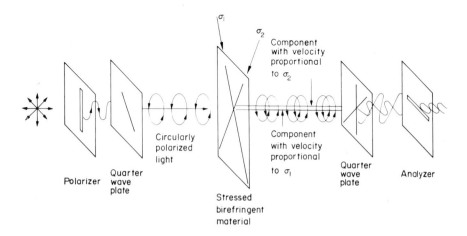

Fig. 7.16. Crossed circular polariscope.

In circularly polarized light the radiations may be imagined to emanate with a helical motion, having a specific velocity in the direction of propagation, that is, along the longitudinal axis of the helix, and at the same time vibrating, or pulsing, radially. The light vector is radial about the longitudinal axis and revolves around that axis, either clockwise (positive) or anti-clockwise (negative), depending upon the direction of alignment of the optical transmission axes of the quarter-wave plates relative to that in the linear polarizer and analyzer.

In Fig. 7.16 the optical transmission axis of the first quarter-wave plate, behind the polarizer, lags $45°$ behind that of the polarizer, while the transmission axis of the second quarter-wave plate, in front of the analyzer, is $45°$ in advance. This places the resultant axes of circular polarization in the analyzer and the polarizer at $90°$ apart, that is, in the circular "crossed" position. Therefore when there is no stress in the birefringent material all light is cut off at the analyzer and the viewed field is dark. But when the birefringent material is stressed an isochromatic field is seen, as was the case with the crossed plane polariscope, in which the relative retardation at any point in the field of view is a measure of the shear stress at that point. For a given stress distribution in the material viewed, the signals displayed by the crossed circular polariscopes will be identical so far as the isochromatics are concerned. The two signals differ, however, in that, with the dark-field crossed-circular polariscope the only completely dark zones are those where no birefringence occurs in the material. All other points will combine to display an isochromatics pattern, which will remain the same no matter at what angular position the crossed polariscope is fixed, relative to the axis of propagation.

The conditions may be pictured if it is imagined that the two light components which determine the relative retardation at any point are lagging in time, one behind the other, both vibrating radially about the axis of propagation, and rotating about that axis, so that the tip of each component describes a helical screw locus in space. No matter at what angular position the crossed polariscope is placed, the receiving end of the system will always receive the same signal from any one point in the field of view.

Viewing Equipment for Photoelastic Transducers

Photoelastic transducers may be used with the circular polariscope. The transducers consist of birefringent materials, in simple geometrical shapes such as strips, discs, and cylinders, which may be either solid or annular. The materials may be either reflectorized or transparent plastics, and glass. Plastics are easy to fabricate and use, but when maximum long-term stability is required the transducers are made of glass.

The equipment that is used for viewing photoelastic transducers varies in character from extremes of simplicity to extremes of sophistication. The arrangements shown in Fig. 7.17 were designed and standardized to render the mode of operation as simple as possible, consistent with the achievement of a degree of precision in measurement that would be likely to produce useful information from an academic viewpoint and at the same time be reasonable and adequate for practical applications in geotechnology.

Fig. 7.17(a) shows the standard filter orientations, relative to the loading direction, for transmitted light viewing, while figs. 7.17(b) and (c) show the standard orientations for reflected light viewing. In the latter two cases the transducer is given a reflective backing. In all the arrangements depicted in Fig. 7.17 the relative orientation of the linear analyzer and its adjacent quarter-wave plate remains the same. Thus, by producing an incident light beam of the correct rotation at the polarizing end of the system, the same analyzer will serve for viewing both transmitted and reflected light transducers.

The effect of the quarter-wave plate at the analyzer is to turn the plane of vibration of the

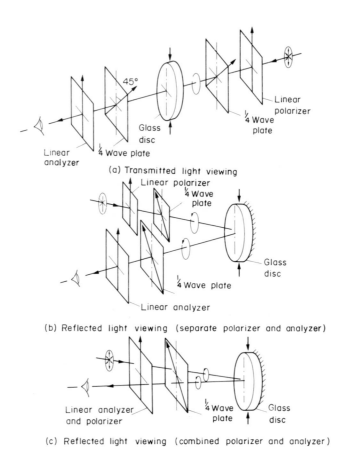

(a) Transmitted light viewing

(b) Reflected light viewing (separate polarizer and analyzer)

(c) Reflected light viewing (combined polarizer and analyzer)

Fig.7.17. Viewing arrangements used with photoelastic transducers (Sheffield system).

beam 90° from its incident direction at the polarizer. The arrangement thus gives a dark field with zero stress in the transducer.

Simple viewer

The simplest form of reflection polariscope consists of a sheet of plane polarizing filter, with a quarter-wave filter attached. These are aligned as shown in Fig. 7.17(c), and the arrangement may be protected by a thin sheet glass, and held in a frame with a handle attached (Fig. 7.18). The same filter combination serves as both polarizer and analyzer when the viewer is held against the transducer with the quarter-wave plate on the side away from the observer, that is, between the plane polarizing filter and the transducer. When in use, the handle of the viewer is aligned with the direction of loading on uniaxial transducers, or with the major axis of symmetry of the signal displayed on biaxial transducers. Sometimes it is convenient to fix the simple viewer system permanently in place, as a covering window on the transducer. The transducer signal is then self-displayed at all times when light falls on it.

Fig. 7.18. Simple viewer for observing photoelastic transducers.

Precision viewers

Precision measurements demand more elaborate viewing arrangements. In the Sheffield system these are provided by making provision for rotating the plane polarizing filter over an angle of 180°, relative to the quarter-wave plate, in the analyzer. This rotation is measured on a semicircular scale, graduated into 100 units. Figure 7.19(a) shows the precision viewer that is used with transmitted light transducers and Fig. 7.19(b) shows that used with reflected light transducers. With transmitted light transducers the polarizer is either carried as a separate unit or else permanently fixed on to the light-incident side of the transducer. The viewer then consists of the analyzer only. The viewer for reflected light transducers incorporates the polarizer and the analyzer, mounted one on each side of the central handle. The polarizer, on the observer's right, carries an attachment into which an electric lamp is inserted. For underground work the light may very conveniently be provided by the headpiece of a miner's cap lamp. The polarizer, together with its light-cup, is hinged so that the beam of light can be directed on to the transducer, which is viewed through the analyzer on the left-hand side of the instrument. The viewer is portable and is hand-held for field use and some laboratory purposes, or it may be mounted on a stand for laboratory bench work.

Photoelastic Transducers

While the detailed and precise interpretation of the birefringence displayed by the materials and models used in conventional photoelastic stress analysis calls for considerable expertise, which can only be gained by study and practice, yet it is possible to employ the basic principles of photoelasticity to bring out useful and accurate information by following very simple rules and operating procedures with equipment that is now commercially available, or which may be constructed with little difficulty. One does not have to be an expert in photoelasticity in order to use photoelastic transducers. At the same time, one should not allow oneself to be deluded,

Fig. 7.19(a) Precision viewers for transmitted light transducers,

by the apparent simplicity of the techniques, into thinking that the problems which are being studied by their aid are equally simple to understand and to solve. It is one thing to make measurements, but another thing entirely to assess their value and meaning in relation to practical problems in geotechnology, either in the laboratory or in the field.

Linear Photoelastic Strain Gauges

A linear strain gauge operating on photoelastic principles was designed and introduced by G. U. Oppel in 1956. It consists of a thin strip of stress-birefringent plastic into which a number of parallel "interference fringes" have been frozen by causing the plastic to polymerize in a condition of strain. To use the gauge it is cemented at its ends to the surface on which strain is to be observed. Any subsequent strain is then displayed as lateral displace-

Fig. 7.19(b) Precision viewer for reflected light transducers.

ment of the fringe pattern, along a scale mounted on the gauge. The gauges are available in a range of lengths and sensitivities, the maximum sensitivity ranging from 15 to 50 micro-strains.

Hawkes also designed a uniaxial photoelastic gauge, which was described in 1968. It consists of a strip of stress-birefringent material with a reflective backing (A, Fig. 7.20). This

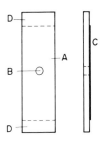

Fig. 7.20. Hawkes' linear photoelastic strain gauge.

strip is drilled with a central hole, the function of which is to produce a stress-concentration around the periphery of the hole, when the gauge is under strain. This increases the fringe orders displayed by the gauge for a given value of linear strain, and thus increases the sensitivity of the gauge. The gauge is then cemented at its ends to the surface under examination, and observed using the precision reflection hand-viewer. The optical patterns displayed by the gauge under strain are shown in Fig. 7.21. The sensitivity of the gauge is directly proportional to the thickness and the "stress-optic constant" of the plastic. With present-day

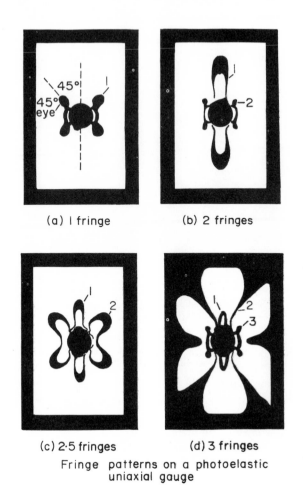

(a) I fringe (b) 2 fringes

(c) 2·5 fringes (d) 3 fringes

Fringe patterns on a photoelastic uniaxial gauge

Fig. 7.21. Optical signals displayed by Hawkes' linear gauge.

materials 3 mm thick, the gauges have a sensitivity of about 345 microstrains per fringe in white light, subdivided and read to 2% fringe order, or 7 microstrains, using the precision hand-viewer.

The Uniaxial Photoelastic Disc Transducer

If a stress-birefringent disc is loaded about a diameter and observed under polarized light, using one of the arrangements depicted in Fig. 7.17, a characteristic isochromatic pattern is seen. Under no load the disc appears to be dark. This is the zero "fringe order". As soon as load is applied light passes through the disc, and, with increase in load, a series of isochromatics is generated, in progressive order, in which specific zones or "tints of passage" can be recognized. The relative retardation between successive tints of passage has a constant value, so that each of these isochromatic bands comprises one unit increment on a linear scale of fringe order. During compression of the disc the isochromatic fringes are generated first at the two loading points, each fringe forming a paired loop, and these grow, to approach the center of the disc from either end of the loaded diameter. The looped fringes merge temporarily at the disc center and they then split into two separate bands. With further increase of load these two bands move, one on either side of the center, along the equator, to approach the disc boundary.

In theory, the locus of the zero fringe, under load, is the circumference of the disc, so that it should disappear from view as soon as load is applied. In practice, however, the zero fringe may remain visible as a dark zone immediately around the periphery of the disc. The successive fringes, however, can never reach the boundary, and they remain visible, stacked along the equatorial axis (see Fig. 7.22).

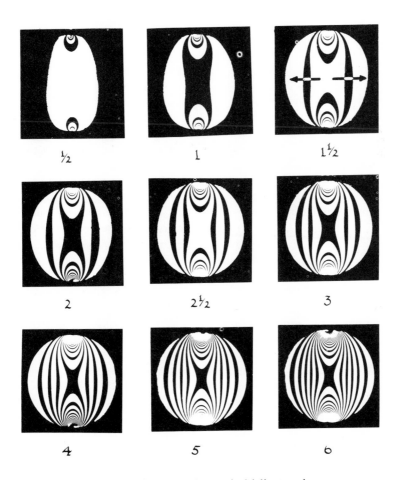

Fig. 7.22. Optical signals on uniaxial disc transducer.

For any disc viewed under polarized light of a given wavelength, the fringe order at the disc center is directly proportional both to the diametral deformation and to the applied load. This makes the arrangement very suitable for use as a photoelastic strain gauge, or as a load cell, because the fringe order at the center of the disc can be accurately measured by a simple count of fringes, coupled with the operation of a precision viewer. However, the main use for this type of transducer in geotechnology, up to the present time, is in the form of glass discs and cylinders which are incorporated into load cells of wide variety in size, range, and purpose.

Biaxial Photoelastic Transducers - the Biaxial Strain Gauge

A cylinder or a disc of stress-birefringent material that is attached continuously around its perimeter to a material that is subsequently placed under load will display birefringence under polarized light. In this transducer the displayed fringe order is directly proportional to the shear stress in the disc. In theory, this has a uniform value over the whole of the area that is enclosed within the cemented perimeter. In practice the fringe order at the center of the disc is likely to be slightly higher than that displayed around the perimeter, that is, there may be a slight radial strain gradient in the disc. Under increasing load the visual effect is that successively higher fringe orders are generated at the center of the disc, and these move radially in all directions, to approach the perimeter. In contradistinction with the line count transducer, however, in this case the fringes do not stack up to remain visible on the disc. Each disappears, to be replaced in its turn by another value on the scale. It follows that, unless the observer is able to recognize specific values of fringe order on the isochromatic scale, it is only possible to make a specific measurement of shear stress if the transducer is continuously observed, either increasing or decreasing, from a datum position on the scale.

But if the disc or cylinder has a central hole, several important effects are produced: (i) the sensitivity of the gauge is increased; (ii) the isochromatic pattern that results is one in which under increasing load, successive fringe orders are generated at the periphery of the central hole and move towards the major principal stress axis, where they stack up and remain visible, to serve as a linear numerical scale; (iii) the characteristics of the isochromatic pattern completely identify the stress field in the plane of measurement. The magnitude and sense (tension or compression) of the major and minor principal stresses can be identified, while the directions of these stresses are self-evident by the axial symmetry of the visual signal.

In Oppel's biaxial "strain compass" the disc is given a circular fringe pattern "frozen" into the plastic, and this pattern is self-displayed through a polaroid window which is integral with the gauge. Oppel-type strain compasses, now available commercially, permit strain magnitudes to be observed to within + or −40 microstrains, and directions to within 5°. Hawkes' biaxial plastic strain gauge, when constructed of current material 3 mm thick and 45 mm outer diameter, and read with the aid of the precision reflection hand-viewer, determines strains to within + or −9 microstrains (2% fringe order).

Photoelastic plastic biaxial gauges are used for the measurement of surface strains, as an alternative to electrical resistance strain gauge rosettes. Biaxial gauges are also used in the form of inclusions, to be inserted and cemented into boreholes drilled into rocks and concrete, or cast into concrete specimens. Such inclusions are of two types: (a) soft inclusions, of plastic, calibrated as strain gauges, and (b) hard inclusions, of glass. The glass inclusions are usually calibrated to read directly in terms of stress, when inserted into materials of known elastic modulus, or when inserted into materials of unknown modulus (but less than half that of glass). When used in this way they are termed "stressmeters".

Reading Photoelastic Transducers using the Precision Viewers

Reading photoelastic transducers consists of a two-part procedure: (i) recognizing the line or fringe count and (ii) measuring any fractional part of one fringe order.

Making the unit count

Recognition of the tint of passage on a photoelastic strain gauge, under white polarized light, depends on the observer's subjective assessment of color. While, in general terms one may define the tint of passage at the first unit fringe as the transition from red to blue, the observer must judge for himself exactly where that transition lies. Also, each successive fringe order has its own distinctive tint of passage and they are not all discerned with equal ease. The second fringe order is the easiest to define with precision, and the others less so, because they become paler, "washed out", and progressively more difficult to pin-point with increase in fringe order. So much, in fact, that for many purposes it is best not to use white light at all, but to insert a filter in the polarizer assembly, so that the fringes appear as dark bands against the background color of the filter. Recognition of the fringe order then becomes simply a matter of counting fringes on the uniaxial line-count transducer, or recognizing optical patterns on biaxial transducers. For practical reasons it is best to limit the range of the scale to 4 units with biaxial transducers and 5 units on the uniaxial line-count transducer. Until practice is gained a pattern comparator is a useful visual aid to recognition of the optical patterns representing 1 to 4 units in various biaxial fields (see Fig. 7.23).

Reading fractional fringe orders

To measure any fractional part of one fringe order, between the unit positions on the scale, the principle of "compensation" is applied. This adds, or subtracts, birefringence to or from the signal, either by sliding a birefringent wedge over the point of observation (Babbinet compensator) or by rotating the plane polarizing filter relative to the quarter-wave plate in the analyzer (goniometric compensation).

The precision viewers, Fig. 7.19, are arranged for the application of goniometric compensation. A clockwise rotation of the polaroid, relative to the quarter-wave plate in the analyzer, reduces the apparent fringe order in a compressional stress field or increases the apparent fringe order in a tensional field. The operation of taking a precise reading is then as follows.

1. Hold the precision viewer in the right hand, with the index of the scale on zero. Align the handle of the viewer with the axis of loading (on uniaxial transducers) or with the major stress axis (on biaxial transducers). Then make the unit fringe count, for example: X units plus or minus some fractional increment.

2. Still holding the viewer firmly in alignment, rotate the index of the viewer over its scale (or in certain viewers rotate the scale relative to the index). In a compressional field this will cause the fringe pattern to shrink inwards to a lower apparent fringe order. In a tensional field the fringe pattern will expand to a higher apparent fringe order. Continue the process of compensation until the viewer displays the next unit fringe order to that observed on the initial fringe count. Note the position of the index on the scale. This now registers the fractional increment which must be added to the initial fringe count of units (compression), or subtracted from the final fringe count (tension).

For example, in Fig. 7.24(a) the initial unit fringe count is 2 plus some fraction. The process of compensation is followed until the pattern shrinks to exactly 2 units, or expands to exactly 3 units, Fig. 7.24(b), and the appropriate fraction (0.38) noted. The final reading would then

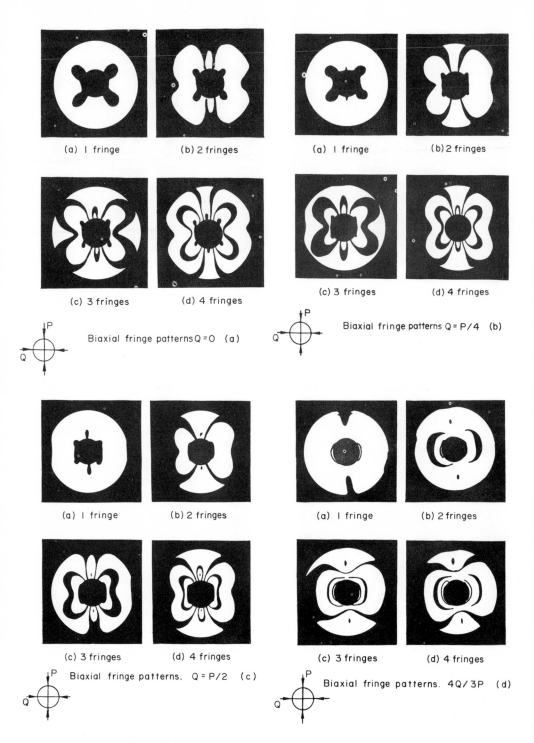

Fig. 7.23. Pattern comparator for biaxial transducers.

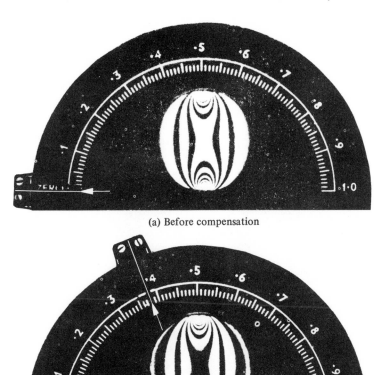

(a) Before compensation

(b) After compensation

Fig. 7.24 The process of goniometric compensation of a photoelastic transducer signal.

be 2.38 (compression) if the fringe orders reduced during the compensation, or 2.62 (tension) if the apparent fringe order on compensation was 3.0.

Locating the optimum point of reference

The observer must choose some point of reference on the displayed pattern. On uniaxial line count transducers each unit position is marked by the appearance of the tint of passage at the center of the disc, in white light, or must be recognized from the symmetry of the dark "caliper cross" at the center of the field in monochromatic light. With practice this can be established to within 0.02 on the goniometric scale. On biaxial transducers the optimum point of reference may not be so obvious. However, the fringe patterns around a hole in a uniaxial stress field give the biaxial transducers some easily recognizable features when the principal stress ratio approaches 1/0. On these transducers the fringes are generated at the periphery of the central hole. With increase of load the fringes move out quickly and they appear to emanate from four round zones, or "eyes" near the central hole on radii aligned at 45° to the principal stress axes of symmetry (see Fig. 7.21 and 7.25). These four "eyes" are convenient reference points at which to read the fringe order with maximum precision

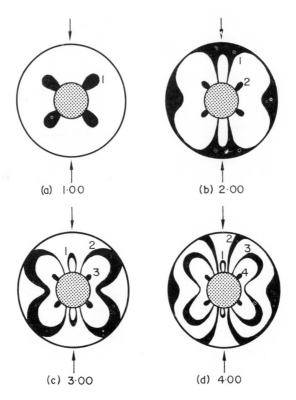

Fig. 7.25. Isochromatic fringe patterns in an annular biaxial transducer under increasing uniaxial stress.

Also, if a strain gradient exists across the transducer this may be measured by compensating to each "eye" separately.

In biaxial stress fields, when the minor principal stress approaches the major stress in magnitude, the 45° "eyes" become progressively less obvious, so that if σ_2 is more than $0.5\sigma_1$ they are not used. In a detailed study of the strain distribution in hollow cylindrical inclusions Hawkes has established reference points for optimum sensitivity and precision in reading, for various values of Poisson's ratio in the gauge material. He shows, for example, that, for a gauge having a central hole 9.5 mm diameter, a thickness of 3 mm, a Poisson's ratio of 0.32, and a strain-optic constant of 0.14, illuminated by white light, the optimum reference point is on the minor stress axis at a distance of 8 mm from the gauge center. He therefore advocates the use of an opaque collar, of the appropriate radius, inserted in the central hole of the transducer. The fringe pattern is then precisely compensated to the edge of this collar, on the minor stress axis, on all occasions, when the 45° "eyes" are not apparent (see Fig. 7.26 and 7.27).

Determination of principal stress ratios on biaxial transducers

An approximate estimate of the overall stress ratio can be made from the visual symmetry of a biaxial signal. A more precise determination may be made by various techniques, one of the simplest being to measure the distance between two dark spots, or "isotropic points"

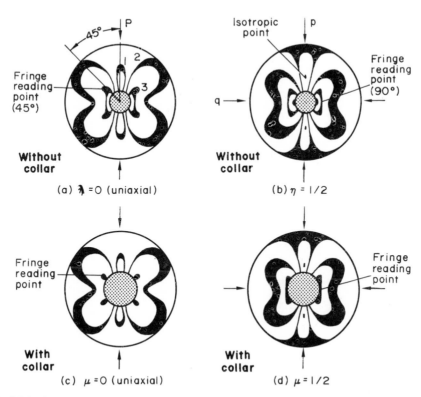

Fig. 7.26. Identification of optimum fringe reading point in a hollow cylindrical inclusion.

that are visible on the major stress axis. This measurement is then expressed as a ratio of the external diameter of the gauge. The isotropic points are points where there is no shear stress in the gauge material, and at these points there is no birefringence in the gauge. They therefore form zero points on the numerical scale that the isochromatic pattern represents. They may be distinguished from the isochromatic fringe orders in that they do not move over the field of view during the process of compensation. Nor are they affected by any change in the stress magnitude, provided that the overall ratio σ_1/σ_2 is maintained constant. But if the stress ratio is changed, then the radial distance of the isotropic points also changes, and this radial movement is directly proportional to the change in the stress ratio (see Fig. 7.27).

Calibration of Strain Gauges

The calibration factors appropriate to bonded electrical resistance strain gauges are supplied by the gauge manufacturers. In case of doubt sample gauges from a batch may be calibrated from observed signals and measured loads on materials of known elastic moduli. Mild steel ($E = 207$ GN/m^2) and aluminium ($E = 70$ GN/m^2) are convenient materials to use in this respect. Proximity gauges, LVD transformers, and the like, together with their respective control circuitry, must be calibrated directly against linear measurements. Photoelastic gauges may be calibrated uniaxially by bonding a representative sample on to a tapered cantilever

G—E

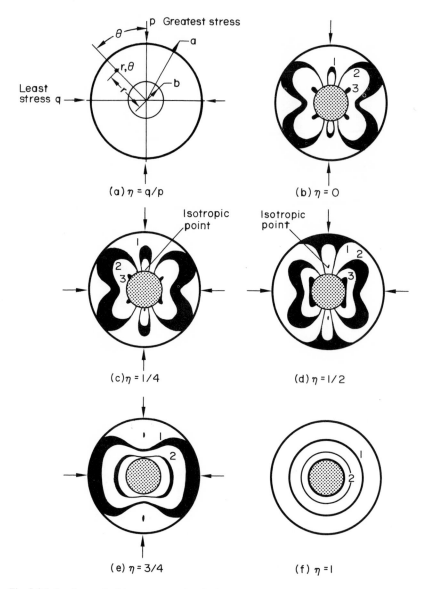

Fig. 7.27. Isochromatic fringe patterns in a hollow cylindrical inclusion. Fringe order 3.0.

bar of steel or aluminium. The bar is then loaded by attaching deadweights to its end. The strain on the upper surface of such a steel bar is given by

$$\epsilon \text{ (calculated)} = \frac{6wx}{30 \times 10^6 \, bh^2}$$

where w = loading weight,

b = width of bar at distance x from the loading axis for a tapered bar x/b is constant), (in.)

h = thickness of bar. (in)

Since the gauge is being calibrated in bending, a correction factor is necessary to compensate for the thickness of the gauge (m), such that

$$\epsilon \text{ (true)} = \epsilon \text{ (calculated)} \ \frac{m + h}{h}$$

Hawkes' biaxial strain gauge is calibrated uniaxially in terms of the major principal strain. The minor principal strain is then deduced from the isotropic point spacing. Alternatively, a biaxial straining cross may be fabricated. Such an arrangement is an invaluable aid to the study of biaxial fringe patterns. Its mode of use is to set up patterns of known strain, as identified by a 90° strain gauge rosette attached at the center of the cross and coaxially with the axes of strain. A biaxial photoelastic gauge is cemented around its periphery, over the top of the electric strain gauge rosette, the leads from which are brought through the central hole of the photoelastic gauge to the control equipment.

Selected References for Further Reading

DAVIDENKOFF, N. The vibrating wire method of measuring deformation. *Proc. ASTM*, vol. 34 (2), pp. 847–860 (1934).

HARDY, H. R., STEFANKO, R. and KIMBLE, E. J. An automated test facility for rock mechanics research. *Int. J. Rock Mech. Min. Sci.*, vol. 8, pp. 17–27 (1971).

HAWKES, I. Theory of the photoelastic biaxial strain gauge. *Int. J. Rock Mech. Min. Sci.*, vol. 5, pp. 57–64, (1968).

HAWKES, I. Moduli measurements on rock cores. *1st Int. Conf. Int. Soc. Rock Mechs., Lisbon*, vol. 1, pp. 655–660 (1968).

HAWKES, I. and MOXON, S. The measurement of in-situ rock stress using the photoelastic biaxial gauge with the core-relief technique. *Int. J. Rock Mech. Min. Sci.*, vol. 2, pp. 405–419 (1965).

HENDRY, A. W. *Elements of Experimental Stress Analysis*, Pergamon Press, London, 1964.

HOLLISTER, G. S. *Experimental Stress Analysis*, Cambridge Univ. Press, 1967.

HOOPER, J. A. Apparatus for applying sustained loads to large specimens. *Int. J. Rock Mech. Min. Sci.* vol. pp. 353–362 (1967).

KOTTE, J. J., BERCZES, Z. G., GRAMBERG, J., and SELDENRATH, R. Stress-strain relations and breakage of cylindrical granite rock specimens under uniaxial and triaxial loads. *Int. J. Rock Mech. Min. Sci.*, vol. 6, pp. 581–595 (1969).

OPPEL, G. U. Photoelastic strain gauges. *Exp. Mechs.*, vol. 1 (3), pp. 65–73 (1961).

SWINDELLS, M. A. and EVANS, J. C. Measurement of load by elastic devices. *N.P.L. Notes on Applied Science*, No. 21, HMSO, London 1960.

CHAPTER 8

The Strength of Rock Materials

Rock Materials and Rock Masses

There is much debate, amongst geotechnologists, concerning the extent to which the engineering strength properties of a rock mass may be assessed from the results of examinations and tests on the rock material. One school of thought, amongst whom engineering geologists are strongly represented, holds that it is futile to attempt to apply fracture theory (as derived from a consideration of idealized rock material) to a rock mass, the behavior of which under load will probably be determined by the nature of major discontinuities such as joints and fissures. Others take an entirely different viewpoint, seeking first to explain the behavior of rock as a material and then to extend their theories to include the influence of discontinuities and non-isotropy. This school includes those who, beginning from a consideration of rock as a brittle material to which the Griffith theory of failure may be applied, have extended this theory to triaxial conditions, taking into account the effects of crack closure in compression. The results enable a criterion of failure to be defined in terms of the internal state of stress, the uniaxial tensile strength, and the pore fluid pressure.

A third school of thought applies statistical theories, not only to take into account such matters as heterogeneity and discontinuities on a macroscopic scale, but also, by applying suitable scale factors, attempting to explain the behavior of the rock mass as well. This approach is particularly useful, for example, when assessing the strength of supporting pillars in a mine. It is also applied to the problem of assessing the relevance of laboratory tests on rock samples to the strength of the same rock *in situ*.

A fourth approach to the problem of assessing the strength of a rock uses energy concepts. The rock mass possesses energy by virtue of its position in space and the existence of gravitational and tectonic forces. The rock material possesses energy by virtue of the forces within the atomic lattices, together with those that control molecular, and inter- and intra-crystalline and granular cohesion. The energy balance in a stable rock mass is disturbed by the creation of an excavation, and in an unbroken rock material it must be disturbed if the rock is to be broken by the application of an external force. In the former case, potential energy released by rock movements, and strain energy stored in the rock mass, will all have obvious implications in relation to rockbursts and to strata control, while in the latter case the efficiency of the rock-breaking process is determined by the characteristics of energy transfer between the mining tool and the rock.

Yet another approach to determine the strength of rocks is empirical. It uses some arbitrary criterion by which various rock materials may be compared for the purpose in hand. The tests may be conducted on samples brought to the laboratory for examination, or they may be conducted on the rock *in situ*.

Failure Mechanisms in Rock Materials

Many research workers have studied the processes of deformation and fracture in rock materials, but in spite of the ever-growing mass of literature on the subject the extent of our knowledge is still far from complete. Probably the most popular approach, in the sense of the number of investigators concerned, is that which uses the Griffith theory of failure in brittle materials, in which it is held that failure originates as a result of stress concentrations around inherent small discontinuities or microcracks within the material. These stress concentrations are the result of the imposed and inherent stress system.

The basic assumption here is that rocks behave as brittle materials. The concept of brittleness may be defined in various ways. Elastic materials such as metals, when subjected to tension tests, demonstrate some ductility before fracture, and it has become conventional to define brittleness as the lack of ductility in a material. A more positive approach, however, defines brittle materials in terms of their behavior under load, as being those which fail by fracture at, or only slightly beyond, their yield stress. Ramsay defines brittle rocks and brittle fracture as occurring in rocks when their internal cohesion is destroyed while they are deforming in the elastic range, and Coates puts a numerical value on brittleness, in terms of the ratio between the reversible elastic strain and the total strain at the point of failure.

Hucka and Das compare various methods of brittleness measurement, the results of which, for any given rock, will depend on the particular definition used.

The Griffith Failure Criterion

In essence the Griffith theory is an energy theory in that stability of the material under load implies that a balance exists between the strain energy and the surface energy within the microcracks and the material structure. If the strain increases to produce an imbalance of energy then the microcracks extend, and fractures are generated. The original Griffiths theory pictured the inherent microcracks as being elliptical in shape, so that, when the material was subjected to compressional loads tension fractures were generated by the stress concentrations at the tips of the microcracks. A failure criterion was thus deduced in terms of the shear stress on the surface of the plane of failure (τ), the normal stress across the plane of failure (σ), and the uniaxial tensile strength of the material (T).

Murrell extended the original Griffith theory, and applying it to rocks, showed that the Griffith failure criterion corresponds with a parabolic Mohr's envelope defined by the equation

$$\tau^2 = 4T^2 - 4T\sigma$$

This assumed that the Griffith cracks remain elliptical up to the point of failure, whereas in practice it is observed that some rock materials demonstrate behavior which indicates that this is not always so. For example, some rock materials under test exhibit volume changes, involving some loss of porosity in the early stages of loading (which indicates a degree of consolidation of the material). This may be followed by dilation (which may be explained in terms of the development of shear dislocation processes in the consolidated material). This implies that the microcracks are closed, and in this event the yield along the incipient fracture surfaces will, in part, be controlled by friction.

Taking all these factors into account, McLintock and Walsh further modified the Griffith criterion to include the effects of friction across closed cracks, and they emerged with a postulated failure envelope for brittle rock similar to the Coulomb–Navier envelope.

Hoek has compared the original and modified Griffith criteria for a number of South African rocks, to produce the fracture data shown in Figs. 8.1 and 8.2. In these diagrams the stress coordinates are plotted as ratios of the uniaxial compressive strength, which thus becomes a basic parameter when describing the strength of a rock material. Hoek describes the rupture

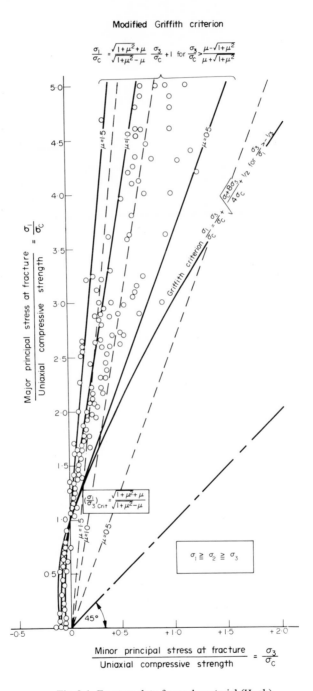

Fig. 8.1. Fracture data for rock material (Hoek).

criteria for homogeneous fine-grained rock to be as shown on Fig. 8.3. The lower part of the envelope defines a region where the strength of the material is determined by its resistance to

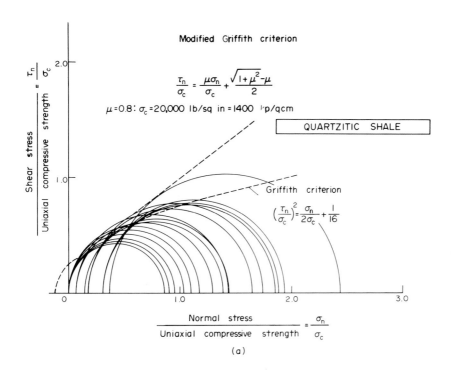

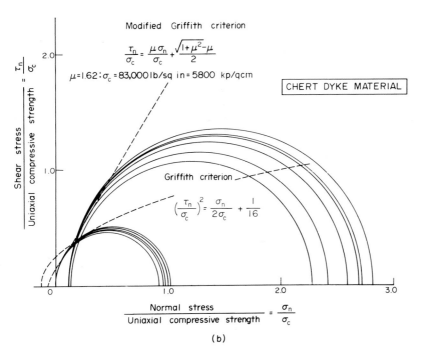

Fig. 8.2(a, b). Fracture envelopes for two South African rock materials (Hoek).

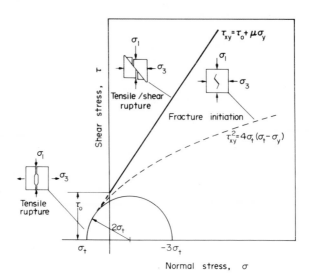

Fig. 8.3. A rupture criterion for brittle rock (Hoek).

tension. In this region a simple tensile rupture occurs when the principal stress σ_3 equals the uniaxial tensile strength of the material. With change in the conditions of load the tension σ_3 decreases, and the strength of the material increases, because the material builds up its resistance to shear as a result of the increasing normal stress across the incipient fracture surface. The material thus demonstrates a change from simple tensile failure to a combined tensile/shear rupture.

At this stage of the failure envelope the compressive stress σ_1 is large enough to force the two sides of the fracture surface together, and to generate friction between them, but the tensile principal stress σ_3 is still strong enough to generate tensile cracks running in the direction of σ_1 from the inclined shear fracture plane. The composite failure surface thus takes a stepped form, in the nature of small faults running through the rock material.

V. N. and E. Z. Lajtai also describe the evolution of stepped fractures in rock specimens under compression, and show how the fracture process may emanate directly from compressive stress concentrations, propagating approximately perpendicularly to the direction of maximum compression, as a normal shear fracture.

Under triaxial confining pressures, with both the principal stresses compressive, the normal stress across the fracture surface reaches a magnitude such that friction between the shear planes dominates the fracture process (although plastic deformation may also occur in the end regions of the microcracks, due to the stress concentrations in those regions). In the lower ranges of triaxial compression the "asperities" on the fracture zone cause some dilation of the rock material as the stepped surfaces ride over one another. At still higher confining pressures these asperities may be crushed, so that the yield of the material will not be accompanied by dilation. The failure process will then be essentially one of shear along an inclined planar fracture. Thus, once past the tensile failure zone the failure envelope maintains a linear envelope defined by the equation

$$\tau_f = \tau_0 + \mu\sigma_y$$

where τ_f is the shear stress at failure,
$\quad\tau_0$ is the shear strength intercept at zero normal stress,
$\quad\sigma_y$ is the normal stress on the closed crack at the start of the fracture process,
$\quad\mu$ is the coefficient of internal friction.

Alternatively, if the rock material is less brittle, so that plastic deformation processes assume more significance, the strength envelope may take the parabolic form described by Murrell, in which the process of failure develops from fractures initiated at the ends of open flaws. In this case the slope of the envelope indicates a transition from brittle towards plastic behavior of the rock material, with increase in confining pressure.

The Brittle-Plastic Transition in Rock Materials

The transition from brittle to plastic behavior in a rock material is mainly of interest in relation to geotectonics and geophysics. In geotectonics it helps to explain why and how exposures of hard, brittle rocks may also exhibit obvious characteristics of plastic flow, because the rock which, at the outcrop, is now hard and brittle, behaved in past geological time as a plastic material, at depths where high pressures undoubtedly exist. It also helps to provide the mechanism whereby lateral translations of segments of the Earth's crust and mantle may be effected, relative to one another. In geophysics it has pronounced significance in relation to seismic phenomena and earthquakes.

In relation to engineering problems the brittle-plastic transition of rocks is probably of lesser significance, since the ambient strata pressures are generally of a much lower order. Nevertheless, there are some important implications in relation to strata control and support in mining and civil engineering excavations, particularly in saline evaporite rocks. The possibility of brittle-plastic transitions and reversions must also be borne in mind when studying such phenomena as rockbursts, rock and gas outbursts, pillar failure in mine support, and the stress distribution around tunnels and shafts.

Yield and Flow of Rock Materials

An apparent anomaly in the behavior of some materials is that they may demonstrate the characteristics of brittle fracture and, at the same time, may possess the fundamental property of a fluid in that they can withstand no shear stress without undergoing permanent deformation. Substances like manufactured glass, natural glasses such as pitchstone, obsidian, and lava flows, and also sand-cement mixtures all exhibit creep phenomena under load that can be described in terms of true viscosity.

TABLE 8.1. *Viscosity of Some Rock Materials (Jaeger)*

Substance	Viscosity (poises)
Glass $(10°C)$	10^{22}
Glass $(575°C)$	1.1×10^{13}
Lava flows	4×10^4

However, most materials behave elastically when subjected to low **stresses**, and they begin to yield only when the imposed loads generate stresses above a finite yield-point. The term "yield" thus implies the onset of plastic deformation. Continued plastic deformation, at a rate controlled by the internal frictional resistance of the material, relative to the generated shear and normal stresses, is termed "flow". "Failure" may thus take the form of flow and fracture in brittle materials or be indicated by yield and flow in plastic materials. The yield-point may be defined in terms of the maximum shear stress. That is, the material will yield if its shear strength is less than the maximum shear stress $(\sigma_1 - \sigma_3)/2$ which acts across a plane whose normal bisects the angle between the major and minor principal stresses σ_1 and σ_3.

Tresca's criterion

Tresca's yield criterion states that yield occurs at a point where the magnitude of the maximum shear stress has a value $\sigma_0/2$, which is a constant of the material.

Hence the criterion may be written

$$\sigma_1 - \sigma_3 = \sigma_0$$

or

$$\frac{\sigma_1 - \sigma_3}{\sigma_0} = 1$$

Tresca's criterion implies that equal yield stresses exist both in uniaxial tension and compression, which is obviously untrue for fracture of brittle materials. It is more feasible for plastic yield of ductile materials.

Von Mises' criterion

A more commonly used concept is Von Mises' criterion, derived from Tresca's, and written

$$\frac{\sigma_1 - \sigma_3}{\sigma_0} = 2(3 + \mu^2)^{-1/2}.$$

Triaxial Compression Tests on Rock Materials

Triaxial tests are required if the complete failure envelope of a rock material is to be determined. As with triaxial tests on soils, a constant fluid pressure is maintained around the cylindrical surface of a prepared core-sample of the rock, but in this case the test cell is constructed of steel, since higher confining pressures are used. The rock specimen must be jacketed, with either a plastic material or metal foil, to prevent penetration of the confining fluid into the pores of the rock under pressure. The ends of the specimen are placed in contact with upper and lower bearing-pistons, which slide inside the pressure cylinder, with *O*-ring seals or chevron packing around the pistons. The bearing-pistons are then held in compression on the testing machine.

With the object of removing the complications generated by the uncertainties of contact between steel loading pistons, spherical seatings, bearing plates, and the rock specimen, Atkinson and Ko have designed a multi-axial cell for triaxial testing. This device utilizes fluid cushions to apply a three-dimensional field of compressive stresses to a 100 mm cube of rock.

In the routine triaxial testing of rock materials it is not usual to measure pore pressures, so that the result is therefore determined by total stresses. Research testing equipment, usually incorporates arrangements for the control and measurement of pore-fluid pressure. and also for observations of longitudinal and diametral strains, on the specimen. Observations of diametral change necessitates instrumentation within the pressure chamber, and since this presents technical problems it is generally not attempted in routine tests.

The routine test then consists of observing the axial compression load at failure, for a given constant value of the confining pressure, from which the axial principal stress and the radial stress are determined, to give the deviator stress $\sigma_1 - \sigma_3$. Upon removing the specimen from the pressure chamber after the test, the mode of deformation and failure of the specimen, and the angle of inclination of the fracture plane, if present, are observed.

The Effects of Pore Fluids in Rocks

In rocks, as in soils, the effects of the pore fluids are extremely important. It is well known that the presence of water weakens a rock, even hard rocks like quartzite and granite, but the extent to which this takes place is not always appreciated. Colback and Wild show the effect of water saturation on the strength envelope of a quartzitic shale, as compared with the same material when dry. Hoek suggests that this reduction in strength is probably due to two causes: (1) a reduction in the molecular cohesive attraction between the rock particles when moisture is present and (2) the pore fluid pressure counteracts the confining pressure so that the effective stress acting across a potential fracture plane is reduced.

The opinion of experts is divided as to the extent to which effective stress concepts, as evolved and proved for soils, should also be applied to rock materials. Skempton advises caution in this respect.

In an examination of the concept of effective stress, as applied to soils, concrete, and rocks, he shows that, while it is commonly assumed that the effective stress controlling changes in shear strength and volume in saturated porous materials is given by

$$\sigma' = \sigma - (1 - a)u_w$$

where σ' = effective stress,
σ = total normal stress,
u_w = pore water pressure,
a = area of contact between the solid particles; this is only approximately true when a is small, as in soils. It is appreciably in error for other materials.

A more correct expression for effective stress in fully saturated materials would be for shear strength

$$\sigma' = \sigma - (1 - \frac{a \tan \psi}{\tan \psi})u_w$$

and for volume change

$$\sigma' = \sigma - (1 - \frac{C_s}{C})u_w$$

where ψ and C_s are respectively the angle of intrinsic friction and the compressibility of the solid substance comprising the particles and φ' and C are the angle of shearing resistance and the compressibility of the porous material.

For fully saturated soils these equations degenerate essentially into Terzaghi's effective stress equation, but for saturated rocks and concrete C_s/C ranges from 0.1 to 0.5, $\tan \psi/\tan \varphi$ ranges from 0.1 to 0.7, and a is not negligible. In these circumstances Terzaghi's equation is not applicable.

Much will depend upon whether the pore spaces are continuous, or whether they are what Bernaix calls "vacuolar". However, Murrell's experiments on the triaxial testing of rocks showed that the effect of pore pressure was to counter the effect of confining pressure, so that if the pore pressure increases to equal the confining pressure then fracture occurs at a constant value of the deviator stress, equal to the unconfined compressive strength. Also, if a rock is not confined by lateral restraints and an increasing pore pressure applied, then the rock fractures when pore pressure equals the unconfined tensile strength.

There is, in fact, ample evidence to show that the law of effective stress may hold for many rocks, and that a wide range of properties depend upon the effective stress, or the effective confining pressure. These properties include the fracture strength, ductility, elastic modulus, sonic velocity, electrical conductivity, and permeability.

Uniaxial Compression Tests on Rock Materials

The uniaxial compressive strength is a basic index in the classification of rock strength (see Fig. 8.4). The test is deceptively easy to perform, but if meaningful results are to be obtained, great care in its performance is required of the investigator. As yet there is no general standardization of techniques and procedures, so that any values quoted in the literature should be treated with considerable reserve, when comparisons are being made. As a general

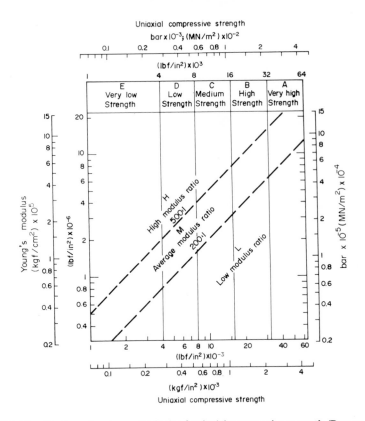

Fig. 8.4. Classification of rocks on the basis of uniaxial compressive strength (Deere and Miller).

rule, the test is made on cylindrical samples of drill-core obtained during field exploration, but precision work in research requires that samples be more carefully prepared in the laboratory. The procedure may involve cutting cores with a laboratory thin-walled diamond core drill on selected blocks of rock. The cylindrical surface of each specimen may be precision-finished by lathe grinding, followed by lapping the end surfaces so that they are perfectly smooth and perpendicular to the specimen axis. The result obtained when such a specimen is placed on a compression testing machine, and loaded in the direction along the cylindrical axis, depends on several factors.

The Effects of End-contact Conditions

One of the major controlling factors in a uniaxial compression test on a rock specimen is the nature of the contact conditions between the specimen and the steel platens, or compression plates, of the testing machine. The purpose of lapping the ends of the specimen before the test is to try to achieve complete planar contact at the ends, and to ensure truly axial loading of the specimen. If the ends of the specimen are not truly flat and contiguous with the compression-plates there will be localized contacts at the beginning of the test, and these will provide points of high stress-concentration that will induce premature fractures emanating from these points. Assuming that the specimen end-surfaces and the machine compression-plates are perfectly smooth, the question of slight obliquity of the ends remains. Such obliquity would result in non-axial loading and a concentration of stress along an edge, to one side of the axis. To prevent this it is necessary that one of the compression-plates of the testing machine should be mounted on a ball-and-socket or a spherical seating. Alternatively an independent ball seat can be attached to the upper end of the specimen during the test.

The Effects of End-restraint Conditions

The frictional contact conditions between the specimen and the compression-plate have a critical effect on the test result. If there is no friction then the specimen will fail under a stress situation that lies at the lower extreme of the strength envelope shown in Fig. 8.3. That is, the specimen will fracture in a direction parallel with the axis of loading, and it will demonstrate its tensional strength in a direction at right angles to that axis. If the ends are not lubricated, then some restraint will be imposed between the compression-plate and the specimen because the two surfaces in contact, steel and rock, have different elastic properties. This means that the vertical compression force generates lateral strains but these will not be identical in magnitude in the two materials. The nature of the end restraint then depends on the frictional properties that exist between the two surfaces in contact and the particular end conditions arranged during the test. The result, on any one material, may vary widely.

To obtain values for the compressive strength of a rock, that will approach a satisfactory consistency, the test conditions must be standardized and various investigators advocate their individual methods that are aimed to achieve this. Such methods include the insertion of thin metal diaphragms between the compression-plate and the specimen the insertion of low-modulus elastic inserts, of plastic or rubber (supported around the periphery of the cylinder so that it does not extrude under load and tear the specimen apart), shaping the specimen larger at its ends than at its middle (with a carefully ground curved profile between the two diameters), or interposing rock cubes or cylinders, of the same material but larger than the test-piece, between the test-piece and the compression-plates.

That there must be considerable variation in possible result, over all these suggested techniques, will be obvious. This being so, probably the most practical suggestion is to arrange complete radial constraint at the specimen ends. This is attained by using hardened steel end-plates on the loading machine, the plates being wider than the diameter of the rock sample.

The Effects of Size and Proportions of the Test-piece

The coarser the grain size of the rock to be tested the larger the diameter of the test-piece should be, if the sample is to be truly representative of the rock material. Several investigators have suggested that the size of the internal "Griffith cracks" from which failure is generated is about the same as the grain size of the material. Leading from this it has been suggested that the ratio test-specimen/grain-size diameter should be at least 10/1, and some investigators recommend that it be as much as 20/1. However, for practical reasons the size of drill-core obtained for testing rarely exceeds 57 mm diameter.

To avoid inconsistent results due to the unknown stress distribution at the specimen ends, caused by the end-restraint conditions, the test sample should have a length not less than two, and preferably three, diameters. If the latter proportions are adopted then the middle third provides the observed test section on which strain measurements may be made for the determination of the uniaxial stress/strain relationships.

The Effects of Pore Fluids and Porosity in the Uniaxial Testing of Rocks

Porous rocks should be tested in their natural environment if a realistic estimate of their strength is to be obtained. This means that they should be tested on the site, using portable equipment. This is not difficult, but it may be inconvenient. If the samples are taken for testing to a laboratory then the residual water that results from the core-drilling process should first be allowed to drain off, before sealing the samples in airtight covers, to await transfer. It is not usual to measure pore pressures, or to make provision for pore-pressure control, in uniaxial testing of rocks, so that here again there is the possibility of some variation in result, depending upon the degree to which the specimen had dried out between the time it was cored and the time of the test. Uncertainty as to the amount of residual water remaining in the rock after the core-drilling operation obscures any assessment of the natural moisture content. Hence some investigators adopt the procedure of drying-out the specimens completely before testing, sometimes using oven-drying at 100°C. With certain types of clay rocks high-temperature drying is not advisable, since it may produce chemical changes and induration of the material. Air-drying is then preferable.

The Effects of the Duration and Rate of Loading

The strength of a rock material is a time-dependent property. If it is stressed above a certain level the material will creep, flow, and crack, until ultimate failure occurs. The time that elapses between the first application of load and the ultimate failure of the specimen will decrease with increase in magnitude of the applied stress. Considering this time factor, at one extreme we have the stress-magnitude, that causes failure, decreasing exponentially to a limiting value as time approaches infinity. At the other extreme we have the capability of the rock to withstand much higher stresses that are imposed only for a short length of time, up to a practical limit, represented by, say, the transient stress-pulse required to fracture the rock by a hammer-blow, or the passage of a stress wave generated by explosive shock. When determining the uniaxial compressive strength of a rock material in the laboratory we are usually concerned with neither of these limiting conditions, but we are concerned with an intermediate situation

The test is conducted by increasing the applied load from zero, but clearly, a number of tests on the same material at different rates of load application will produce different values for the compressive strength. Here again, a standardized procedure is required. This could be arranged on the basis of "rate of loading" or "rate of strain". In a linearly elastic material both these rates are the same, but for most rocks this cannot be safely assumed. With rocks the test-piece may still be deforming even if the rate of increase of load is brought to zero, at the maximum value of resistance to load that is demonstrated by the rock.

The USBM standardized procedure for rock tests quotes loading rates between 689–2758 kN/m^2 per sec, while ASTM 6170–50 specifies a loading rate not more than 689 kN/m^2 per sec, or alternatively a rate of deformation (measured by the rate of lowering the compression-head of the testing machine) not greater than 1.3 mm/min. If the purpose of the test is solely to determine the long-term or "static" compressive strength of the rock then a slow loading rate is best. However, the same will not be true if we are also determining the deformation modulus and the rock is susceptible to creep. In the latter case the test should be conducted at a rate of loading for which creep is negligible. It is obvious therefore that the rate of loading is a factor that should be specified and incorporated into the reported results of compression tests on rock materials.

The Effects of the Testing Machine

When a brittle material is placed in compression on a testing machine of conventional design, or what is sometimes termed a "soft" testing machine, such as is used in a materials laboratory for testing metals, then elastic strain energy is stored in the testing machine itself, as the load on the specimen is increased. When the specimen reaches the limit of its resistance to load it begins to yield and at this point the energy stored in the testing machine is suddenly released. The compression-plates of the machine then move rapidly towards one another, and the material under test is shattered with explosive violence. While brittle failure of this character does occur sometimes in nature, it is generally the result of dynamic shock loads, and seldom is it caused by a slow build-up of compression. Therefore, if we want to study the behavior of rock materials in response to static loads, or to loads that increase in magnitude only slowly, then the test on a soft compression machine does not tell us the whole story. In general, rock materials retain some strength beyond the limit of their maximum resistance to load, and the complete stress-strain characteristic of a rock under compression takes the form shown in Fig. 8.5.

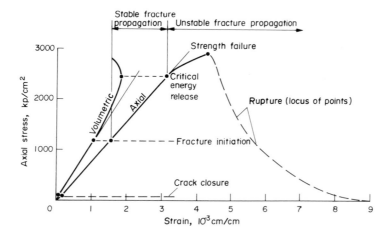

Fig. 8.5. The process of deformation and failure of a brittle rock material under compression (Bieniawski, Denkhaus, and Vogler).

In order to preserve the specimen for observation during the period when the strain is continuing under decreasing load, beyond the limit of strength failure, the testing machine must function as a "stiff" machine. The movement of the compression heads in such a machine is controlled, so that, upon reaching the strength limit of the test specimen, they do not come together at a rate greater than that at which the specimen deforms on the downslope side of the stress-strain characteristic. The strain energy in the machine is then not released suddenly. This requires that a stiffening element be incorporated into the test system, in parallel with the rock specimen. Such an element can be provided by steel bars, or by hydraulic jacks, or by a feed-back servo-control to prevent acceleration of the compression plates as the specimen fails.

The Effects of Non-isotropy of the Rock Materials

An isotropic material displays the same physical properties irrespective of the direction in which those properties are measured. Rocks are complex assemblages of granular and crystalline materials in which the question of isotropy must be approached in a statistical sense. In such a material, if the particles, grains, crystals, pore spaces, and "Griffith microcracks" all have a random orientation, a section cut through the material would expose a similar assemblage, independent of direction.

Many rocks, however, contain "preferred" orientations of their constitutents, caused by the geological processes of sedimentation and metamorphism which they have undergone during their past history. These materials are anisotropic, and the physical properties displayed by such rocks will vary, depending on the direction of measurement. Sometimes the anisotropy is displayed in what appear to be "planes of weakness" in the rock material. Rocks such as slates display a pronounced cleavage or foliation. At first sight one tends to assume that the direction of cleavage represents the weakest planar direction in the rock. But this may not always be so. It depends on the character of the forces that are applied, in order to break the rock. If we are attempting to open-up the rock, by hammer and wedge, then the cleavage planes represent major planes along which the rock may be split by forces of tension. But if the rock is loaded by forces of compression, a very different situation exists.

Figure 8.6 shows the influence of the discontinuities upon the uniaxial compressive strength of a foliated rock specimen. It is apparent that the compressive strength of this rock varies when measured in different directions, from a maximum of 68.95 MN/m² to a minimum of 17.24 MN/m². But the rock displays approximately the same compressive strength when loaded at right angles across the planes of foliation as it does when loaded along the direction of the planar discontinuities. The minimum value of compressive strength is displayed when the direction of loading in compression is at such an angle that the planar discontinuities coincide with the direction of shear fracture, as determined by the failure criterion for the rock. In this particular case that angle is approximately 30° from the axis of loading.

The Effects of Heterogeneity

A material is homogeneous if its constituents are so distributed that small specimens are truly representative of the whole. So far as earth and rock materials are concerned, the question of homogeneity is principally one of scale. The coarser the grain size of the material and the larger specimen must be if its measured properties are to be truly representative (cf. "Effects of size and proportions of the test-piece", p. 124).

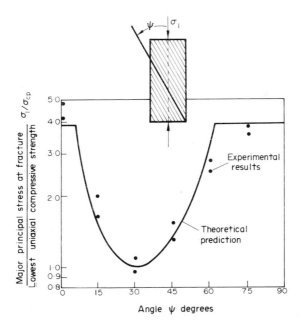

Fig. 8.6. Relationship between bedding plane orientation and the compressive strength of a slate (Hoek).

The Number of Samples Tested

It will be apparent from the foregoing discussion that a realistic determination of the uni-axial compressive strength of a rock material is a statistical problem. One or two measurements on random samples are likely to be quite meaningless, no matter how carefully the test is performed. Even if the most rigorous standardization of the methods of specimen preparation, and testing technique, are employed, several samples of the same material should be tested. The USBM proposed standard procedure specifies at least ten air-dried specimens. The custo-mary measure of variability among a group of tests is the standard deviation, and this provides an index of reliability for the test. Obert and Duvall quote attainable values of deviation ranging from 3.5% to 10%, depending on the rock type, while Hoskins and Horino found that the standard deviation may go up to around 27% if the ends of the specimen are not carefully lapped.

Compression Tests on Irregularly-shaped Rock Samples

The time, trouble, and expense involved in making uniaxial compression tests on rock materials by the proposed standard laboratory procedures, if the results are to be acceptably reliable, leads many geotechnologists to seek other methods of approach, in the assessment of rock strength. Some engineers go so far as to discount entirely the usefulness of what they term "laboratory rock squashing" as an aid towards the solution of field problems in rock mechanics. Where comparative assessment of rock strength has to be made it is sometimes possible to dispense with the conventional uniaxial compressive test, in favor of some other form of test, specifically designed to suit the particular circumstances concerned. The information that is so obtained can seldom be used in an absolute sense, however.

The uniaxial compressive strength is a basic rock property, and its determination is often necessary before theoretical concepts of design can be applied, either in relation to the rock itself or to rock excavation processes. At the same time there are some rocks that are so brittle, or friable, or so weakened by the circulating fluids used in core-drilling, that satisfactory core-samples are difficult, if not impossible, to obtain. In such circumstances one may make an assessment of compressive strength from tests on irregularly shaped rock fragments.

The technique, originally devised by Protodiakonov, consists of selecting samples of rock debris, of irregular shape, measuring the individual volume and weight of each fragment specimen, and then crushing the specimens in a compression press. Protodiakonov quoted the mean breaking force P as being related to the specimen volume C, in the form

$$\log P = \log f + 0.63 \log V$$

and

$$f = 0.19 f_c$$

where f_c is the uniaxial compressive strength.

The method requires that a large number of specimens be tested, the International Bureau for Rock Mechanics recommending a standard procedure involving the selection of from fifteen to twenty-five fragments, irregular in form but roughly egg-shaped, having a ratio of the smallest to the largest diameter about 1.5 to 1 and a volume about 100 cm^3. The fragments should have mass differences less than 2%, and should be crushed parallel to their longest axis and perpendicular to any obvious planes of lamination.

A modification of Protodiakonov's technique, recommended by Hobbs, crushes irregularly shaped rock specimens between the jaws of a 10-ton hydraulic press. In this case the specimens are loaded parallel to their smallest dimension, and perpendicular to any laminations. The maximum height of each fragment, in the direction of loading, is measured before each test. After the fragment has been loaded to failure the area of contact of the ends of the fragment on the loading plates is also measured. This is done by placing a piece of carbon paper and a piece of graph paper between the specimen and each of the loading plates. After the test the area of the carbon imprint on each of the graph papers is measured, and the mean of the two areas taken as the area of contact of the plates with the specimen. The results are analyzed statistically, and plotted to determine the regression line of the compressive strength f_c, relative to the average applied stress at fracture. This was observed by Hobbs to be

$$f_c = 0.91 I_a - 3180 \text{ psi}$$

when the compressive strength f_c was determined on cylinders of mudstone and siltstone measuring 1 in. long by 1 in. diameter.

Determination of Compressive Strength from Drilling Tests on Rock

Tsoutrelis describes a technique to determine the compressive strength of rocks from the results of drill tests. The method is based on the relationship

$$U_0 = K_0 (P - P_0)$$

where U_0 is the penetration rate of a rotary diamond drill, measured as penetration distance per revolution of the bit at the start of drilling, that is, when the bit is sharp.

P is the thrust applied to the drill, while P_0 and K_0 are constants.

$$K_0 = a.z - b$$

where $z = 10^4/f_c$,

f_c = uniaxial compressive strength,

a and b are drilling constants, depending upon the geometry and type of bit used.

The relationships between the constant K_0 and the uniaxial compressive strength of a number of rock materials, using Tsoutrelis' experimental drill, are shown in Fig. 8.7. Any type

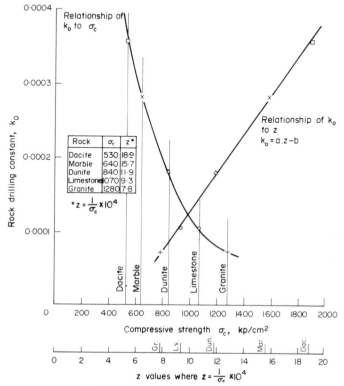

Fig. 8.7. Relationship of Tsoutrelis drilling constant to the compressive strength of rocks (Tsoutrelis).

of bit may be used, although hard, abrasive, materials will necessitate the use of a diamond rotary bit. The constants a and b must be determined for the particular drill used and the drilling technique employed. This is done by test drilling into a "reference standard" material (such as massive limestone or marble) whose compressive strength is known.

The Tensile Strength of Rocks

Rock materials are relatively weak in tension, compared with their compressive strength. Brittle failure theory predicts a ratio compressive-strength/tensile-strength of around 8/1. In practice, however, the ratio is not easy to determine, principally because of the difficulties that are experienced when trying to establish a reliable value for the tensile strength of a rock material.

When he is designing structures in reinforced concrete the structural engineer may proceed on the basis that his concrete will be well able to take compressional loads, but any tensions will be taken by the reinforcement. That is, it is assumed that the concrete has no strength in tension. There are occasions when the reinforcement of rock masses by steel components can be effected, to take care of tensional forces, but such occasions are rare. In general we must assume that the tensional strength of rock is very limited, so that if stress concentrations in tension occur at any rock exposure, then failure is likely to occur at those points. At low confining pressures, such as exist around the peripheral walls of an excavation, the ability of the strata to withstand tensional forces is no higher than that represented by their uniaxial tensile strength.

The tensile strength of rocks is also fundamentally important in that it determines the response of the material to dynamic forces, in drilling, blasting, and breaking ground by means of wedge-penetration devices, ploughs, and rock rippers.

Determination of Uniaxial Tensile Strength

In spite of the importance of tensile strength in relation to the ability of a rock to resist dynamic attack, or to support static loads, the concept is little used as a design factor in rock mechanics. The main reason for this is that too little reliance can, as yet, be placed on the results of laboratory tensile tests on rock materials. Because most rocks are more or less brittle, when unconfined, the test samples cannot yield plastically to relieve the stress concentrations that are produced at localized points around the specimens, where these are gripped to be pulled apart by the testing machine. Consequently, premature failures are generated from these points. Difficulties in ensuring truly axial loading also exist, so that the specimen is liable to be twisted or bent when gripped and pulled from either end.

Various testing methods have been designed, in attempts to resolve these difficulties. The methods include the preparation of precisely cut and ground cylindrical samples, fastened at each end by epoxy resin or other strong cement, to end-caps of steel, or aluminium, which are gripped by the machine. Alternatively the ends of the rock cylinder are cemented into metal collars. Tension is applied to the end-caps through the medium of light steel cables, sometimes attached through swivels and ball-joints, to eliminate twist on the specimen. Another alternative to using cemented end-caps is to grind the rock cylinder in a lathe, to a dumbbell shape, which can then be held in a claw-grip at each end. This is, however, a tedious and expensive procedure, so that dumbbell-shaped rock specimens are more likely to be used only for research and triaxial tests and not for routine uniaxial tensile tests.

Indirect Tensile Tests on Rock Materials

An alternative to the direct tensile test is the indirect "Brazilian Test" to determine the uniaxial tensile strength. The method is to place a disc, or a cylinder, of the rock material under compression across a diameter. In theory, such a loading situation generates a uniaxial tensile stress in the material at right angles to the compressed diameter, and the cylinder should fail when this tension reaches the uniaxial tensile strength of the material.

Hence $$T = \frac{2W}{\pi dL}$$

where W is the applied load,
d is the diameter,
L is the axial length of the cylinder or disc,
T is the uniaxial tensile strength.

In practice it is found that the failure of such specimens of rock is very often generated by localized crushing under the diametral compression-pads, along the axis of loading, and not by diametral tension at right angles to that axis. Curved-jaw loading rigs may be used in an attempt to improve the loading conditions, so that they may more closely resemble the theoretical condition (Figure 8.8). Uncertainties associated with the premature generation of failure under

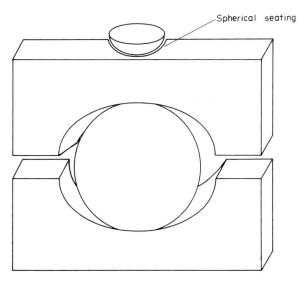

Spherical seating

Fig. 8.8. Curved-jaw loading rig for the "Brazilian" indirect tensile test on rock discs (Hawkes and Mellor).

the compress-pads are sometimes removed by drilling a central hole in the disc, and then testing the annulus so constructed. This is sometimes termed the "ring-test".

The stress distribution in such an annulus under diametral compression takes a complex form, but stress concentrations in tension are produced along the axis of compression, in a direction at right angles to that axis, at two positions on the perimeter of the central hole. In theory, therefore, failure should be generated from these points, in tension, and the hope of the investigator is that this failure should precede the effects of crushing under the compression-pads. However, the value for the tensile strength obtained in such tests is greatly dependent upon the geometrical proportions of the ring. Too small a central hole and crushing under the compression-pads generates failure, too large a central hole and the annulus flexes, so that bending stresses are introduced. Also, as with the uniaxial compression test, the result of a tensile test is also dependent on the rate of loading, the extent of this dependence varying in different rock materials, as shown in Fig. 8.9.

Opinions as to the value of the Brazilian and ring tests are varied. On the basis of their test experience using a wide variety of materials and five different testing machines, Mellor and Hawkes state that the Brazilian test is useful for brittle materials but for other materials the test may give wholly erroneous, although consistent, results. The same investigators discount the value of ring tests because the results they yield are generally greatly in excess of the direct uniaxial tensile strength when the hole is small, while they approach the flexural strength as the diameter of the central hole increases, relative to the outside diameter of the ring. The ring test, too, is not self-consistent, in that the effective stress concentration factor for a given specimen size varies with the rock type and its physical condition.

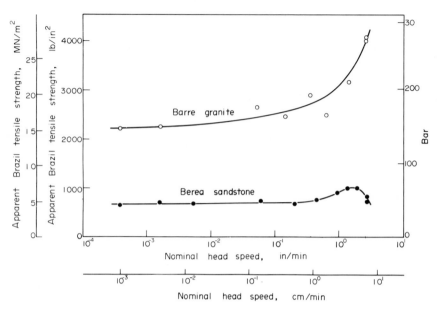

Fig. 8.9. Apparent Brazilian tensile strength as a function of loading rate, for two rock types
(Hawkes and Mellor).

The Tensile Strength of Rocks in Bending

The strength of rock materials spanning an excavation, such as a tunnel of rectangular cross-section, leads to interest in the extreme skin stress that is generated on the lower surface of a beam, loaded from above, in bending. The tensile strength observed on rock materials loaded in this way is usually somewhat greater than the direct uniaxial tensile strength. The USBM standardized test procedures for rock materials include the determination of the flexural strength, or "modulus of rupture", on 150mm lengths of drill-core. The flexural strength of such a rock sample is R_0

where
$$R_0 = \frac{8F_t L}{\pi D^3}$$

and F_t is the applied load at failure
   L is the horizontal distance between the lower support knife edges
   D is the diameter of the core sample.

For a specimen beam of rectangular cross-section

$$R_0 = \frac{3F_t L}{2a^2 b}$$

where a is the depth and b the width of the beam section.

In the USBM test the tensile stress is a maximum along a line contact directly underneath the applied load. Only a very small proportion of the beam is therefore under test. If a four-point bending rig is used, as shown in Fig. 8.10, then a uniform bending moment and a uniform tensile stress are generated on the lower surface of the beam.

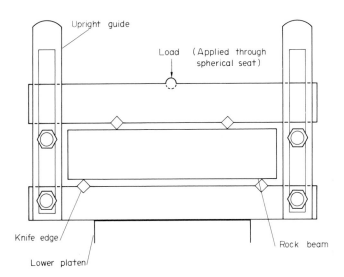

Fig. 8.10. Four-point loading arrangement for tensile tests on rock beams in bending.

The Unconfined Shear Strength of Rocks

If a reliable measure of the uniaxial tensile strength can be effected by direct or indirect tests, together with a measure of the unconfined compressive strength, then the shear strength of a rock material when unconfined can be estimated from the Mohr envelope construction. The degree of reliance that could be place on such an estimate will depend upon the accuracy with which the locus of the Mohr envelope can be established. Since there is often considerable doubt as to the linearity of the envelope at low confining pressures and in the vicinity of the uniaxial condition, particularly on the tensile side, an observed direct shear test may be preferred.

Questions of interpretation apart, while shear box tests on soils are now well-established and standardized procedures, there is no comparable standard laboratory test for rocks. There are several types of shear test used on rock materials. The tests may be conducted on prepared slabs of rock on which the vertical force required to shear the slab is measured, either in single shear (Fig. 8.11(a)) or in double shear (Fig. 8.11(b)). Another type of shear test is conducted on a compression testing machine, with the rock sample prepared in prismatic form and held between compression-jaws that are shaped to hold the specimen across a pair of diagonal angles (Fig. 8.12(a)). Alternatively, using irregularly shaped samples, as collected in the field, the rock specimen may be held inside a split, hollow, steel box, as shown in Fig. 8.12(b). The sample is contained within a cast of cement poured around it, and is orientated so that the split in the box lies at an angle to the axis of loading. In both the arrangements depicted in Fig. 8.12 the shear strength of the rock material is measured along an impressed plane of fracture.

Another type of shear test is conducted by forcing a cylindrical punch into the rock specimen, as shown in Fig. 8.13, and yet another form of test is to twist a cylindrical core-sample between torsion grips. In the punch shear test thin discs of rock, sliced from core sample, are placed on a die, and a load is then applied through a cylindrical punch. The shear strength is then given by

$$S_S = W/\pi dt$$

where W is the punch load applied to cause failure,

   d is the punch diameter,

   t is the thickness of the rock disc.

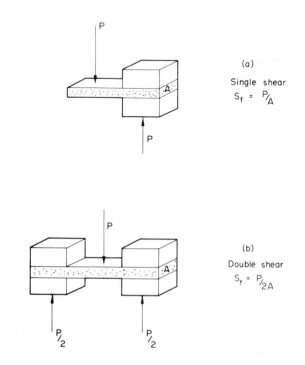

(a)

Single shear

$$S_f = P/_A$$

(b)

Double shear

$$S_f = P/_{2A}$$

Fig. 8.11 (a) Test of a rock beam in single shear. (b) The double shear test.

In all these shear tests the results obtained are determined by the testing arrangement and technique applied, as well as by the rock material. The tests convey no information that is descriptive of the inherent properties of the rock material in an absolute sense, or if they do, at the present state of knowledge we are unable to make such an interpretation. Yet the tests are useful in a laboratory that applies its own standardized procedures for comparative purposes. In this sense the tests serve a similar purpose for rock materials to that served by many forms of field tests for assessing the strength properties of rock masses.

In discussing the several varieties of shear tests on rock materials Everling suggests that the shear strength should only be derived from a series of compression tests under triaxial loading. In particular he makes a clear distinction between the strength of a material in pure shear, and the shear stress that is required to cause failure when the normal stress on the plane of fracture is zero. For most rocks there is considerable difference between these two values.

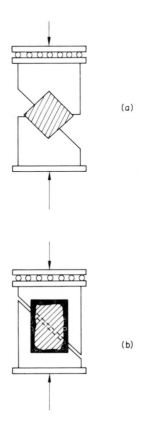

(a)

(b)

Fig. 8.12 (a) and (b) Arrangements for shear tests in compression loading.

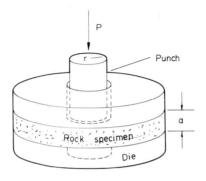

Fig. 8.13. The punch shear test.

Geotechnology

Comparative Tests of Rock Strength

Moh's Scale of Hardness

Moh's scale of hardness has long been used by mineralogists. It grades minerals according to their relative hardness as shown by a scratch test, in which minerals of a higher number on the scale are able to scratch materials of a lower numerical scale order. Thus, including, for comparison, some common materials:

1. Talc.
2. Rock salt, Gypsum. (Fingernail)
3. Calcite. (Brass pin)
4. Fluorspar
5. Apatite. (Glass)

6. Felspar. (Mild steel)
7. Quartz. (Hardened steel)
8. Topaz. (Tungsten carbide)
9. Corundum.
10. Diamond.

This scale is not an absolute measure, nor is it a satisfactory index of rock strength, since the abrasive qualities of a material, and not only its hardness, will determine its order as the result of a scratch test. Also, when two rocks of similar strength but different mineral content are to be compared, the scale must be suspect. Nevertheless, although crude, it is a simple test which in many cases may serve as a useful qualitative basis, for reference.

Hardness Tests

Hardness tests are sometimes used as an aid when assessing the comparative response of rocks to various forms of mechanical attack, such as drilling and blasting, or rock cutting. Two general methods are employed — measurement of indentation hardness and observation of rebound hardness.

Indentation hardness

The Vicker's indentation test has been used on rocks, although it is a test which is more generally applied to metals. The method is to force the tip of a square-based pyramid, of diamond hardness, into the prepared smooth surface of the material to be tested. The "Vickers hardness" is then defined as the ratio of the applied load to the surface area of the impression.

$$H_v = W/A = 2W/d^2 . \sin \theta/2$$

or

$$H_v = kW/d^2$$

where H_v is the Vickers hardness
W is the applied load
A is the surface area of the impression
d is the length of the diagonals of the permanent impression
θ is the angle between opposite faces of the pyramid
k is a constant

While the concept of Vickers hardness is thus simple and explicit, so far as the surface hardness of a homogeneous material is concerned, its connection with the response of a rock material to mechanical attack is obscure, to say the least. Nevertheless, it can be argued that a test, similarly conducted, could form the basis for a comparative classification of rock strength, if the equipment be suitably modified for rock materials. It is with such an argument that the use of such equipment as the Franklin point load apparatus has been advocated.

136

The point load index

The Franklin point load index apparatus consists of a small hydraulic pump and ram, coupled to a compression loading frame. A portion of a rock core-sample is loaded between pointed jaws, and the force P that is required to break the specimen is read from a calibrated pressure gauge in the hydraulic circuit. The distance D between the jaw contact points is also read from a graduated scale on the loading frame, from which the point load index

$$I_s = P/D^2.$$

It is claimed by its originators that the point load index is a measure of tensile strength. However, it is different from the indirect Brazilian and ring tests (in which crushing under the point loads destroys the validity of the theoretical concept of tensile failure at the center of the core sample) because in this case it is failure by crushing under the point loads that is directly observed. The specimen fails at a load lower than that which it could sustain under the conventional uniaxial compression test, and the relative contributions of the processes of shear and tensile failure are, no doubt, complex. Nevertheless, some of the uncertainties associated with the direct compression test, notably the variation in platen contact control, are eliminated, because failure of the specimen is deliberately induced by splitting under the point loads. Also, it is to be expected that the geometry of the specimen will have a much smaller influence on the results of a point load test than is the case with conventional compression testing. For all these reasons it may well be that the index provided by indentation tests may prove to be a rock strength parameter that can be more easily and more consistently observed, and hence of greater practical value, than either the uniaxial tensile or the uniaxial compressive strength.

Rebound Hardness Tests on Rock Materials

As with the Vickers hardness test, rebound testing on rock materials was first performed by the techniques designed for testing metals. The techniques were subsequently extended to the use of equipment specifically designed for testing concrete, and then to rock materials.

Shore scleroscope hardness test

The Shore scleroscope test consists of dropping a small weight, the underside of which is in the form of a pyramid of diamond hardness, on to the surface of the material to be observed. The height of fall is maintained constant and the height of rebound is observed for a number of tests, from which the results are treated statistically. The test measures the hardness of a very small area of the rock surface, on which the existence of very hard crystals such as quartz contributes to high values, and soft matrixes to low values, of rebound. Various investigators have attempted to correlate average rebound values or maximum rebound values to the compressive strength of rock materials. There appears to be a degree of correlation sufficient to encourage the evolution of rebound hardness testing for routine application to rocks, both in the field and in the laboratory.

Schmidt rebound test-hammer

The Shore scleroscope is a laboratory instrument, but the Schmidt rebound test-hammer was designed specifically to test concrete at the field site. The Schmidt hammer is also used extensively on rocks, in connection with mining and civil engineering operations, particularly on the Continent of Europe. The operating principle of the hammer is to release a steel mass, which, on the release of a catch, is projected a fixed distance by the action of a spring, to impact upon an anvil which is held against the surface under observation. The hammer is designed to develop a kinetic energy of 9.0 J upon striking the anvil. When in use the instrument is directed so as to be normal to the rock surface, the gravitational effect of any inclination of the tool being allowed for in the observation of rebound distance, which is measured on a scale.

The velocity of impact of a hard elastic body upon another body is dependent upon the coefficient of restitution of the impacted material, or the "dynamic elasticity", such that

$$k = V_2/V_1 = h/H$$

where k is the coefficient of restitution,
     V_1 is the velocity of impact,
     V_2 is the velocity of rebound,
     h is the height of rebound,
     H is the height of impact.

The velocities of impact and rebound are both functions of the hammer design, and H is constant. h is observed to be used as an approximate indication of the strength of the rock

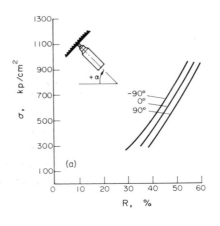

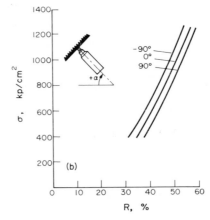

Fig. 8.14. The height of rebound R as a function of compressive strength for (a) argillaceous shale and (b) siltstone and sandstone (Hucka).

against which the hammer is applied. Hucka quotes the results for rebound hardness tests on three sedimentary rock types. It is apparent that there was an approximately linear correlation with compressive strength, so far as these materials were concerned.

Impact Tests on Rock Materials

The response of rocks to impact loads is a matter which takes us into the realms of rock dynamics which, so far in this text, has not been discussed. We have, however, looked at some, and we will have mention of other, comparative tests, which are used to assess both the "hardness" and the compressive strength of rock materials. The geotechnical engineer is often concerned with matters in which the combined static and dynamic properties of rock materials are relevant — for example, the study of rock fracture as a result of attack by mining tools. For the sake of completeness in our catalogue of tests of rock strength we should include at this point some mention of impact tests. The ability of a rock material to withstand impact is sometimes designated the "toughness" of the rock. Two methods of impact testing are generally employed — the split Hopkinson Bar technique and the vertical drop-hammer.

The Hopkinson Bar

The Hopkinson Bar technique is primarily a research tool, and is used to study the propagation of stress pulses generated by a mechanical hammer, or by a small explosive charge, impacted against one end of the suspended bar system. Attewell's arrangement of the apparatus, as applied to rock testing, is shown in Fig. 8.15. The arrangement consists of two steel bars, 25mm diameter, suspended horizontally by piano wires. The specimen to be tested is prepared either in cylinder or disc form, to the same diameter as the steel, and is held axially in contact between the two steel bars. Suitable sensing devices, to measure the strain and stress pulses, are placed along and at one end of the bar. Impact from a mechanical hammer, or a small explosive charge, is applied at the other end of the bar. Typical dynamic stress curves for a sandstone, observed by Attewell, are shown in Fig. 8.16.

The ASTM Rock Toughness Test

The vertical drop impact test has been recommended as a standard "toughness" test for rock materials in the U.S.A. In such a test the toughness of the rock is expressed in terms of the height of drop, required of a standard hammer, to produce fracture of a test specimen of standard dimensions. The ASTM equipment consists of a 2-kg hammer mounted in a vertical guide-sleeve, and able to fall freely to impact upon a 50-kg, cast iron anvil. The test specimen is prepared in the form of a cylinder 25mm in length by 24 to 25mm diameter, from selected blocks of rock free from flaws or incipient fractures. Each specimen is placed, end down, on the anvil and then subjected to repeated blows from the drop-hammer, starting with 1 cm fall for the first blow, 2 cm fall for the second, and increasing 1 cm fall for each successive blow, until the rock fractures.

The height of blow at failure is reported as the toughness of the test specimen. At least six specimens of the same rock material are tested, three being prepared so that the direction of impact is perpendicular, and three parallel, to the plane structural weakness, should any such plane be apparent. The individual and the average toughness of the three specimens in each set are separately identified and reported.

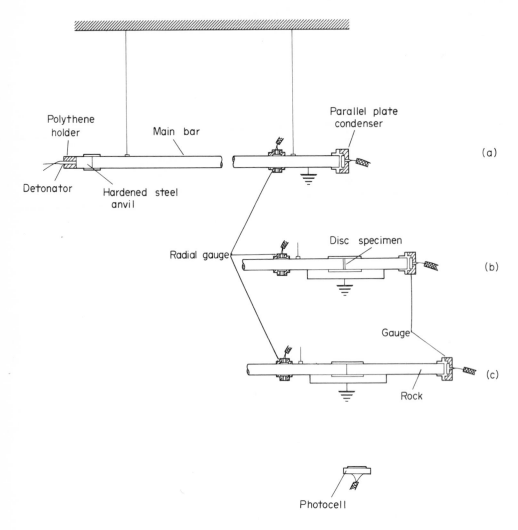

Fig. 8.15. Arrangements of Hopkinson Bar equipment for the observation of dynamic stress–strain behavior of rock materials (Attewell).

Protodiakonov Impact Toughness

A technique that is applied to determine the strength of coal forms the basis of design for coal-mining machinery in the U.S.S.R. This is the index of impact toughness, evolved by Protodiakonov, in which a number of samples of the rock material under test, of a graded sieve size, are placed at the bottom of a metal pipe, blanked off at the base. A standard weight is dropped from a fixed height in the pipe, after which the contents of the pipe are again sieved. The amount of degradation of the rock material provides a measure of its "toughness" for which the "Protodiakonov Number" serves as an index. This is derived from the volume of material passing through a given sieve size after the test.

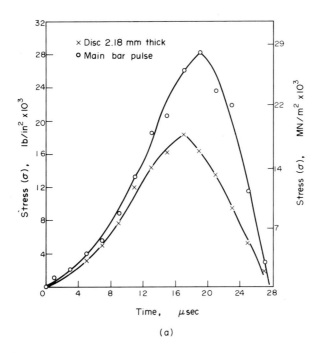

Fig. 8.16 (a) Dynamic stress-time curve for a sandstone (Attewell).

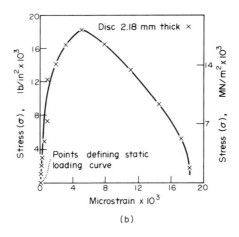

Fig. 8.16 (b) Dynamic stress-strain curve for the same sandstone. (Points at 2 microsec intervals). (Attewell).

The Expanding-bolt Rock Tester

In Great Britain the toughness of coal seams has been assessed by an expanding rock-bolt method. The test consists of drilling a hole, about 150mm deep and 37mm diameter, into the rock face, and then inserting an expansion-bolt into this hole. The end of the bolt that protrudes from the rock is then gripped and pulled by a hydraulic ram. As a result of this pull, the wedge-end anchor of the bolt is expanded sideways into the rock. With increasing ram-pull the strength of the rock in which the bolt is embedded is ultimately overcome, and failure of the rock ensues. A conical segment of rock is thus torn from the side of the rock face.

Early applications of this technique, devised by Foote, related the dimensions of the cone of fracture to a strength index, possibly allied to the shear strength, using the formula

$$S = \frac{3F}{\pi h (R + 2r)}$$

where F is the breakage force required,
  h is the cone depth,
  R is the cone base radius,
  r is the radius of the central bolt-hole,
  S is the strength index.

However, Evans has suggested a tensile failure theory, according to which

$$F = 3.5T (h^2 + rh)$$

where T is the tensile strength.

Since this testing technique is mainly applied in connection with the assessment of the probable ability of coal ploughs to cut into the seam concerned a (mechanism in which tensile failure of the rock is more apparent than shear) the tensile theory would appear to be the more realistic. In this connection the method has added interest as a means of determining the tensile strength of friable rock materials such as coal, which often presents practical problems when the conventional techniques, previously described, are attempted.

The CERCHAR Rock Resistance Index

The rock mechanics laboratories of CERCHAR, France, have a technique to assess both the hardness and the abrasive characteristics of rock materials. Both these properties are significant in relation to any assessment of the relative ease by which the material may be drilled or cut by mining tools. The hardness of the rock is measured by drilling into it, using a tungsten-carbide drill-bit. An index of hardness is then defined as the time taken to drill a hole 1 cm deep, at a constant speed of rotation, and using a standard thrust on the drill. This hardness test permits of a certain degree of strength classification of rock materials (cf. Tsoutrelis, p. 128) but CERCHAR holds that it is not enough in itself, because drill tests carried out on rocks with similar hardness values can produce very different rates of wear at the cutting-tip of the tool, other things being equal. Consequently, an abrasivity test is added, in which the wear on a pointed steel rod is measured, after dragging it across the surface of the rock. As a result of this, rocks of the same hardness may be further classified in order of abrasion quality.

The Specific Energy Concept

The concept of specific energy is applied to processes of rock excavation by drills and rotary boring and tunneling machines. Specific energy is defined as the energy required to excavate unit volume of rock. It is therefore a direct measurement of the rock strength in relation to the effectiveness of any rock-cutting process. For example, it may be used to establish the relative efficiency of various tools, machines, or cutting processes in a given rock material. Conversely it can establish the relative order of resistance of various rock materials to a given machine, or to a specific mode of excavation. It has yet to be shown, however, whether the performance of a given tool in various rocks can be quantitatively related to the performance of other tools in the same rocks.

The specific energy is determined by using a rotary drill on the rock material. In this operation work is done by the thrust F and by the torque T. If the speed of rotation of the drill bit is N rev/min, the total work done in 1 min is $(Fu + 2\pi NT)$, where u is the penetration rate.

The volume of rock excavated in 1 min is (A_v), and the specific energy (E_s) = work/volume.

$$E_s = F/A_v + (2\pi/A_v) (NT/u).$$

The term (F/A_v) is the "thrust" component of the specific energy, and the term $(2\pi/A_v)$ (NT/u) is the "rotary" component.

If p is designated the penetration per revolution, in rotary drilling, then $p = u/N$, and the rotary component of specific energy becomes $(2\pi/A_v)$ (T/p) where T is the torque required to move a layer of rock of depth p, in one revolution of the drill.

The ratio T/p may itself be used as a comparative index of specific energy in rock materials.

Gaye has extended Teale's concept of specific energy, to scale-up small-hole laboratory drill tests to field-scale machine cutting. He found that in a given cutting process the size of debris produced is a function of the specific energy, and also a function of the ratio

$$\frac{\text{Compressive strength}}{\text{Specific energy}} = f_c/E_s$$

This ratio f_c/E_s, Gaye terms the "rock number" (N_r). N_r is approximately constant for a given cutting process, irrespective of scale, provided that the tool is being operated under normal working conditions. By using the rock number, and observing the results of small-scale drill tests, it is possible for a laboratory "back-up" service to give field engineers a sound qualitative estimate of the resistivity of various rocks to a particular mode of attack.

Selected References for Further Reading

ASTM Standards Part 2. *Toughness of Rock,* Section D3-18, p. 427 (1942).

ATKINSON, R. H. and KO, H. Y. A fluid cushion, multiaxial cell for testing cubical rock specimens. *Int. J. Rock Mech. Min. Sci.,* vol. 10, pp. 351-361 (1973).

ATTEWELL, P. J. The response of rocks to high velocity impact. *Trans. Inst. Min. Met. London,* vol. 71, pp. 705-724 (1962)

BIENIAWSKI, A T., DENKAUS, H. G. and VOGLER, V. W. Failure of fractured rock. *Int. J. Rock Mech. Min. Sci.,* vol. 6, pp. 323-341 (1969)

BRACE, W. F. and MARTIN, P. J. A test of the law of effective stress for crystalline rocks of low porosity. *Int. J. Rock Mech. Min. Sci.,* vol. 5, pp. 415-426 (1968).

COATES, D. F. and PARSONS, R. C. Experimental criteria for classification of rock substances. *Int. J. Rock Mech. Min. Sci.,* vol. 3, pp. 181-189 (1966)

DENKHAUS, H. G. Strength of rock materials and rock systems. *Int. J. Rock Mech. Min. Sci.,* vol. 2, pp. 111-126 (1965).

EVANS, L. The expanding bolt seam tester — a theory of tensile breakage. *Int. J. Rock Mech. Min. Sci.,* vol. 1, pp. 459-473 (1964).

EVERLING, G. Comments on the definition of shear strength. *Int. J. Rock Mech. Min. Sci.,* vol. 1, pp. 145-157 (1964).

FOOTE, P. The expanding bolt seam-tester. *Int. J. Rock Mech. Min. Sci.,* vol. 1, pp. 255-260 (1964).

FRANKLIN, J. A., BROCH, E. and WALTON, G. Logging the mechanical character of rock. *Trans. Inst. Min. Met. London,* vol. 80, section A (1971).

HAWKES, I. and MELLOR, M. Uniaxial testing in rock mechanics laboratories. *Engng. Geol.,* vol. 4, pp. 177-284 (1970).

HETENYI. M. *Handbook of Experimental Stress Analysis,* p. 15, J. Wiley & Sons, New York, 1966.

HOBBS, D. W. A simple method for assessing the uniaxial compressive strength of rocks. *Int. J. Rock Mech. Min. Sci.,* vol. 1, pp. 5-15, (1964).

HOEK, E. Rock mechanics — an introduction for the practical engineer. *Mining Magazine, London,* 1960.
HOEK, E. The brittle failure of rock. In *Rock Mechanics in Engineering Practice,* Stagg and Zienkiewicz (Eds.), pp. 99-124, Wiley, 1968.
HUCKA, V. A rapid method for determining the strength of rock *in situ. Int. J. Rock Mech. Min. Sci.,* vol. 2, pp. 127-134 (1965).
HUCKA, V. and DAS, B. Brittleness determination of rocks by different methods. *Int. J. Rock Mech. Min. Sci.,* vol. 11, pp. 389-392 (1974).
LAJTAI, V. N. and LAJAI, E. Z. Fracture from compressive stress concentrations around elastic flaws. *Int. J. Rock Mech. Min. Sci.,* vol. 10, pp. 265-284 (1973).
MELLOR M. and HAWKES, I. Measurement of tensile strength by diametral compression of discs and annuli. *Engng. Geol.,* vol. 5, pp. 173-225 (1971).
MORLEY, A. *Strength of Materials,* p. 35, Longmans Green & Co., London, 1944.
MURRELL, S. A. F. A criterion for the brittle fracture of rocks and concrete under triaxial stress, and the effect of pore pressure on the criterion. *Proc. 5th Rock Mechs. Symp., University of Minnesota,* pp. 563-577, Pergamon Press, 1963.
OBERT, L. and DUVALL, W. I. Mechanical property tests. In *Rock Mechanics and the Design of Structures in Rock,* pp. 318-354, Wiley, 1967.
OBERT, L. and DUVALL, W. I. *Rock Mechanics and the Design of Structures in Rock,* p. 728, John Wiley & Sons, New York, 1967.
RALEIGH C. B. and PATERSON, M. S. Experimental deformation of serpentinite and its tectonic implications. *J. Geophys. Res.,* vol. 70, pp. 3865-3885 (1964).
RAMSAY, J. G. *Folding and Fracture of Rock,* p. 289, McGraw Hill Book Co., London, 1967.
SKEMPTON, A. W. Effective stress in soils, concrete, and rocks. In *Pore Pressure and Suction in Soils,* Butterworth, London, 1960.
TEALE, R. The concept of specific energy in rock drilling. *Int. J. Rock Mech. Min. Sci.,* vol. 2, pp. 57-73 (1965).
TSOUTRELIS, C. E. Determination of the compressive strength of rock *in situ* or in test blocks, using a diamond drill. *Int. J. Rock Mech. Min. Sci.,* vol. 6, pp. 311-322 (1969).

CHAPTER 9

The Engineering Properties of Rock Masses

Correlation of Material Properties with Mass Strength

The extent to which the strength properties, and the load-deformation characteristics, of a rock mass may coincide with the result of tests made on laboratory specimens of the rock material is a matter of major concern in geotechnology.

As with rock materials in the laboratory, so with rock masses in the field, the engineering strength properties must be evaluated in terms of measured, deduced or assumed changes in load, stress, and strain. The situation is more difficult in the field, however, since it is not always possible to control the stresses, strains, and loads to which the specimen — the rock mass — is subject. While mass deformations can be measured with relative ease, and loads may sometimes be imposed by controlled increments, or sometimes fairly accurately deduced, on other occasions the applied loads may only be crudely estimated. The detailed distribution of the stresses produced by those loads is, more often than not, a matter of pure conjecture.

Measurements and qualitative observation together must provide the evidence from which analytical and intuitive deductions must be made. But measurements in themselves do not provide solutions to practical problems in geotechnology. Interpretation is the key. The measurements must be interpreted and assessed in relation to all other available evidence and they must then be applied in the light of the engineer's own background, education, training, practical experience, and expertise in the particular field concerned.

Natural and Induced Mass Deformations

Deformations of large masses of earth materials may be the result of natural phenomena, or they may be induced by some other means.

Natural deformations

Among natural mass deformations we should include the steady creep of earth slopes, possibly culminating in landslides, under the influence of gravitational loads. Natural deformations also include the build-up of strain in the vicinity of fault zones within the Earth's crust and mantle. In this case tectonic forces are in evidence. Ultimate failure occurs in the form of an earthquake, in the process of which impulsive shock loads are generated, with rapid release of the energy stored in the strained rock.

Induced deformations

Changes in load produced by operations such as the construction of structures, tower blocks, dams, and refuse tips, will deform the earth materials upon which they stand. In the case of dam construction a further heavy load is imposed by the impounded reservoir. For example, it is estimated that the construction of the Hoover Dam which impounds Lake Mead created a load due to the weight of water, of around 40 million tons. The resulting deformation in the underlying strata extends some considerable distance beyond the lake boundary. Deformation in rock masses may also be produced as a result of the relief of load, and the removal of constraint caused by extensive excavations, open-pits, and earth and rock cuts in roadway and canal construction.

In the case of dam and reservoir construction, impounding the water behind the dam raises the water table in the underlying strata. This alters the water-seepage pressures in the dam foundations and in the rock slopes bordering the reservoir, causing deformation of these materials, the extent of which should be observed and controlled by the engineers in charge. Underground excavations create intensified load in their supporting pillars which will yield to a greater or lesser extent, determined by their size and strength. The effects will extend through the overlying rocks to appear as surface ground subsidence. The construction of underground excavations can produce widespread surface subsidence and damage to surface structures.

Competent and Incompetent Rock

In reviewing the correlation of material properties with mass strength it is useful to classify rock masses as being either "competent" or "incompetent". A competent rock mass is one which is able to sustain the applied loads without mass failure. The strength properties of the rock material together with the bond strength between any discontinuities are able to maintain equilibrium of the rock mass as a whole.

It is only when the rock is competent, homogeneous, and elastic that the results of laboratory tests on small specimens may safely be translated to the mass. In general the discontinuities existing in a rock mass reduce its uniaxial compressive strength, as compared with the value determined in laboratory tests, but in a homogeneous rock mass only comparatively minor discontinuities are likely to be present. The strength in the interior of large supporting rock pillars in a mine, or within the walls of an excavation, can be expected to be higher than the uniaxial compressive strength of laboratory specimens, depending upon the degree of confinement at the point concerned. However, if the width of the pillars is small, in proportion to their height, their uniaxial compressive strength will probably be less than that demonstrated by the rock material in the laboratory.

The walls of excavations in certain types of rock, such as brittle, elastic, fine-grained igneous rocks and massive limestones, are prone to demonstrate localized failure in the nature of "spalling", in which thin shells of rock are detached from the mass, sometimes with explosive violence. The phenomenon is probably associated with high stress gradients in the immediate walls of the excavation and relaxation in the surface layers of rock, in circumstances where the *in situ* stress level is high.

Some rock masses may include major discontinuities, such as laminations, in the nature of sedimentary layers and joint partings parallel with the sedimentation planes. Others may be jointed, with discontinuities forming three or more planes of weakness in intersecting directions, so that the rock mass is divided into separate and discontinuous blocks. The "competency" of these rock masses is primarily dependent upon the tensile, shear, and frictional strength properties inherent in the joint infilling material, and the orientation of the discontinuities relative to the *in situ* stress directions.

It is possible to make laboratory direct shear tests and triaxial tests on rock specimens to include a typical discontinuity or plane of weakness. The effects of such planes can also be included in the determination of failure envelopes for the rock. Considerations of size, however, limit the extent to which the direct testing of laboratory specimens is possible, and the translation of analytical treatments to full-scale situations in which all the relevant parameters (stress and strain distribution, etc.) may be unknown must be questionable.

Where the rock mass is competent and laminated, as is the case with many sedimentary deposits, the laboratory tensile strength of a rock beam, in flexure, may be relevant, in relation to the strength of a rock layer spanning the roof of an excavation. However, the geotechnical engineer must exercise caution here. It must not be forgotten that, while the two-dimensional representation of rock excavations in cross-section may show geometrical distributions of load, stress, and strain very similar to those that exist in the elastic prototypes used by the structural engineer, representing beams and plates in bending, and the like, the real situation in a rock mass, in three dimensions, will probably be far more complex than those represented in these diagrams.

Some rock masses, such as rock salt and the evaporite rocks, may be competent and homogeneous, but inelastic. They may demonstrate brittle fracture under uniaxial load and at low confining pressures, but the effects of increased lateral confinement will change the failure characteristic to one of plastic flow. In these circumstances an analytical approach to the design of excavations is difficult, as elastic theory is irrelevant here and the application of other approaches is hampered by lack of basic information, particularly that regarding the stress distribution within and the deformability of the rock mass.

When the rock mass is "incompetent" it is implied that the rock cannot withstand the imposed loads and remain intact. The rock mass is broken or becomes plastic and its equilibrium is destroyed. The occurrence of rock and earth slides, subsidence, and caving are the result. So far as rock excavations are concerned, techniques of reinforcement and support of the rock walls are required to restrain movement and, if possible, to restore equilibrium. If the total restoration of equilibrium is not possible, then rock reinforcement and support techniques must be devised so as to control the rate and extent to which the rock mass moves in its tendency to close the excavations. This is the practice of "strata control" in mining, in which intuition, developed as a result of practical experience, plays a major role.

While analytical approaches to design in the saline evaporite non-elastic rocks, and in incompetent rock masses, are of little assistance, so that qualitative assessments must take precedence, some help may be gained from scale-model studies in simulated rock materials, or "equivalent" materials, and sometimes in the rock material itself.

The Effect of Scale on Rock Strength

The strength of a material is a statistical concept if it is assumed that failure originates as a result of the progesssive extension of microcracks. The chances of a sample of the material containing a microcrack of critical size and orientation from which failure will develop relative to the imposed stress field will diminish with decrease in size and increase with increase in size of the specimen. This is the essence of Weibull's "weakest-link" theory, in which it is held that the relative strength of different volume samples of a material is governed by a relationship

$$\log \frac{V_2}{V_1} = m \log \frac{\sigma_2}{\sigma_1}$$

where σ_1 and σ_2 are the strengths of specimens of volume V_1 and V_2, respectively, and m is a constant.

Protodiakonov also studied the size effect on the strength of rocks and derived the formula

$$\sigma_d = \frac{d + mb}{d + b} \sigma_m$$

where $m = \sigma_0 / \sigma_m$

σ_0 is the strength of a cubical rock specimen of side length d,

σ_m is the strength of the rock mass ($d = \infty$),

σ_0 is the strength of the specimen when $d = 0$,

b is the distance between discontinuities in the rock mass.

However, there is considerable doubt as to the extent to which these relationships may safely be applied. Some investigators assert that Weibull's theory is applicable to rocks in tension, but not in compression. Others have found that neither Weibull's nor Protodiakonov's formulae agree with experimental results beyond a limited size range. The matter is important both in its implications regarding the effects of scale on the strength displayed by laboratory rock specimens, and in its effect on the significance of field tests as a measure of rock mass properties.

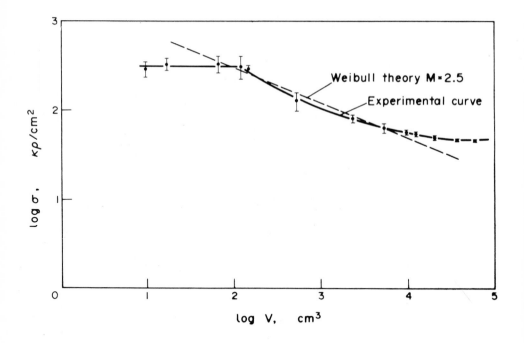

Fig. 9.1. Strength-specimen volume relationship for coal.
Experimentally and according to Weibull's formula (Bieniawski).

Bieniawski determined the relationship between strength and specimen size of coal pillars loaded in uniaxial compression to be that illustrated in Fig. 9.1. From this it would appear that the strength demonstrated in his experimental tests approached a constant value when the coal pillars under test reached around 2m width. Pratt *et al.*, quoting the results of field and laboratory tests on quartz diorite, agree with Bieniawski that there is relatively little change in strength for specimen sizes greater than a few metres. The deformation moduli for a quartz diorite and a granodiorite decreased very little, with increase in specimen size. The size at which constancy of strength is approached will undoubtedly be related to the rock type and its geological structure. Pratt *et al.*, summarize available data collected from various sources in Fig. 9.2, but much more

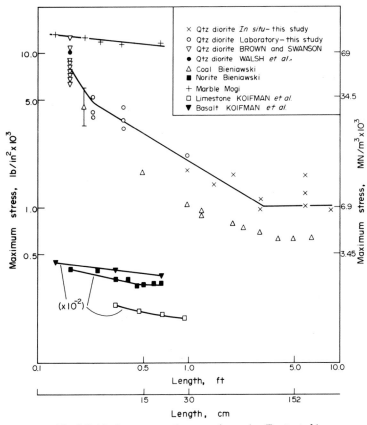

Fig. 9.2. Maximum strength vs. specimen size (Pratt *et al.*).

research is required before the matter can be satisfactorily resolved. Since field tests on a large scale are expensive, and present many practical problems, the subject is of considerable importance. In the absence of firm information, the prudent engineer should make his field tests on as large a scale as his budget, and the practicalities of the situation, will allow.

Field Tests on Soils and Rocks (Static Loading)

Various field tests may be applied to obtain information as to the strength of soils and rock masses.

Bearing Capacity

The bearing capacity of an earth material is important in that it determines the maximum value of contact pressure that can be applied to the material without causing shear failure. When discussing failure criteria it has so far been assumed that the major mechanism of yield is elastic, but when considering the settlement and yield of material, such as the foundation strata beneath a wall or a dam, or the soil upon which a heap of debris has been dumped, plastic yield may be equally, if not more, important. While the failure characteristics of a soil foundation under an imposed load are often analyzed as a process of shear failure on a curved slip surface, there is sometimes to be seen a "heave" or plastic yield of the soil around the foundation, accompanied by a sudden settlement of the structure concerned.

Plastic yield of rock materials in the mass is also important in relation to hard rock boring and tunneling machines operating at high values of thrust unaccompanied by impact in cutting (although sometimes assisted by wedge penetration at the point of application of the thrust). It is important, too, in connection with the operation of mine-strata control systems in which the interposition of a rigid support component may be required in order to form a strong "break-off" line to induce caving of the roof beds. The rigidity of such a support system is determined by the ability of the roof and floor strata to take the pressures generated by the imposed load, without yielding in such a way that the supports penetrate the rocks with which they are in contact.

Theories of bearing pressure

Theories describing the plastic yield of material subjected to a bearing pressure have been developed by several researchers. Prandtl described the mode of failure illustrated in Fig. 9.3,

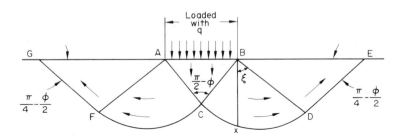

Fig. 9.3. Mode of failure under a bearing load, as assumed by Prandtl.

for which it was assumed that the material is homogeneous, isotropic, and weightless, and that its Mohr failure envelope is linear. The material in the wedges AFG, ABC, and BED was assumed not to deform, while wedges AFC and BCD deform plastically. In those plastic regions the stresses vary with direction, but along any radial direction such as BX they are constant.

Prandtl showed that the ultimate bearing capacity of the foundation material is

$$q = \frac{c}{\tan \varphi} \left[\frac{1 + \sin \varphi}{1 - \sin \varphi} e^{\pi \tan \varphi} - 1 \right]$$

where c is the apparent cohesion of the material and φ the angle of internal friction, in the Coulomb equation $s = c + \sigma_n \tan \varphi$, and s = shear strength.

According to Prandtl's formula, a soil without cohesion, such as a dry sand, would have no bearing capacity either, which is patently untrue. Corrections to take into account the weight of the material are also required.

For shallow foundations and most soil conditions, coefficients determined by Terzaghi's bearing capacity analysis are commonly applied. In this the mode of failure is assumed to be as shown in Fig. 9.4 in which a uniform strip load is applied to a homogeneous material. The bearing capacity is then given by

$$q = \frac{by (1 - \tan^4 \beta)}{2 \tan^5 \beta} + \frac{2c}{\tan \beta \sin^2 \beta}$$

where y is the bulk density of the material,
         $2b$ is the width of the loaded strip,

$$\beta = \left(\frac{\pi}{4} - \frac{\varphi}{2} \right)$$

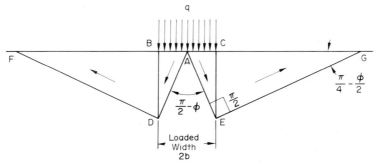

Fig. 9.4. Mode of failure under a bearing load, as assumed by Terzaghi.

If it is assumed that the foundation material behaves elastically then the pressure distribution in the foundation strata can be deduced by using Boussinesq's analysis, which describes the distribution of vertical pressure on a horizontal plane subjected to point loading. Using the appropriate mathematical treatment the analysis may be extended to determine the theoretical stress distribution for all points below the loaded surface. That stress distribution may then be represented graphically by "pressure bulbs", such as depicted in Fig. 9.5.

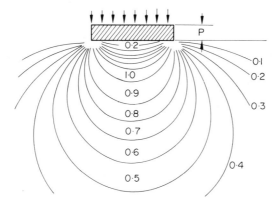

Fig. 9.5. Maximum shear stress under a uniform strip load.

Safe bearing capacities

Safe bearing capacities for various earth materials are quoted in national codes of practice, in foundation engineering. Typical figures, extracted from Civil Engineering Code of Practice No. 4 (Inst. C.E. London), are given in Table 9.1.

TABLE 9.1. *Maximum Safe Bearing Capacities for Shallow Foundations (2 ft below ground surface) on Horizontal Strata under Vertical Static Load (Inst. C.E. London)*

	Type of material	Max. safe bearing capacity (tons/ft^2)		
Rocks	Igneous. Gneiss (Sound)	100		
	Massive bedded limestones and hard sandstones	40		
	Schists. Slates	30		
	Hard shales, mud-stones, soft sandstones	20		
	Clay shales	10		
	Hard, massive chalk	6		
		Dry	Saturated	
Non-cohesive soils	Compact, well-graded sands and gravel-sand mixtures	4–6	2–3	
	Loose well-graded sands and gravel-sand mixtures	2–4	1–2	
	Compact uniform sands	2–4	1–2	
	Loose uniform sands	1–2	½–1	
Cohesive soils	Stiff boulder clays and hard clays with shaley structure	4–6		
	Stiff clays and sandy clays	2–4		Susceptible to long-term consolidation settlement
	Firm clays and sandy clays	1–2		
	Soft clays and silts	½–1		
	Very soft clays and silts	½–nil		

Field Determination of Bearing Capacity

The plate bearing test on soils

The determination of the bearing capacity of a foundation material in the field is performed by means of a bearing plate, of steel, of a size ranging from $0.1 m^2$ to a disc up to 1m diameter. The plate is forced against the surface under examination, by a hydraulic jack abutting against a temporary bridge structure which provides the necessary reaction point. The load of the jack is measured by a pressure gauge in the hydraulic system, and any movement or settlement of the plate is observed by a number of dial gauges mounted around the periphery of the plate. The average settlement recorded by these gauges is plotted against the applied load, to give a curve similar to one of those shown in Fig. 9.6.

On stiff clays, and well-compacted, dense sands and gravels, the curve is likely to resemble line AB, which depicts an initial linear relationship with a well-defined "break-point". The safe bearing pressure may be specified as q/f, where q is the bearing pressure at the break point and f the "factor of safety". But if the foundation material consists of soft clay or loose sand the results of a plate bearing test would be more likely to follow a curve such as AC, on which no definite failure point is apparent. The interpretation of such a result calls for great care, because the test reflects only the properties of the ground immediately beneath the applied load, to a depth approximately equal to the width of the bearing-plate. The extent to which the results may safely be extrapolated to a full size foundation is, as yet, a matter of considerable doubt.

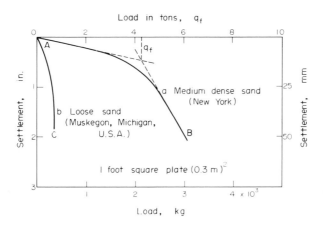

Fig. 9.6. Typical results for plate-bearing tests on various soils (after Terzaghi).

Plate bearing tests on rocks

Ward and Burland describe a site investigation to determine the deformation properties of chalk foundations for a large engineering structure. The investigation was comprised of three stages. (1) A detailed visual examination of the geological structure, with particular attention to patterns of jointing, lithology, and any petrological features likely to influence the relative deformability at the points of examination. (2) Measurement of the deformation characteristics, as shown by plate bearing tests made at successive depths at the base of auger holes approximately 0.3m diameter. (3) A large-scale surface deformability test, termed a tank test.

The plate loading tests were made using a circular steel plate 0.86 m diameter and 0.13 m thick, loaded centrally through a tubular loading column, which was pressurized by hydraulic jack. This was braced against a cross beam, anchored by a pair of concrete piles. The gross deflections were measured at the center of the plate, by a rod extensometer located inside and independently of the loading column. A thermally stable reference beam, also separately supported, provided the measurement datum. Load was applied in increments over three cycles to 4.5 kg/cm^2, and then up to a maximum of 16.7 kg/cm^2.

Plate-bearing tests are sometimes conducted on rocks underground, for example within the site-investigation tunnels of the foundation rocks and the abutment zones of a proposed dam, or in connection with mining and tunneling operations. The primary object here is to determine the load-deformation characteristics within the rock mass, and these are sometimes expressed in terms of a "deformation modulus". Heuze and Goodman describe such a test in a limestone mine.

The relatively low values for the deformation modulus (E), recorded by this investigators, are primarily determined by the closure of discontinuities in the zones of influence of the jack loads, and not by the strength properties of the limestone mass material. Jaeger and Cook point out that while the stresses and displacements are theoretically related to the applied load by a Boussinesq relationship, in practice the boundary conditions beneath the bearing plate must lie between the extremes posed respectively by constant displacement with varying stress, and uniform stress with varying displacement. If the bearing plate does not yield, but is completely rigid during the test, the rock must fail around the perimeter of the plate, because the stress

153

there is, theoretically, at an infinite value. In practice this is not likely to be so, for the bearing plate will be flexible, to some extent. Some degree of uncertainty as to the bearing plate contact conditions must therefore exist.

It is customary to insert a flexible layer between the bearing pad and the jack, the purpose of the layer being to try to achieve a uniform distribution of pressure over the loaded surface. The layer may take the form of a rubber pad, or sometimes a hydraulic "cushion" or "flat-jack" will be used to apply the load. This is an envelope of sheet metal containing oil. Another layer of cement mortar is usually placed below the bearing pad, to level-off any uneven parts and interstices on the rock surface.

Details of a series of plate loading tests made in an underground gallery at a depth of 30 m, in directions perpendicular and parallel to the stratification, in schist, are given by DeBeer, Delmer, and Wallays. These investigators used loading plates measuring 1 m^2, and a maximum applied pressure of 34.5 kg/m^2. The underground gallery tests were supplemented by a surface plate loading test using a circular plate 4 m^2, with 5 kg/cm^2 maximum applied pressure, obtained by deadweight loading.

Strata Penetration Tests

Tests to determine the strength of roof and floor strata in mines are sometimes conducted by means of hydraulic dynamometers. In this case the object of the test is to determine the maximum resistance that the strata can afford, in the support-unit contact zones. This determines at what value of hydraulic pressure the supports must yield, if a continuity of support is to be maintained in moving ground, or at what distance the supports must be spaced, if they are to provide a rigid "break-off" line along which the roof strata can be induced to fracture on a caving system.

The dynamometer prop is, in effect, a hydraulic jack fitted with a pressure gauge, which displays the load on the prop at any time. Various sizes and shapes of bearing-plate may be fitted at the base, and at the head, of the prop, to control the areas of contact. Penetration of the ends of the prop into the strata is measured on a vertical scale, and, when necessary, observed by telescope from a distance. Another instrument, used for a similar purpose, is the "penetrometer", on which the load is measured by proving ring, for greater accuracy.

Borehole rock penetrometer

Stears has described a hydraulically operated penetrometer which was developed by the U.S. Bureau of Mines in an attempt to measure the capacity with which the strata around mining excavations can provide secure anchorage for rock-bolt reinforcement. Essentially the device consists of a closed hydraulic system containing two movable pistons. These pistons, the pump piston and an indenting piston, are forced sideways into the wall of a borehole, when pressure is applied through a screw pump. Tests on the penetrometer showed that it was impossible to make any absolute deductions of the anchorage capacity of the rock, from the observed penetration-bolt tension relationships. Nevertheless the device is useful in that it can identify the relative hardness of the various strata that may be encountered along the length of a borehole. Hence, it can locate the best anchorage position in that particular hole. The device can differentiate between homogeneous materials of different hardness, and repetitive penetrometer readings in the same materials are reproducible to within approximately 5.5%.

Soil penetrometer

Penetration tests of a different kind are made in foundation engineering, to assess the penetration resistance of soils at depth. The tests are varied in character, ranging from the use of a simple probe (the purpose of which is to locate a resistant zone or to determine the "bottom" of a weak stratum), to more elaborate cone-penetration tests (intended to determine the characteristics of side-friction and point-resistance in piling). A typical representative of these tests is the Dutch cone-penetration test, in which a metal cone having an end-bearing area of 10 cm^2 is used. The cone may be forced ahead by an internal push-rod passing through an external cylinder. The external wall of the cylinder can thus measure side-friction independent of end-resistance. The tests may be conducted so as to observe the resistance of a soil to a continuously applied load, or sometimes to observe its resistance to penetration by impact. In the latter case, the distance penetrated by a specific number of measured blows on the penetrometer is used as a resistance criterion.

In situ Deformability Tests

It is customary to measure the deformability of rock masses that are to form the foundations and abutments of large civil engineering structures, such as dams. The tests may be made in exploration galleries that are driven into the rock walls in the abutment zones, or driven laterally from shafts sunk into the foundation rocks. The deformability tests usually include plate-bearing tests, sometimes repeated in different directions in order to study the effects of non-isotropy in the rock mass. The general procedure is to apply cyclic loading and unloading, with four or five cycles, increasing in successive increments of 20% to 25% of full load. The rock mass usually displays considerable hysteresis and, depending upon the requirements of the investigation, a "deformation modulus" may be deduced from measurements of gross deformation with increasing load or, alternatively, an "elastic modulus" may be determined by taking into account only that proportion of the deformation that is seen to be recovered after full load is released.

Flat-jack deformability tests

Continental European practice favors the use of hydraulic flat-jacks, so as to extend the area of contact on the rock surface, over which pressure may be directly applied by the jack. The jack may then take the form of a pair of circular or square metal plates, measuring up to about 2 m across, and these are welded together around their periphery to enclose hydraulic fluid. They are cemented by concrete into slots cut into the floor of an exploration tunnel, as shown in Fig. 9.7. Hydraulic pressure is applied by means of a hand-pump. In Kudjundzic's system the insertion of an expansion chamber and volume-measuring gauges into the hydraulic system enables any change in volume of fluid (due to expansion of the jack under pressure) to be measured. The load-deformation characteristics are thus observed, making corrections for compression of the contained jack-fluid and also for the deformation of the concrete in which the jack is set. Similar tests should be made at four or five measurement stations, to obtain an overall assessment of the deformability of the rock mass.

There are several variants of the test. One variant employs two annular flat-jacks, suitably arranged and supported so that pressure can be applied against the opposite walls of an exploration tunnel, as shown in Fig. 9.8. The annular shape of the jacks, in this case, enables the deformations to be measured at the center of the jacks, as well as around the edges. Another variant of the flat-jack test is that employed by Bieniawski, to measure the strength-to-failure of coal

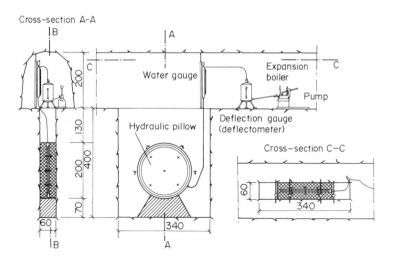

Fig. 9.7. *In situ* deformability test, using hydraulic flat-jacks (Kudjunzic).

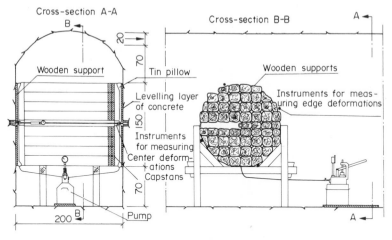

Fig. 9.8. *In situ* deformability tests using two flat-jacks (Kudjunzic).

pillars *in situ*. In this case rectangular flat-jacks are used, inserted and cemented into horizontal slots cut into the rock. The jacks are pressurized and the load-deformation characteristic is observed. Ultimately, taking the load up to the point of pillar failure, a measure of the uniaxial compression strength *in situ* is obtained.

In all the flat-jack tests so far described considerable uncertainty exists as to the pressure distribution imposed by the jacks, and the compressibility of the concrete in which the jacks are set is a major factor which affects the overall load-deformation relationship. This complicates and obscures any interpretation of the observed measurements. These complications have been reduced in the methods devised by Rocha *et al.*, of the National Civil Engineering Laboratory at Lisbon, Portugal, in which the jacks are inserted directly into slots, cut into the rock by precision rotary diamond-wheels and rock-saws, so that the minimum clearance required to insert the jack is obtained. Thus, when the jack is pressurized it quickly makes contact with the rock and no cement is required.

Kudjundzic describes an arrangement of sixteen flat-jacks, applying pressure radially against the surface of a circular rock tunnel (Fig. 9.9). The pressure contact area extends over a prepared length of tunnel measuring a little less than 2 m axially, with a diameter about 2.5 m.

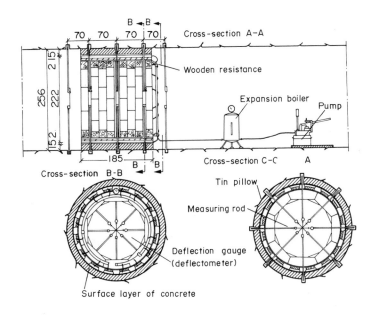

Fig. 9.9. Radial jacking test for rock deformability (Kudjunzic).

Pressure-chamber tests

The radial pressure test may be made directly on the rock mass, over a prepared length of tunnel, by constructing a pressure chamber which is loaded by water, under pressure and retained by bulkheads and manhole covers (see Fig. 9.10). The diametral deformation over the central cross-section of the test chamber is measured by radial extensometers, and averaged, from which

$$E = \psi \, \frac{(1 + v)}{\mu} \, d.p.$$

where E = deformation modulus,
 v = Poisson's ratio for the rock material,
 d = diameter of the test section,
 μ = measured diametral deformation at the center of the test section,
 ψ = is a constant factor, the magnitude of which depends on the proportions length/ diameter of the test section. ($\psi < 1.0$ and approaches 1.0 as the axial length of the loaded section approaches $2d$.)

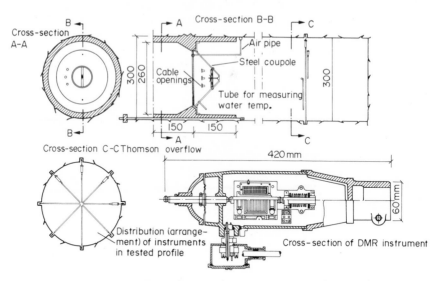

Fig. 9.10. Pressure chamber test for rock deformability (Kudjunzic).

Since the rock is more or less porous and permeable there will be some loss of water from the chamber, under pressure. This can be measured by pumping, to maintain a constant chamber pressure, so that an indication of the permeability is thereby gained, under the pressure applied. Tests at the same site may be repeated before and after grouting, to check the effectiveness of the grout in sealing-off the rock. If permeability tests are not to be made, the walls of the pressure chamber may be sealed by an impervious sheet, attached as a lining, to retain the fluid without loss of pressure during the test.

Borehold Rock Deformability Tests

The cost of *in situ* rock deformability tests increases progressively with their size and degree of elaboration. The expenditure involved in one large-scale test could exceed that of a score or more small-scale tests, and still leave unresolved the question of correlating the test result with the real situation. This being so, the possibility of making *in situ* deformability tests from boreholes drilled into the rock mass is attractive, since many boreholes must necessarily be drilled into the strata for the purpose of detailed geological site investigation. The possibility of using the same boreholes for a subsequent exploration of the engineering qualities of the strata offers obvious economies, when limited finances are available. Several alternative devices have been designed for insertion into boreholes, in such a way that the walls of the borehole can be subjected to load, and some effect of the rock's response to that load measured. In broad terms they may be classified as being either borehole dilatometers or borehole jacks.

Borehole dilatometers

These instruments are essentially borehole pressure cells, which apply pressure radially in all directions to the wall that encloses them. The devices are cylindrical, and incorporate some means of measuring changes of diameter, or volume, with variation in pressure. The change in diameter may be deduced and averaged from measurement of the volume change, or it may be measured across specific directions. Typical of the latter, more elaborate, device is the LNEC (Lisbon) borehole dilatometer. This is 76 mm diameter and 870 mm long. It can operate submerged in water, to depths exceeding 100 m, and it applies radial pressures to the borehole wall, up to 150 kg/cm². Diametral deformations are measured by linear differential transformers, aligned in four directions 45° apart.

Pressure is applied by pumping water into the annulus between the inner steel cylinder and the rubber jacket. This jacket is deflated to allow the device (which weighs 12 kg) to be raised or lowered, by cable and winch, to any desired measurement station along the length of the borehole. Typical results from a LNEC dilatometer test are shown in Fig. 9.11, in relation to which

$$E = \frac{2a\,(1+v)\,p}{\Delta}$$

where $2a$ is the initial borehole diameter,

$\quad\Delta\quad$ is the change in diameter, when pressure p is applied,

$\quad v\quad$ is Poisson's ratio for the rock material

The mean value of the slope of the stress-strain characteristic during the second and third

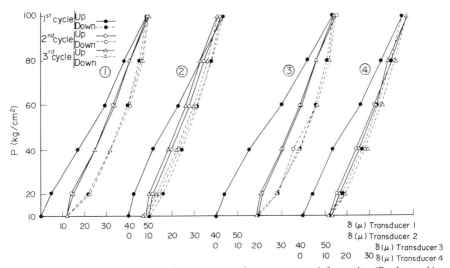

Fig. 9.11. Typical LNEC borehole dilatometer results, pressure vs. deformation (Rocha *et al.*).

load cycles is measured, to give E at the measurement site. Observations on four diameters enable the deformation ellipses and coefficients of anisotropy to be deduced.

Other borehole dilatometers include the Menard Pressuremeter, developed in France, and the U.S. Bureau of Mines cylindrical pressure cell. The Menard instrument is designed for pressure up to 10.34 MN/m² and, although primarily intended for use in soils, may also be applied in soft rocks (up to $E = 6895$ MN/m²). It consists essentially of a cylinder containing hydraulic fluid, to which are attached a meter to measure volume change and a pressure gauge. A compressed gas (nitrogen or carbon dioxide) cylinder is connected to the metering system and also to the pressure cell, by means of coaxial tubing and appropriate valves.

Geotechnology

The pressure cell is in the form of a probe, and comprises three independent sections of which the central section is the measurement cell. The pressure-volume relationships of the probe are observed, with the probe inserted into a borehole, and these observations are correlated to an instrumental calibration in terms of compressibility and inertia. The compressibility calibration determines the volume of fluid required to compress the instrument at various pressure levels, while the inertia calibration determines the pressure required to expand the probe to various sizes. The pressure required to expand the probe to any given volume may then be deducted from the pressure-gauge reading in a field test at that volume.

The USBM cylindrical pressure cell is a device for determining the modulus of rigidity of rock by direct measurement *in situ,* at pressures up to 68.95 MN/m^2. The cell may be constructed in various sizes, but most reported work refers to tests in 38 mm diameter boreholes, using pressure cylinders 200 mm long. The cell is first calibrated by expanding it in two metal test cylinders. These have different expansion characteristics, either because they are constructed of different materials, or if of the same material they have different wall thickness. The dual calibration tests enable two calibration constants to be determined, in a similar fashion as was described for the Menard meter. Thereafter, when the pressure cell is expanded inside a borehole in rock, the change of diameter of the cell may be determined from the volume of fluid pumped into it. The basic measurement equipment is similar to that on the Menard pressure cell, namely, a volume-metering fluid pump and a Bourdon pressure gauge, although at a higher pressure range.

Borehole jacks

The borehole jack type of deformability apparatus applies load at a higher stress level than the dilatometers, but only to a limited area of contact, by means of hydraulic pistons, wedges, or jacks. Directional load is applied over opposite sections of a cylinder wall. One of the most recent introductions of this type of instrument is the Goodman jack, which forces two steel plates, covering a radial arc of 90° each, over a length of 200 mm, into opposite sides of a 57 mm -diameter borehole. The plates are forced apart by twelve hydraulic pistons, while two retracting pistons close the plates together when the device is to be moved along the borehole. Two linear transformers measure the diametral deformation at either end of the plates. A hydraulic pump generates up to 68.95 MN/m^2, and produces a uniaxial pressure on the borehole wall of 64.12 MN/m^2, from which

$$E = 2.40k\,(v)\,\frac{\Delta\,Qh}{\Delta\,U_d}$$

where E is the deformation modulus,
ΔU_d is the average diametral displacement, for a measured increment of pressure, ΔQh.

Values of $k(v)$ range from 1.09 to 1.38.

Comparison of Borehole Deformability Meters

In a review of rock modulus-measurement devices Hall and Hoskins quote comparison data to represent the range over which the respective instruments may be regarded as capable of giving reliable estimates of deformation modulus. However, the data should not be taken too literally. It probably gives a fair estimate of the relative suitability of the various instruments for a specific range of conditions, but a less-certain indication of absolute reliability. There is, in fact, considerable doubt as to what is actually measured by the instruments. In general, it is felt that the modulus observed when using one of the high-pressure instruments, such as a borehole jack, is likely to be 30 or 40% low, due to fractures caused by the jack in the strata around the jack-rock contacts. The borehole dilatometers, such as the Rocha instrument and the USBM cylindrical pressure cell, are less prone to this defect, but there are considerable uncertainties associated with the correlation of their calibration conditions with the actual *in situ* measurement conditions. Indeed, there is, as yet, no satisfactory instrument that is capable of providing a reliable measurement of the *in situ* deformability modulus of a rock mass. Probably the best indication, at the present state of knowledge, would be that obtained from the results of laboratory tests on samples taken from rock cores from boreholes drilled at several locations, in various directions, distributed over the rock mass.

Methods of Applying Load in Field Tests

The application of load in field tests sometimes presents problems. Deadweight loading is sometimes possible, to a limited extent, but a more common procedure is to apply thrust from hydraulic jacks. The necessary reaction base may be provided by a box structure, loaded with scrap metal or rock. Alternatively the test site may be spanned by an arched beam, the ends of which are secured by deadweight, or pinned down by concrete piles, or anchored into the

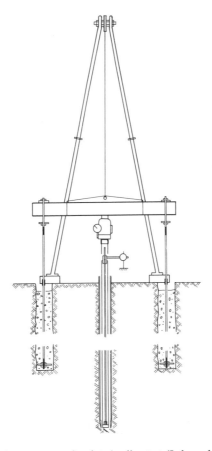

Fig. 9.12. Arrangement of a plate-loading test (Lake and Simons).

bedrock by bolts or cables (Fig. 9.12). In Ward and Burland's field study of the deformation characteristics of the Munford Chalk, deadweight loads up to 1.83 kg/cm² were applied by a circular water-filled steel tank, 18.3 m diameter and 18.3 m high. The tank was built directly on to the chalk surface, after removing about 1.7 m of silty sand. The deformation of the ground underneath the tank, and in the surrounding area, was monitored by displacement transducers placed at various depths in a number of vertical wells. These same wells also provided the locations for plate loading tests.

In underground galleries, load can be more easily applied by jacking from reaction bases thrusting against the opposite side of the gallery. Gallery tests are, however, expensive to perform. An alternative is to jack against the reaction provided by cables anchored at depth in the rock. Such tests may be carried out at the surface, or from underground galleries.

Cable Jacking Tests

A typical arrangement for a single-cable test is illustrated in Fig. 9.13. A minimum depth of anchorage of from 6 to 8 times the bearing-pad diameter is recommended by Zienkiewicz and

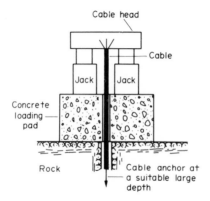

Fig. 9.13. Single-cable jacking test (Zienkiewicz and Stagg).

Stagg, for this type of test. A single cable can provide a load of up to 1000 t. The double cable arrangement, Fig. 9.14, provides a facility for generating still higher loads, and also permits investigation of the directional variation of deformation characteristics in non-isotropic rock masses. Using assumed values for Poisson's ratio and the shear modulus, Zienkiewicz and Stagg show that if the elastic modulus in the direction perpendicular to the rock surface is E, and if nE is the elastic modulus in the plane parallel to the rock surface, then, for a square loading pad with a side length a, the average displacement of the pad in a direction perpendicular to the rock surface is w,

$$\text{where } w = 2.97A_1 . P/a$$

and the average displacement of the pad in a direction tangential to the rock surface is u,

$$\text{where } u = 2.97B_1 . Q/a,$$

A_1 and B_1 are functions of E and n.

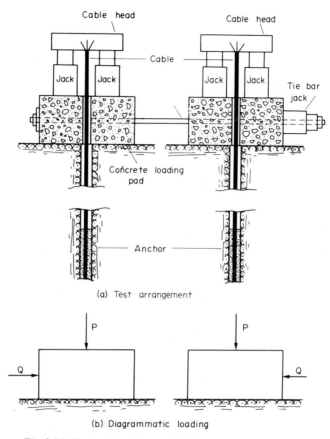

(a) Test arrangement

(b) Diagrammatic loading

Fig. 9.14. The double-cable test (Zienkiewicz and Stagg).

The Direct Measurement of Strength of Rock Masses

It can be argued that, as the strength of a chain is no more than that of its weakest link, so the strength of a rock mass is no more than the bond strength of its weakest discontinuity. Therefore, no important foundation or structural problem involving discontinuous rock masses should omit some direct investigation of bond strength, if the critical factor can be isolated. This is not always possible. For example, a rock slope stability problem may exist in circumstances where movement is general within a discontinuous and fragmented mass of material, and not located on a specific slip or fracture surface, or surfaces. There are other occasions, however, where the risk of potential failure can be attributed, with reasonable confidence, to specific fracture planes, such as bedding, cleavage, joints, fissures, and faults. In these circumstances some direct measurements of bond strength may be made, to include a range of typical discontinuities.

In situ Shear Tests

Shear tests on selected samples of the rock mass, each sample including a discontinuity the bond strength of which may be in question, may be conducted by means of portable equipment carried on to the site. The equipment incorporates a hydraulic ram, operated by hand-pump, to apply a force normal to the plane of the discontinuity, and a similar hydraulic system to apply the shear force along the plane of fracture. For this purpose the sample, as broken from the mass, must be suitably orientated and held in place, usually in a cement-mortar cast, within a shear box. The size of sample that can be handled in this way is necessarily limited to no more than about 28 dm^3, and is sometimes as small as 3.5 dm^3. Direct shear tests on specimens with test surface area measuring up to 300 cm^2 may be made on apparatus such as the Lombardi Direct Shear Test Apparatus, an outline diagram of which appears in Fig. 9.15. This apparatus is

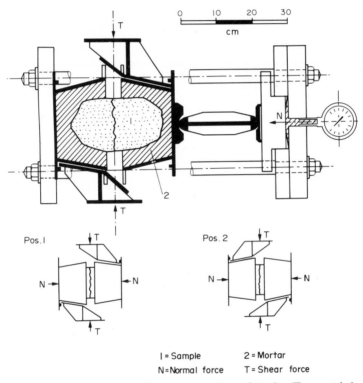

Pos. I   Pos. 2

I = Sample   2 = Mortar
N=Normal force   T = Shear force

Fig. 9.15. The Lombardi direct shear apparatus for rock testing (Terrametrics).

used with a laboratory compression-testing machine, which applies the shear force, while a horizontal press applies an adjustable normal pressure on the specimen.

Hoek's portable shear box is described by Walton. It is designed to take joint samples with a contact area not larger than 0.12 × 0.15 m.

While the shear-box test can be applied to many small samples, selected from various locations over the site, such as an open-pit, it is more usual, in the case of large foundation investigations, to make the tests on a smaller number of typical discontinuities *in situ,* and not detached from the mass. The tests are usually made in underground exploration galleries. Excavations are made to expose a block of ground with the discontinuity as a base. Hydraulic jacks are positioned, to react against the roof and against one side of the gallery, with cushions of concrete mortar to distribute the jack pressure on the rock surfaces under load. The vertical jack or jacks are

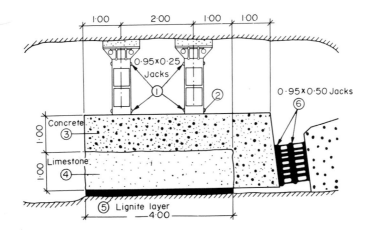

Fig. 9.16. Arrangement of the direct shear test at Mequinenza on 4 × 4-m samples (Jimenez Salas).

pressurized to provide a constant normal force, while the horizontal jack pressure is slowly increased up to the point of failure. Typical arrangements for such a test are shown in Fig. 9.16. The shearing-jack system is inclined slightly, to eliminate any turning moment that would tend to counteract the normal force on the plane of shear.

The size of block so tested customarily ranges from 0.5 to 1 m^2, in plan, but occasionally may be as large as 4 m^2. A notable example of a large-scale *in situ* shear test was that conducted in the foundation site investigations for the Mequinenza Dam, in Spain. This investigation is of special interest, as it included a range of shear tests, on three sizes of block.

Torsion Shear Test for Rocks

The method by which the *in situ* shear test is conducted in rock must inevitably raise some questions as to the validity of the results gained thereby. The pre-test preparation of the block disturbs the original equilibrium of, and is likely to weaken, the material tested. There is some uncertainty as to the stress distribution imposed by the jacks, and as to the manner in which this may be related to the natural stress field in the rock mass. Alternative methods of testing may therefore be considered. One such alternative is the torsion shear test for rocks. An outline arrangement for such a test is shown in Fig. 9.17. A hole drilled into the rock mass gives an anchor location for a rock bolt. The depth of this anchor goes beyond that of the discontinuity, whose bond strength is to be tested, and its orientation is normal to that test surface. A rock annulus is drilled around the central bolt, by means of a large-diameter core drill. This isolates the core-segment of rock for the torsional shear test.

The test is conducted by placing a normal compressive load on the rock by tensioning-up the rock bolt, and this load is measured on the central load cell. Torque is then applied by lateral hydraulic jacks, operating on a torque tube inserted into the rock annulus, the necessary reaction being provided by restraining rock anchor-bolts. Extensometers, fastened to the rock anchors, measure torsional strain, and the point of failure is detected by means of shear strips bonded to the rock and to the torque tube. Torque is applied by loading–unloading cycles, at increasing increments of load, and the strain rate is observed, up to the point of failure.

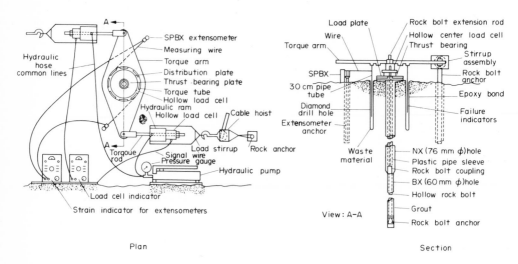

Fig. 9.17. Plan and section of torsional shear test arrangement for rocks (Terrametrics).

In situ Rock Compression Tests

The difficulty of correlating the results of laboratory compression tests with the strength of the same rock *in situ* leads to interest in the possibility of making the test on the rock in its natural state, and not on specimens transported to a laboratory. One such arrangement is illustrated in Fig. 9.18. Here, at Mequinenza, the main object of interest was the strength of the lignite, which was too friable to withstand core-drilling, and which, in large specimens, disintegrated during transport. The results of an *in situ* compression test of the Mequinenza pattern are dependent on the deformability of the material between the face plate and the anchor, and, ultimately, on the anchor strength of the stress-transmission column. The test then becomes a larger-scale version of that devised by Foote to provide a strength index for coal *in situ* – the expanding-bolt seam tester.

The Determination of Pillar Support Strength in situ

Various attempts have been made to determine the strength of small rock pillars, *in situ*, by direct loading. These have mainly been conducted in sedimentary deposits such as coal and salt. The method is to isolate the pillar to be tested, on four sides by vertical cuts using a rock saw or shearing machine, and on one face by a horizontal cut into which a vertical jacking system is inserted. The largest pillars that have been so tested are about 2 m² in plan. Theoretical considerations, applying the "weakest-link" argument, predict that the strength of pillars tested in this way should decrease with increase in size (see Fig. 9.1). Bieniawski's experimental results, however, indicate an approach to a constant value when the specimen rock cube reaches about 2 m side length. If this is so, then Bieniawski's observations can be summarized by the statement that, for coal pillars 2 m or more in width,

$$\sigma = 400 + 22\,w/h$$

where σ = specimen strength
      w = width of pillar
      h = height of pillar

and the ratio w/h is greater than unity.

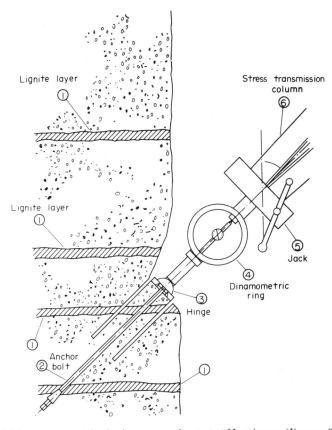

Fig. 9.18. Arrangement for *in situ* compression test at Mequinenza (Jimenez Salas).

While, at the present state of knowledge, the general application of Bieniawski's results should only be advocated with some reservations as an absolute indication of the strength of large pillars, it is possible that they may be applied more confidently to provide a comparative index of pillar strength. For example, comparing South African coal seams, if Bieniawski's determined strength of the Witbank coal is given a reference strength index of 100, then the strength indices for other South African coals are as detailed on Table 9.2.

Residual Strength of Rock Masses

While the tests so far described are intended to observe the resistance characteristics of rock masses to forces of tension, compression, and shear, up to the point of failure, in practice, very often it is not the maximum resistance of the rock mass which is important. Where, for example, the stability of a rock slope is in question, or when the walls of an excavation must be supported

TABLE 9.2. *Comparison of Coal Strength Data for Collieries representing all Major Coalfields in South Africa (Bieniawski)*

Coalfield	Colliery	Locality	Coal density (kg/m³)	Uniaxial compressive strength (MN/m²)		Number of specimens tested	Number of batches	Strength index
				Mean	Standard deviation			
C	Coalbrook North	Second seam	1514	40.82	14.1%	54	3	103.9
C	Cornelia	Bertha Section	1434	43.78	16.9%	60	3	111.5
B	Durban Navigation	Section 5	1435	34.48	10.0%	52	2	87.7
B	Durban Navigation	Section 40	1355	38.44	16.7%	72	3	97.8
A	Kendal	—	1346	44.13	22.6%	40	2	112.4
C	Sigma	—	1517	37.92	14.3%	39	2	96.5
D	Springfield	—	1466	40.51	16.9%	100	5	103.0
A	Witbank, Wolvekrans	No. 4 seam	1453	39.30	18.1%	35	6	100.0
A	Witbank, Wolvekrans	No. 2 seam	1491	42.75	27.4%	78	9	108.9
A	Witbank, Wolvekrans	No. 1 seam	1488	50.54	16.3%	49	4	143.9

Note: (i) All specimens were 25 mm cube in size and were loaded normal to the bedding planes at a constant rate of 7 kg/cm² /sec in a standard laboratory testing machine.
(ii) Coalfields: A – Witbank–Breyten Coalfield, Transvaal
B – Klip River Coalfield, Natal
C – Vereeniging Coalfield, Orange Free State
D – Balfour Coalfield, Transvaal.

in moving ground, the rock mass is already fractured and broken. It has passed the point of maximum resistance, but the fractured material still possesses the capacity to resist load, to a limited extent. It has residual strength, and it is this residual strength which is of major importance from a practical engineering viewpoint. This is the effective strength of the rock mass which must be considered in relation to matters of slope stability, pillar support, and the reinforcement of rock walls and excavations in moving ground, or in "incompetent rock". Shear tests on rocks should therefore be conducted, not only to determine the fracture strength, but also the residual strength after fracture. In general the value of residual strength increases with increase of normal stress across the fracture surface (see Fig. 9.19).

Indirect Observation of the Strength of Rock Masses

The modulus of elasticity of a rock mass may be determined indirectly by seismic methods, in which the velocity of sound waves traveling through the rock is measured. The matter is discussed in the section "Strength of Rocks Under Dynamic Loads".

The "Hydrofrac" technique

Hydraulic fracturing is a technique used in the petroleum industry to stimulate production from a depleted oil well. It consists of sealing-off a section of a borehole in the rock, then introducing fluid pressure into this section and increasing the pressure until the wall of the borehole is caused to fracture. The technique is, therefore, an *in situ* test of the strength of the rock mass,

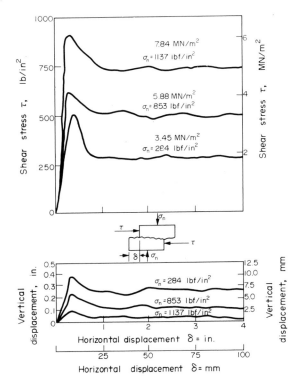

Fig. 9.19. Relationships between shear stress and displacement and vertical and horizontal displacements for porphyry joints tested in a shear machine at different normal stresses (Hoek and Pentz).

in this particular context. One of the implications of the extension of brittle fracture theory to triaxial stress conditions in a rock, taking into account the effects of pore pressure, is that when the pore pressure equals the confining pressure the cohesive power of the rock may be reduced to no more than its unconfined tensile strength. In hydraulic fracturing it is assumed that the borehole wall cracks when the maximum induced stress around the pressurized section of the borehole reaches the tensile strength at some point on the wall, when the tensile stress induced by the pressurized fluid overcomes the compressive stress induced around the wall by the action of drilling the borehole into the regional stress field of the rock mass.

In an analysis of theoretical concepts concerning hydraulic fracture, Fairhurst shows that if the regional stress field is defined by three orthogonal principal stresses, one of which is assumed to coincide with the axis of the borehole, then fracture will occur at the wall in a direction normal to the maximum induced tensile stress, when this reaches the tensile strength of the rock. The fracture will propagate in a plane perpendicular to the least compressive principal stress, and the fluid pressure required to propagate the fracture, once initiated, will be equal to that stress.

Observation of acoustic emission from stressed rock

The release of kinetic energy during the internal processes of deformation within rocks and rock masses generates noises. These noises follow an impulse pattern that is indicative of the impulsive deformations in the rock material. Consequently, the acoustic emission may be used to observe and monitor the process of deformation, and the approach to failure of the earth material. Although it is simple in its basic principle, the technique is complicated by several factors. For example, rock deformation-impulses generate several wave components, each with different characteristics, and which travel at different velocities. Not only do the wave fronts separate, with increasing distance from the source, but the overall spectrum of frequencies is selectively absorbed, some wavelengths more rapidly than others. The interpretation of the signals received by a geophone, which may be placed at some distance from the impulse source, therefore calls for considerable expertise. Nevertheless the technique is widely used, not only in the laboratory to study the failure mechanism of rocks, but also in the field, to observe rock-burst, gas outbursts, and earthquake phenomena, and to monitor the stability of earth and snow masses and rock slopes.

Selected References for Further Reading

ASTM. *Testing Techniques for Rock Mechanics,* Spec. Tech. Pub. No. 402, New York, 1965.

BIENIAWSKI, Z. T. The effect of specimen size on the compression strength of coal. *Int. J. Rock Mech. Min. Sci.,* vol. 5, pp. 325-375 (1968).

DE BEER, E., DELMER, A. and WALLAY, M. Large-size plate loading tests in a gallery and on the rock surface. *Int. Symp. on Rock Mechs.* (Editorial Blume), Madrid, pp. 13-33, 1968.

FAIRHURST, C. *Measurement of in situ Rock Stresses with Particular Reference to Hydraulic Fracturing,* Univ. Minnesota, School of Mineral and Met. Engng., Res. Rept. Minneapolis, 1964.

GOODMAN, R. E., VAN, T. K. and HEUZE, F. E. The measurement of rock deformability in boreholes. *10th Symp. on Rock Mechs., Austin, Texas,* AIME, New York, 1969.

HALL, C. J. and HOSKINS, J. R. *A Comparative Study of Selected Rock Stress and Property Measuring Instruments,* Tech. Rept. No. U1-BMR-2, Advanced Research Projects Agency, Washington, DC, 1972.

HEUZE, F. E. and GOODMAN, R. E. Mechanical properties and *in situ* behavior of the Chino Limestone. *Proc. 9th Symposium on Rock Mechs., Colorado,* AIME, New York, 1968.

HOEK, E. and PENTZ, D. L. *The Stability of Open-pit Mines,* Imperial College London, Rock Mechs. Unit Res. Rept. No. 5, 1968.

KNILL, J. L., FRANKLIN, J. A. and MALONE, A. W. A study of acoustic emission from stressed rock. *Int. J. Rock Mech. Min. Sci.,* vol. 5, pp. 87-121 (1968).

KUDJUNDZIC, B. Experimental research into mechanical characteristics of rock masses in Yugoslavia. *Int. J. Rock Mech. Min. Sci.,* vol. 2, pp. 75-91 (1965).

LAKE, L. M. and SIMONS, N. E. Investigations into the engineering properties of chalk at Welford Theale, Berkshire. *Conf. on in situ Site Investigations in Soils and Rocks,* British Geotechnical Soc., pp. 23-30, 1970.

PRATT, H. R., BLACK, A. D., BROWN, W. S. and BRACE, W. F. The effect of specimen size on the mechanical properties of unjointed diorite. *Int. J. Rock Mech. Min. Sci.,* vol. 9, pp. 513-529 (1972).

ROCHA, M., DA SILVEIRA, A., GROSSMAN, N. and DE OLIVEIRA, E. Determination of the deformability of rock masses along boreholes. Memo. No. 339, Lab. Nac. Engng. Civ. Lisbon, 1969.

SALAS, JOSE A. J. Mechanical resistances. Introductory lecture. *Int. Symp. on Rock Mechs.* (Editorial Blume), Madrid, pp. 115-129, 1968.

STEARS, J. H. *Evaluation of Penetrometer for Estimating Rock Bolt Anchorage,* U.S. Bur. Mines Rept., Invest. No. 6646 (1965).

WALTON, G. Discussion of Hoek, E. Estimating the stability of excavated slopes in opencast mines. *Trans. Inst. Min. Met. London,* vol. 80, p. A75 (1971).

WARD, W. H. and BURLAND, J. P. Assessment of the deformation properties of jointed rock in the mass. *Int. Symp. on Rock Mechs.* (Editorial Blume), Madrid, pp. 35-40, 1969.

ZIENKEWICZ, O. C. and STAGG, K. G. Cable method of *in situ* testing. *Int. J. Rock Mech. Min. Sci.,* vol. 4, pp. 273-300 (1967).

CHAPTER 10

Determination of the State of Stress in Rock Masses

In situ Stress Measurement

As the science of rock mechanics has evolved during the years, no problem has been more elusive of solution than the determination of the *in situ* state of stress in a rock mass. Yet few problems can have been given more attention. Instruments and techniques, devised with this end in view, are legion, and each rock mechanics conference appears to produce more. That this should be so is indicative of the fact that no single technique has, as yet, proved to be generally acceptable, either from a practical or from a theoretical viewpoint. However, some investigators have been successful in making measurements of stress *in situ,* and although the validity of their results may be questioned in precise quantitative terms, from a general and qualitative viewpoint they are often of great interest and value. The subject is not an easy one to investigate. While the necessary equipment is sometimes commercially available, for the most part its use is expensive, and some of the best equipment is either exclusively owned or otherwise difficult to obtain. The techniques of application are highly specialized. They require practice and perseverance on the part of the operators, if useful results are to ensue. Any investigator venturing into *in situ* stress measurement without fully appreciating what is involved is very likely to be frustrated and disappointed as a result of his early efforts. Many instruments and techniques have failed to progress far beyond the prototype stage, and not a few investigators have expressed the view that time and money can be much more profitably spent in other ways than in the attempt to measure *in situ* stresses. They suggest that we should think in terms of deformations, rather than stress, since these can more easily be observed and measured. They say we should ask ourselves at what degree of distortion will the material lose its stability, and under what deformation will it rupture?

Nevertheless, so long as we are unable to determine the state of stress in the Earth's crust, and in the rocks around our excavations, engineering in rock masses will remain an art. If rock mechanics is to develop as a theoretical science with practical applications in engineering then we must devise acceptable methods for *in situ* stress measurement. Without the facility for making such measurements much of the work that has been done in recent years, for example, the development of failure criteria for rocks, and the evolution of rational methods of design for underground excavations, will remain of academic interest only. And without basic information on such matters as *in situ* deformation moduli, and *in situ* stress values, any analytical approach to geotechnology can only begin from uncertain premises and doubtful assumptions.

Stresses in the Earth's Crust

The state of stress in the Earth's crust, at a given time and locality, is the resultant of forces of various origin and character. It is conventional to regard the stress state which exists before any engineering work begins as the virgin stress. This stress state is disturbed by the construction of excavations, earthworks, and engineering structures, which produce a new distribution of induced stress within the rocks in their vicinity. The virgin stresses include components of gravitational stress, due to the weight of the overlying rocks, and the affects of lateral constraint within them, and there are also components of inherent or latent stress, some of which originate in processes such as crystallization, metamorphism, sedimentation, consolidation, and dehydration, depending upon the particular rocks concerned, while other virgin stress components originate in tectonic forces and crustal movements.

The concept of gravitational stress conventionally assumes that the rock mass behaves as an elastic material under complete lateral constraint, in which case the stress state at a depth H is defined by

$$\text{Vertical principal stress } \sigma_1 = Hw$$

where w = weight per unit volume.

$$\text{Lateral principal stress } \sigma_2 = \sigma_3 = \frac{v}{1-v}Hw$$

where v = Poisson's ratio.

The ratio Lateral/Vertical principal stress is then C where

$$C = \frac{v}{1-v},$$

but when the lateral constraint is not completely rigid, C will attain higher values. If the rock behaved as a perfectly plastic material a hydrostatic condition would exist, in which C = unity. Horvath postulates that, for a rock with certain characteristics of yield strength, weight per unit volume, and Poisson's ratio, there is a limit depth above which the lateral stresses may be computed on the basis of elastic theory, but below which the horizontal principal stress may be derived from a plastic yield criterion, so that

$$\sigma_2 = H_w - \sigma_F$$

where σ_F = yield stress.

Seagar also describes a situation in which a rock might behave elastically, provided the difference of the principal stresses is below a value corresponding to the yield shear stress, and thereafter deforms plastically under further loading. On relief of load the material would behave elastically again when the stress difference falls below the yield stress. Such a cycle could be imagined to occur if the strata had been deeply buried and then partly uncovered. If adjacent strata have different yield stresses, Seagar's calculations suggest that they may well be under different lateral pressures, even though they have experienced the same loading history. A strong rock stratum might not have yielded under load, while a weaker stratum would yield more and generate a greater lateral pressure. According to Seagar's hypothesis the lateral pressure is always greater than that which is calculated on the basis of pure elasticity, and it can quite easily exceed the vertical pressure in magnitude.

If the rock mass is subject to inherent stresses in addition to those that originate from superincumbent load the lateral stress may well exceed the vertical stress, and it could also be expected to be highly directional in character. Some authorities, however, maintain that creep under sustained loads on a geological time scale is likely to have relieved stress differences and so produced a hydrostatic stress system. Only actual *in situ* stress measurements can resolve these specula-tions, and Herget has plotted such observations, as reported from a variety of world sources, in Figs. 10.1 and 10.2. Figure 10.1 shows the increase of vertical stress with depth, which on reg-ression analysis gives

$$\sigma_V = (272 \pm 178) \text{ psi} + 1.154 \pm 0.123) \text{ psi/ft depth.}$$

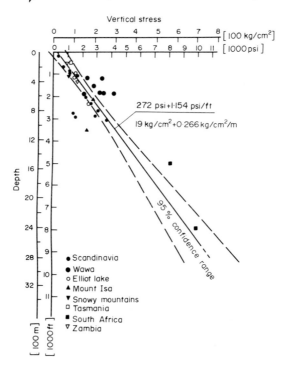

Fig. 10.1. Measured increase of vertical strata pressure with increasing depth from the surface (Herget).

Figure 10.2, which plots average measured horizontal stress, shows three groupings with general relationships

$$\text{(i)} \quad \sigma_H < \sigma_V$$

$$\text{(ii)} \quad \sigma_H = \sigma_V$$

$$\text{(iii)} \quad \sigma_H > \sigma_V$$

About 75% of the total number of observations collected by Herget fall into category (iii), for which a regression analysis yields

$$\sigma_H = (1184 \pm 78) \text{ psi} + (1.842 \pm 0.101) \text{ psi/ft depth.}$$

This confirms the general relationship, reported some years earlier by Hast, as a result of his own observations in Scandinavia.

Stress Distribution Around Excavation in Rock

When considering the virgin state of stress it is convenient to think in terms of principal

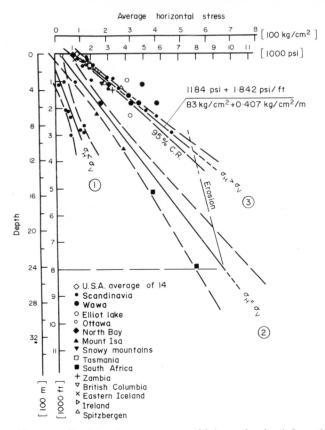

Fig. 10.2. Measured increase of horizontal strata pressure with increasing depth from the surface (Herget).

stresses in three orthogonal directions, one of which is vertical. These are, in fact, secondary principal stresses which serve to identify the primary resultant stress, which is unknown both in magnitude and direction. Identification of the resultant stress is possible if the nine components of stress acting upon an elemental cube are known. These components consist of six quantities, all of which are independent variables.

The virgin state of stress is, however, disturbed in the rocks around excavations and earth-works, and in the vicinity of structural foundations. A redistribution of stress is produced, in which new stress concentrations, some compressive and others in tension, are induced at various localities. If structural stability is to be maintained between the extremes of overdesign on the one hand and an adequate factor of safety on the other hand, it is important that the engineer should know the locations, magnitudes, and directions of these induced stress concentrations, relative to the original stress state. He may obtain some knowledge of this, in idealized materials with specific properties, by methods of calculation or by photoelastic model studies. Similar analyses can also be extended to models made of real or simulated rock materials. To the extent that the models can reproduce the characteristics of the real situation, they are useful, but some-times the models fall far short of reality, which is three dimensional and inconstant. When the ground yields, the stress concentration zones may be deflected to an unknown depth within the rock walls. Yield may be elastic and the walls remain intact up to the point of failure under increasing stress, or it may be time-dependent and plastic, so that the excavation steadily closes under constant load. In hard elastic and brittle rock, failure may originate in stress concentrations around discontinuities and cracks within the interior of the mass, at some distance from the walls, and not only around the periphery of the excavation. During the progress of engineering, in rock excavation, tunneling, and blasting, the induced stresses will change as the work proceeds, while over and above all, the rocks of the Earth's crust are subject to the diurnal variations of stress generated by earth tidal forces. The engineer concerned with the determination *in situ* stress is therefore required to determine a quantity which cannot be predicted confidently by theory, and which is known to vary both in space and time.

Principles of Rock Stress Measurement

We are concerned with two types of measurement: (a) determination of the absolute stress state and (b) measurement of relative stress or change of stress. Absolute stress measurement in rocks which demonstrate elastic behavior may require the application of a "stress-relief" technique, in which the rock element containing the measuring device is relieved from the stress that is generated by the confinement of the surrounding rock. The strain which results from this stress-relief is measured, so that a conversion to the relief component of stress can be made on the basis of known, or assumed, stress-strain relationships for the rock concerned. Relative stress change at a locality may be determined by measuring absolute stress at each end of a time interval, but this is not always necessary. Wherever possible, stress-relief techniques, which can be costly and time-consuming, are not applied in the measurement of relative stress. Generally speaking, the instruments used in both types of measurement are similar, but although any instrument that is designed to measure absolute stress will also measure relative stress, some instruments which are of simple design, and intended for relative stress measurement, cannot be applied to absolute measurements without further modification.

The stress-relief technique

The stress-relief technique is illustrated diagrammatically in Fig. 10.3. The measuring instrument may be attached to the surface of the rock which forms the wall of the excavation. Then the rock to which the gauge is attached is relieved from the confinement of its surroundings either by cutting slots on four sides of it, using a rock saw, or by drilling a ring of overlapping holes around it. Alternatively, the rock and gauge is overcored by a hollow rotary core-drill of a suitable diameter. The strains that are registered on the detached rock, as a consequence of the relief of stress upon it, are then measured. Instruments used in this way include linear extensometers, strain-gauge rosettes, and photoelastic biaxial gauges. The result identifies the secondary principal stress components in the plane of the wall of the excavation. The third principal stress in this case is zero.

Exploration of the virgin stress state requires that the measurement be extended beyond the zone of influence of the walls of the excavation. This may be attempted by drilling a borehole into the wall and then setting a gauge in, or at the back of, the borehole. The process of overcoring then follows, and the procedure may be repeated at successive depths into the rock wall. The result again identifies the secondary principal stresses in a plane at right angles to the axis of the borehole. Since we are now concerned with a triaxial stress field, at least six strains, two in each of three orthogonal directions, are required to provide a solution. Until comparatively recently this was seldom attempted and it was more usual to make measurements only in one

G—G

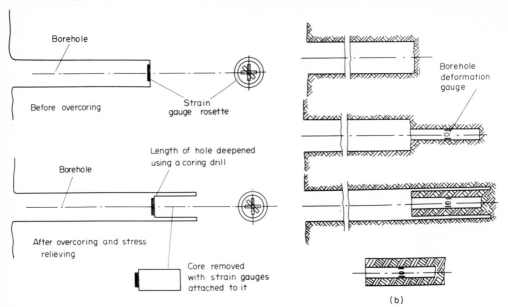

Before overcoring

Borehole

Strain gauge rosette

Length of hole deepened using a coring drill

Borehole

After overcoring and stress relieving

Core removed with strain gauges attached to it

Borehole deformation gauge

(b)

Fig. 10.3. The stress-relief technique. (a) Using borehole strain cell. (b) Using borehole deformation gauge.

borehole. With the instruments that are generally available this requires that certain simplifying assumptions be made about the direction of the third principal stress. Two such commonly made assumptions are that one principal stress lies in a vertical direction (in which case the measurement borehole is drilled horizontally into the rock wall) or (if the borehole is not horizontal), the third principal stress lies along the central axis of the borehole.

Three distinct types of borehole strain-measuring instruments are used. They may be classified as borehole deformation meters, borehole strain cells, and borehole inclusion stressmeters.

Borehole Deformation Meters

A borehole deformation meter is a device which measures changes in the cross-sectional dimensions of a borehole in rock, when the borehole is deformed as a result of stress change. The stresses are calculated by elastic theory.

The general equation for plane strain is:

$$\Delta D = \frac{\alpha_1 D}{E} \left\{ (1 + K) - v\,L + 2\,(1{-}K)\,(1 - v^2)\cos 2\theta_1 \right\}$$

or

$$\Delta D = \frac{D}{E} \left\{ (\sigma_1 + \sigma_2) - v\,\sigma_3 + 2\,(\sigma_1 - \sigma_2)\,(1 - v^2)\cos 2\theta_1 \right\}$$

and for plane stress:

$$\Delta D = \frac{D}{E} \left\{ (\sigma_1 + \sigma_2) + 2(\sigma_1 - \sigma_2)\cos 2\theta \right\}$$

where ΔD = change in diameter inclined at θ_1, to the direction of principal stress σ_1,

$\quad D$ = original diameter,

$\quad K$ = σ_2/σ_1,

$\quad L$ = σ_3/σ_1

If the deformation is measured across three different diameters and the modulus of elasticity and Poisson's ratio are known, the magnitude and direction of the stresses σ_1 and σ_2 can be computed.

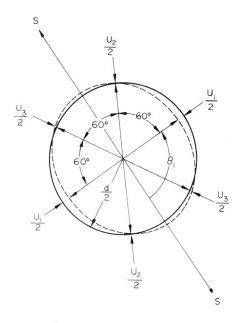

Fig. 10.4. Deformation of a hole in a biaxial stress field, 60° rosette.

In Figure 10.4

$$\sigma_1 + \sigma_2 = \frac{E}{3D(1 - v^2)} (U_1 + U_2 + U_3)$$

and

$$\sigma_1 - \sigma_2 = \frac{\sqrt{2}S}{6D(1 - v^2)} \left\{ (U_1 - U_2)^2 + (U_2 - U_3)^2 + (U_1 - U_3)^2 \right\}^{\frac{1}{2}}$$

where U_1, U_2, and U_3 are the measured deformations.

The angle between σ_1 and the direction in which the deformation U_1 is measured is:

$$\tan 2\theta_1 = \frac{\sqrt{3}(U_2 - U_3)}{2U_1 - U_2 - U_3}$$

Where the measured deformations are 45° apart:

$$\sigma_1 + \sigma_2 = \frac{E(U_1 + U_3)}{2D(1 - v^2)}$$

$$\sigma_1 - \sigma_2 = \frac{E \left\{ (U_1 - U_2)^2 + (U_2 - U_3)^2 \right\}^{\frac{1}{2}}}{2D \sqrt{2} (1 - v^2)}$$

$$\tan 2\theta_1 = \frac{-2U_2 - U_1 - U_3}{U_1 - U_3}$$

Merrill and Peterson give the following rules for determining θ_1:

For a $60°$ rosette:

1. If $U_2 > U_3$, θ_1 lies between $+90°$ and $+180°$ or between $0°$ and $-90°$.

2. If $U_2 < U_3$, θ_1 lies between $0°$ and $+90°$.

3. If $U_2 = U_3$ and if
$\qquad$ (a) $U_1 > U_2$ $\quad \theta_1 = 0°$,
$\qquad$ (b) $U_1 < U_2$ $\quad \theta_1 = 90°$.

and for a $45°$ rosette:

1. If $U_2 > \dfrac{U_1 + U_3}{2}$ $\quad \theta_1$ lies between $+90°$ and $180°$ or between $0°$ and $-90°$.

2. If $U_2 < \dfrac{U_1 + U_3}{2}$ $\quad \theta_1$ lies between $0°$ and $+90°$.

3. If $U_2 = \dfrac{U_1 + U_3}{2}$ $\quad$ and if

$\qquad$ (a) $U_1 > U_3$ $\quad \theta_1 = 0°$,
$\qquad$ (b) $U_1 < U_3$ $\quad \theta_1 = 90°$.

The USBM single-component borehole deformation gauge

The original instrument of this design was introduced by Merrill in 1962. It is constructed for insertion in a 38 mm diameter drillhole. The sensing element is a beryllium copper cantilever on which four electrical resistance strain gauges are bonded, in a wheatstone bridge circuit. The instrument has a sensitivity of about 20 microstrains, which corresponds to a stress sensitivity of approximately 93 KN/m² for a rock in which $E = 20.685$ GN/m². The installation and measurement procedure with this gauge involves the following sequence of operations.

A 150 mm diameter hole is drilled into the wall, to a distance sufficient to extend beyond the fracture zone near the rock face. The core is removed, and guides are placed in the hole to center a 38 mm diameter core bit and barrel. A 38 mm diameter hole is then drilled into the center of the end of the 150 mm hole, to a further depth of 3 m or so. The borehole deformation gauge is then inserted, to a depth of from $150 - 250$ mm, in the central hole. The gauge is orientated to measure along the vertical diameter of the borehole, and the leads from the gauge are brought through the drill-rods to the strain-gauge measurement box. An initial reading is taken, then the overcoring drill is started. As overcoring proceeds, readings of deformation are taken at regular intervals, as indicated by the bridge-box. If the procedure is successful the deformation will be seen to occur more or less gradually as the overcoring drill-bit passes over the contact pins which transmit pressure from the borehole wall to the cantilever in the gauge. This exerts a force of from 45 to 130 N. When the drill has passed 25 to 50 mm beyond this point, stress relief is usually completed. A final reading is taken and then the rock core containing the gauge is removed from the hole for subsequent tests to determine the deformation modulus for the rock.

The instrument is calibrated by depressing the lever a known amount, observed by micrometer, and noting the corresponding signal on the strain-gauge bridge. Since successive measurements

U_1, U_2, and U_3 are not made in the same plane it is necessary to interpolate between the readings observed in the field. This is a disadvantage which limits the usefulness of the instrument in localities where high strain gradients can be expected, as for example, in and adjacent to the stress concentration zones around an excavation. Subsequent developments in borehole deformation meters have therefore been directed towards the construction of multiple-component instruments, measuring simultaneously along three directions 60° apart. These measurements are in a single plane, normal to the borehole axis.

The USBM three-component borehole deformation gauge

The arrangement of Merrill's three-component borehole deformation gauge is shown in Fig. 10.5. Six beryllium copper cantilever strips are arranged hexagonally to operate six pistons,

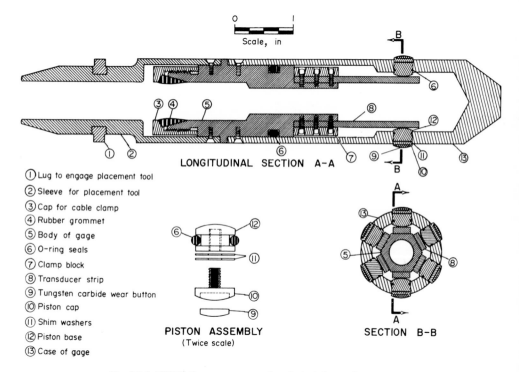

① Lug to engage placement tool
② Sleeve for placement tool
③ Cap for cable clamp
④ Rubber grommet
⑤ Body of gage
⑥ O-ring seals
⑦ Clamp block
⑧ Transducer strip
⑨ Tungsten carbide wear button
⑩ Piston cap
⑪ Shim washers
⑫ Piston base
⑬ Case of gage

LONGITUDINAL SECTION A–A

PISTON ASSEMBLY
(Twice scale)

SECTION B–B

Fig. 10.5. USBM three-component borehole deformation gauge

making contact with the wall of the borehole around one diametral plane. Foil resistance strain gauges are bonded on the upper and lower surfaces of each cantilever, and each pair of opposing transducers constitute a measurement component of the gauge. Each component is connected to a strain indicator. The transducers are calibrated before assembly into the instrument so that six transducers with matched characteristics may be selected. After assembly, the completed instrument is again calibrated in a cylindrical steel jig, so that the borehole deformation-strain gauge characteristic may be measured by six micrometers in one plane around the circumference of the jig.

Geotechnology

Borehole Strain Cells

By applying strain-gauge rosettes to the back of a borehole, the overcoring tool can be the same drill that is used to make the borehole itself. This considerably extends the possible range of measurement into the rock walls.

The CSIR "doorstopper" borehole strain cell

Attempts to overcore electrical resistance strain gauges set at the back of a drill hole sometimes meet with problems, mainly due to the difficulty of insulating the gauges and contacts from the water that must be circulated around the core-bit during drilling in hard rock.

Leeman solved this problem very effectively by encapsulating the electrical connections to the gauges in a silicone rubber moulding 34 mm diameter, on the front face of which a rectangular strain gauge rosette is mounted. The leads from the rosette are connected to four copper pins in an insulated connector plug (Fig. 10.6). The equipment can be used in a 60 mm diameter diamond-drill hole.

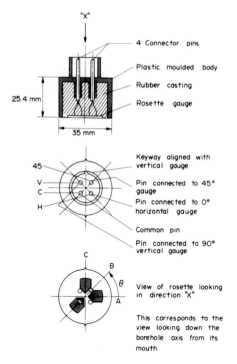

Fig. 10.6. The Leeman (CSIR) "doorstopper" borehole strain cell.

A manually operated setting tool is used to install the instrument which is orientated to measure strains in the vertical, horizontal, and 45° directions. The setting tool remains in place while the strain cell is cemented to the rock and is then withdrawn so that overcoring may proceed (Fig. 10.7).

If the difference in the readings of the strain gauges in the vertical, 45° and horizontal directions, before and after overcoring, is, respectively, ϵ_v, ϵ_{45}, and ϵ_H. Then the principal strains ϵ_1 and ϵ_2 in the rock at the end of the borehole are:

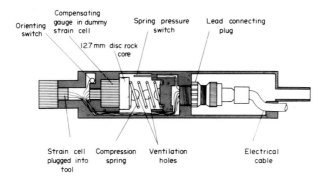

Fig. 10.7. Setting tool for the "doorstopper" strain cell.

$$\epsilon_1 \text{ and } \epsilon_2 = \left\{ (\epsilon_H + \epsilon_v) \pm \sqrt{2\epsilon_{45} - (\epsilon_H + \epsilon_v)^2 + (\epsilon_H - \epsilon_v)^2} \right\}$$

The directions of ϵ_1 and ϵ_2 are θ_1 and θ_2 measured anti-clockwise from the direction of ϵ_H, and are given by

$$\tan \theta_1 = \frac{2(\epsilon_1 - \epsilon_H)}{2\epsilon_{45} - (\epsilon_H + \epsilon_v)},$$

$$\tan \theta_2 = \frac{2(\epsilon_2 - \epsilon_H)}{2\epsilon_{45} - (\epsilon_H + \epsilon_v)}.$$

The principal stresses in the rock at the end of the borehole are

$$\sigma_1 = \frac{E}{1 - v^2} (\epsilon_1 + v\epsilon_2),$$

$$\sigma_2 = \frac{E}{1 - v^2} (\epsilon_2 + v\epsilon_1).$$

The photoelastic plastic biaxial strain gauge

Hawkes and Moxon describe the use of Hawke's biaxial plastic strain gauge for *in situ* stress determination in rock, applying the stress-relief technique. This technique, the Leeman "doorstopper", and the direct application of electrical strain gauge rosettes, all provide an essentially similar function and are conducted in much the same way. All are entirely dependent on the investigator securing a good cement bond between the gauge and the rock. This presents difficulty in wet holes, so that some method of drying is necessary. Free water must be drained out, the rock surface cleaned, and then an acetone spray over the back of the hole will dispel water from the rock surface for a sufficient length of time (10 to 15 min) to allow the cement to polymerize. Considerable care must be taken to guard against temperature effects. The sensitivity of the photoelastic gauge is not significantly affected by temperature over the range $0 - 150°C$, and a dummy gauge in the circuit will compensate for temperature effects on resistance strain gauges. However, differences in the coefficient of expansion between the gauge, its cement, and the rock can give rise to errors unless due precautions are taken. Readings should be made when the gauges attached to the overcored rock are within $2°C$ of the original temperature at the back of the hole. During the drilling process it is important to use water that is not at a widely different temperature from that of the rock. Drilling should proceed slowly, using good, sharp, bits.

The interpretation of the measurements made by any of this class of instrument must take into account the effect of stress concentration at the back end of the borehole. The results of an experimental investigation by Leeman gave a stress-concentration factor of 1.53, while Hawkes and Moxon found the factor to be 1.58. These figures agree closely with the theoretical analysis of Galle and Wilhoit, but Hoskins points out that the factors are valid only when the borehole axis coincides with one of the principal stress directions. Van Heerden found errors of nearly 30% introduced by assuming the principal stress direction to be coincident with the borehole axis. Coates and Yu, in an analytical solution of the problem, using the finite element method, quote concentration factors, dependent on Poisson's ratio, according to the equations:

$$a = 1.366 + 0.0250v + 0.502v^2,$$
$$b = -0.125 + 0.154v + 0.390v^2,$$
$$c = -0.520 - 1.331v + 0.886v^2$$

where a is the stress concentration factor to be applied to the observed stress in the radial direction, b is the factor in the tangential direction, and c the factor in the axial direction, respectively, v is Poisson's ratio.

Hoskins also maintains that the effect of the stress acting along the borehole axis is not negligible unless that stress is also negligible. He found that the effect of the longitudinal stress along the borehole axis is always to reduce the estimate of compressive stress acting in the plane of the gauge, and it may falsely indicate the presence of tensile stresses in that plane when none actually exist. This, together with Van Heerden's findings on tensile fractures, may explain the apparently high values of tensile stress components that are sometimes indicated by strain cells situated at the back of a borehole.

When any of the borehole strain cells is being used it is essential to prepare the end of the borehole to receive the cell. There must be no annular socket remaining at the end of the borehole when the strain cell is inserted. The general practice is to use a flat diamond or hard-surfaced "bull" bit to grind the back of the hole and remove all traces of socket. Some investigators grind the back of the hole to a hemispherical profile, claiming that not only does this remove the risk of leaving an annular socket, but it also gives a stress concentration factor of unity. Van Heerden, on the results of photoelastic studies and the application of a brittle fracture criterion for the rock, showed that fractures may initiate at relatively low strata pressures, near the flat end of a borehole drilled into rock, but there was no particular advantage, so far as the extent of the potential fracture zone around the borehole end was concerned, in using a spherically shaped borehole end in preference to a flattened one.

Borehole Inclusion Stressmeters

A borehole inclusion stressmeter differs from a borehole deformation meter in that it may be calibrated directly in terms of stress, even though its response to stress is a measured strain, or other effect which results from that strain. "Stressmeters" are, in fact, hard, rigid, or near rigid, inclusions, whereas deformation meters and strain cells may be either soft inclusions offering little resistance to borehole deformation or simply profile measuring devices that offer no resistance at all.

Rigid inclusions

It can be shown that when an elastic circular inclusion is situated in an elastic host material under uniform uniaxial loading, and the inclusion is welded around its boundary into the host material:

If a change of stress σ occurs in the host material a uniformly distributed stress σ' will be set up in the inclusion such that:

$$\frac{\sigma'}{\sigma} = (1-v^2)\left\{ \underbrace{\frac{1}{(v-1) +\frac{E}{E'}(v'+1)(1-2v')}}_{1} + \underbrace{\frac{E}{E'}(v'+1) + (v+1)(3-4v)}_{2} \right\}$$

where E = Young's modulus of host material,
E' = Young's modulus of the inclusion,
v = Poisson's ratio of the host material,
v' = Poisson's ratio of the inclusion.

The ratio σ'/σ approaches a limiting value when E'/E is infinite but is virtually constant at 1.5 when $E' > 5E$.

The shear stress induced in the inclusion is:

$$S = \frac{\sigma_1 - \sigma_2}{K - k}$$

where σ_1 and σ_2 are the principal stresses in the host material and K and k are two constants,

$$K = \frac{(1-v')\,(3-4v')}{8(1-v)\,(1+v)} \times \frac{E}{E'} + \frac{5-4v}{8(1-v)}$$

and

$$k = \frac{(1+v')\,(1-4v)}{8(1-v)\,(1+v)} \times \frac{E}{E'} + \frac{4v-1}{8(1-v)}$$

The numerical values of K and k are therefore functions of the ratio between the elastic moduli of the inclusion and the host material. The characteristics of their variation show increasing slope as E'/E approaches zero, i.e. when the inclusion is increasingly soft, compared with the host, but a rapidly decreasing slope with increase in hardness of the inclusion, compared with that of the host. This decrease in variation is such that the values of k, K, and the ratio σ_1/σ are, for all practical purposes, constant when E' is equal to or more than $2E$.

This means that if a hard-inclusion stressmeter is firmly cemented or fixed in contact with the walls of a borehole in rock, changes in rock stress will produce a change in the stressmeter that will have little dependence on changes in the modulus of elasticity of the rock. That is, it is not necessary to have precise knowledge about the rock modulus. The more rigid the meter the less important it becomes to know what is the deformation modulus of the rock.

Meters which are applied on this principle, have been designed by Hast, Wilson, Potts, May, and Hawkes. All these devices can be "prestressed" after insertion so that they can be applied to measure absolute stress by the overcoring technique or to monitor relative stresses above and below the initial prestress level. All these instruments operate on the same basic principle. They differ in the method by which the borehole wall is loaded from within, or "prestressed", after the meter is inserted in the borehole and before it is overcored. They differ also in the type of transducer that is incorporated into the inclusion, to measure its deformation when the rock containing it is overcored. Hast's stressmeter may be taken as representative of its class.

Geotechnology

Hast's stressmeter

This meter employs a magnetostrictive gauge in the form of a nickel-alloy spool on which is wound a coil protected by a permaloy cylindrical screen. The spool is loaded through platens forced against the borehole wall by a multiple wedge system (Fig. 10.8).

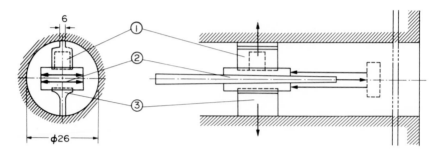

Fig. 10.8. Hast's stressmeter. (1) Magnetostrictive armature. (2) Prestressing wedge mechanism. (3) Loading platens.

Deformation of the spool under load alters the magnetic permeability of the system and the impedance voltage drop across the coil, which is measured. The width of the loading platens in contact with the wall can be selected to suit the site conditions, to ensure that the effective modulus of the cell is higher than that of the rock.

Calibration of rigid inclusion stressmeters is usually performed by inserting the meters in blocks of rock and loading these in the laboratory testing machine. This involves the transport of large rock specimens, which is not always convenient. An alternative is to calibrate in materials of known elastic moduli which are assumed to be similar to those of the rocks in the field. Hast calibrates in steel blocks in the laboratory and applies correction factors intended to suit the particular properties of the rocks on the site.

The calibration characteristics of all the prestressed borehole stressmeters are dependent upon the level of prestress as well as the relative elastic properties of stressmeter and rock. In some of the meters the prestress level is not easily controlled and in all the meters it is necessary to select a level of prestress to suit the rock before commencing the field tests. As the properties of the field rock are usually unknown to begin with, the optimum choice of prestress level becomes very largely an intuitive procedure. In general, the use of these meters presents special problems both in the field and in the laboratory. Since they are somewhat larger than the borehole deformation meters and borehole strain cells, overcoring is usually more difficult and costly, while in the laboratory there are difficulties in calibration. It is understandable, therefore, that the use of the prestressed rigid insertion stressmeters has almost invariably been limited, in each case, to the designer.

Photoelastic glass-insertion stressmeters

The idea of using the birefringent properties of glass under load, as an optical indicator of stress in a solid structure, has been put forward many times since Brewster first suggested it in 1816. However, the first extensive and successful application of the principle in rock mechanics was reported from Japan, by Hiramatsu *et al.*, in 1957. This consisted of the use of solid cylindrical and prismatic inclusions of glass, set in shallow holes in concrete mine shaft and tunnel

184

walls, and observed by reflection polariscope. A subsequent development in the former Post-graduate School of Mining at Sheffield University extended the technique to facilitate the determination of stress in boreholes at various depths within the rock walls.

The photoelastic glass-insertion stressmeter may be in the form of a solid inclusion, in which event the observed birefringence is a measure of the shear stress in the glass, and hence, by cali-bration, of $(\sigma_1 - \sigma_2)$ in the rock in a plane normal to the longitudinal axis of the borehole. When the glass insertion has a hole drilled along its central axis it forms a biaxial gauge which is cali-brated in terms of the major principal stress in the plane of measurement (N/m^2 per fringe). The device is primarily intended to instantaneously measure and display any increase of stress that may occur around it after the time of its insertion in the host material. That is, it measures rela-tive stress above the initial setting value, and it is not overcored. However, to a limited extent, depending upon the strength of its cement bond to the host material, it can respond to a relative decrease in the ambient stress level, after the time of its insertion. In visco-elastic and in plastic materials, such as rock salt and ice, it will gradually "pick up" the absolute stress in the host, through a load-transfer process resulting from creep of the host material in the stress-concentra-tion zone around the inclusion.

The photoelastic stressmeter in its simplest form consists solely of the annular glass cylinder. This is inserted in the rock or concrete wall, and a circularly polarizing light probe is inserted down the central hole by the operator whenever a reading is taken. Observation is through a small hand-viewer, of a size suitable for carrying in the pocket, using also a telescope for distant viewing when required (Fig. 10.9(b)).

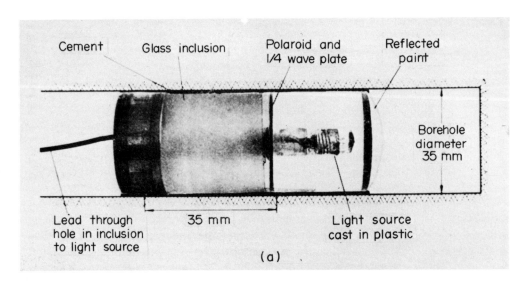

Fig. 10.9. The photoelastic glass insertion stressmeter. (a) The stressmeter arrangement in a borehole.

The form of stressmeter used for deep insertion in a borehole is shown in Fig. 10.9(a). This includes an integral circularly polarized light source, current for which is carried through leads taken through the central hole.

The sensitivity of the photoelastic glass stressmeter is dependent on the length of the light path through the instrument, the modulus of elasticity (effective) of the rock into which it is inserted, the principal stress ratio in the biaxial stress field, and the point of reference on the optical signal at which the reading is taken. Hawkes has detailed the effects of all these factors,

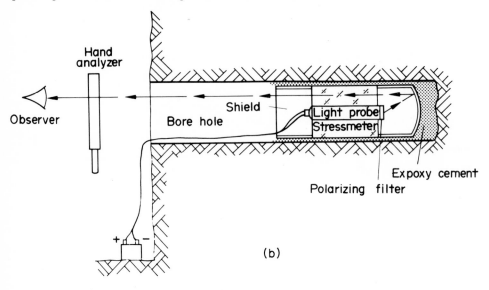

Fig. 10.9. The photoelastic glass insertion stressmeter. (b) Viewing system.

which are summarized in the diagram (Fig. 10.10 (b)). This shows the calibration constants for a glass stressmeter 1 in. long in host materials of Young's modulus ranging from 5×10^5 to 3×10^7 psi, in all biaxial stress fields, ranging from the uniaxial ($\mu = 0$) to hydrostatic ($\mu = 1$). Alternative calibration curves for the $45°$ and $90°$ reference points are shown, the former being that used in a uniaxial or near-uniaxial stress field. This $\mu = 0$ ($45°$ point) calculated calibration curve is also compared with the experimentally observed curve. It can be seen that, while the calculted calibration sensitivity is constant at approximately 450 psi/fringe-inch (122 kN/m² per fringe-mm) in materials whose E value extends up to 3×10^5 psi, (2068.5 MN/m²) — in practice the sensitivity is observed to be constant, at the same figure, in all host materials up to $E = 5 \times 10^6$ psi (34475 MN/m²). This means that the glass stressmeter functions as a rigid inclusion in a wide range of sedimentary rocks, including most porous sediments, shales, sandstones, and medium-strength concretes and limestones. For materials whose E value is higher than 34475 MN/m² the deformation modulus of the host material must be determined before the appropriate calibration constant can be supplied.

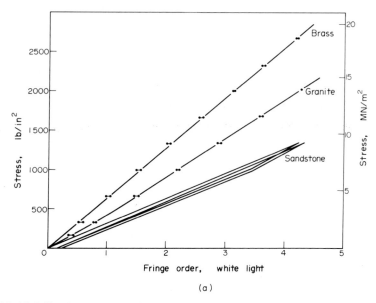

Fig. 10.10. (a) Calibration characteristics of a 1.5-in. long photoelastic stressmeter in three materials (uniaxial stress field).

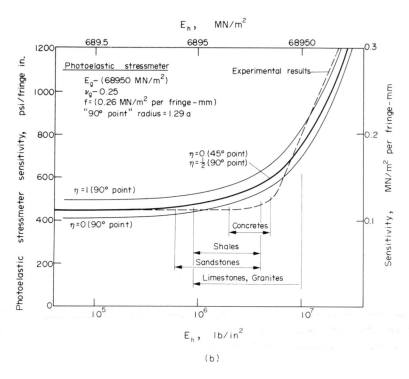

Fig. 10.10. (b) Calibration characteristics (observed and calculated) of photoelastic stressmeters in uniaxial and biaxial stress fields, in a range of host materials (Hawkes).

The difference between the observed and the calculated calibration curves of the glass stress-meter in porous sediments and non-elastic materials is due to the fact that the effective modulus for these materials, in the stress-concentration zone around the rigid inclusion, is the initial tangent modulus, which is very much lower than the average nominal value or the secant modulus of the material. It means that the glass insertion can be used as a stress-indicator, without reference to strain or modulus of deformation, in a wide range of earth materials, elastic and non-elastic rocks, saline evaporites, permafrost, and ice.

Various setting techniques are used, depending upon the depth of insertion and the site conditions. The setting-tool head for deep insertions is shown in Fig. 10.11. The stressmeter is

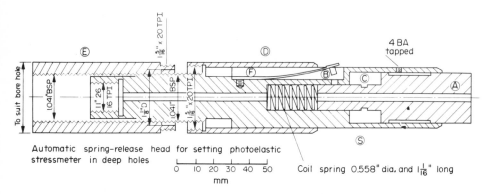

Automatic spring-release head for setting photoelastic stressmeter in deep holes

Coil spring 0.558" dia. and $1\frac{1}{16}$" long

Fig. 10.11. Deep-hole setting tool for photoelastic stressmeter.

cemented into the rock by a thin annulus of epoxy-resin cement, to which has been added a proportion of carborundum filler, which gives the cement high rigidity after it has polymerized.

Determination of the Complete State of Stress by Measurements in a Single Borehole

Resolution of the three-dimensional resultant state of stress within a rock mass may be facilitated by the use of computers, employing equations into which are fed the observations made in three mutually perpendicular boreholes. The problem is complicated by the fact that the rock mass is not likely to be either homogeneous or isotropic. The stress field too, will, in all probability, neither be homogeneous nor constant. The compilation of a precise solution, from at least three biaxial, and sometimes nine uniaxial, measurements, separated in time and space, is obviously impossible. The best that can be hoped for is an approximate solution, and even to obtain this much hard and expensive field work, involving core-drilling into the rock, must be done. Hence there is much interest in the development of instruments and techniques that will yield a complete solution from multiple measurements made simultaneously in one borehole.

The Leeman (CSIR) three-component gauge

A further development of the borehole strain cell technique is reported from South Africa, where Leeman has produced a multi-component cell to measure nine strains, three each at three locations in a borehole; (i) in the "roof", (ii) in the "sidewall", and (iii) at an intermediate point making an angle of $7\pi/4$ from the horizontal diameter (see Figs. 10.12, 10.13). By attaching

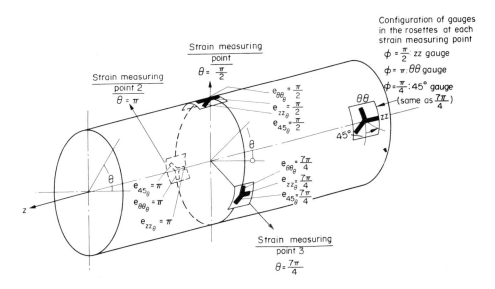

Fig. 10.12. Arrangement of gauges for determination of the complete state of stress from measurements in a single borehole (Leeman).

Fig. 10.13. The Leeman multi-component borehole strain cell.

strain-gauge rosettes to the rock at each of these positions the magnitudes (ϵ_1 and ϵ_2) and directions (ϕp) of the principal strains are found. From which, knowing Young's modulus (E) Poisson's ratio (v) for the rock, the stress components are calculted. Referring to Fig. 10.12:

$$P_{\theta 1} = E/2 \left(\frac{\epsilon_1 + \epsilon_2}{1-v} + \frac{\epsilon_1 - \epsilon_2}{1+v} \cos 2\,\phi p \right),$$

$$P_{zz} = E/2 \left(\frac{\epsilon_1 + \epsilon_2}{1-v} - \frac{\epsilon_1 - \epsilon_2}{1+v} \cos 2\,\phi p \right),$$

$$P_{\theta z} = E/2 \; \frac{\epsilon_1 - \epsilon_2}{1+v} \sin 2\,\phi p$$

for each value of θ, namely $\theta = \pi/2$, $\theta = \pi$, and $\theta = 7\pi/4$.

By suitably orientating three rectangular rosettes, as shown in Fig. 10.13, nine strains are measured, of which only six are required from which to compute the stress components σ_x, σ_y, σ_z, τ_{xy}, τ_{yz}, and τ_{xz}, using the equations

$$\sigma_x = 1/8 \; (3P_{\theta\theta} \underset{\theta=\pi/2}{} + P_{\theta\theta} \underset{\theta=\pi}{}),$$

$$\sigma_y = 1/8 \; (3P_{\theta\theta} \underset{\theta=\pi}{} + P_{\theta\theta} \underset{\theta=\pi/2}{}),$$

$$\sigma_z = P_{zz} \underset{\theta=\pi}{} + v/2 \; (P_{\theta\theta} \underset{\theta=\pi/2}{} - P_{\theta\theta} \underset{\theta=\pi}{}),$$

$$\tau_{xy} = -1/8 \; (P_{\theta\theta} \underset{\theta=\pi}{} + P_{\theta\theta} \underset{\theta=\pi/2}{} + 2P_{\theta\theta} \underset{\theta=7\pi/4}{}),$$

$$\tau_{yz} = -1/2 \; P_{\theta z} \underset{\theta=\pi}{},$$

$$\tau_{xz} = -1/2 \; P_{\theta z} \underset{\theta=\pi/2}{}$$

The instrument is inserted into a 38 mm diameter borehole, and the strain gauge rosettes are carried, cast in rubber plugs, to be forced out against the wall of the hole by pneumatic pressure at about 413.7 kN/m², after smearing the face of each plug with glue. The setting-tool contains a temperature-compensating gauge, cast in rubber and glued to a rock disc which is not loaded. Leads from the gauges are carried through the setting-tool and then through a control-box to a strain-gauge measurement-bridge. A mercury switch, built into the setting-tool, facilitates orientation of the rosettes in the borehole, the correct alignment being indicated when a lamp on the control-box lights up. After allowing time for the glue to harden, the setting-tool is removed and the mouth of the hole is then plugged, so as to protect the cell from the water circulated around the bit of the 89 mm overcoring drill which follows. Overcoring proceeds to about 50 mm beyond the end of the gauge in the central hole. The core, containing the gauge, is then broken off and removed intact. The central hole is then unplugged, so that the setting-tool can be reconnected and readings of relief-strain taken.

Leeman describes the results of initial laboratory and field tests of his instrument, while subsequent experience with the equipment is reported by Herget. The tests show that the equipment, and the available analysis (for which Herget also includes a computer program), can obtain what are claimed to be realistic estimates of the ground stress, provided that the elastic constants of the rock are carefully determined, and the ground at the test site fits the assumption of a homogeneous elastic body. The success of the method depends on obtaining intact rock core cylinders, so that the requirements for rock quality are higher than those that apply when uniaxial or biaxial strain cells are used.

The LNEC (Lisbon) single borehole stress-gauge

Rocha and Silverio describe a multi-component borehole strain cell, for the complete determination of the state of stress in a rock mass, by means of measurements carried out in a single borehole. The cell consists of an epoxy-resin cylinder, into which ten strain gauges are embedded, distributed along the middle part of the cylinder. Four of the gauges serve to provide check measurements, over the minimum requirement of six observed strain components. The gauges are connected by cable, one end of which is cast into the cylindrical strain cell, so as to preserve a complete seal against the ingress of water. This cable is also connected to a dummy-gauge embedded in epoxy-resin but carried in a capsule at the upper end of the cell. The overall dimensions of the cell are 440 mm long by 35 mm diameter.

The strain cell is inserted into a vertical or near-vertical down-hole. Epoxy-resin cement is first poured into the hole, or, alternatively, carried in a flexible plastic bag, to the bottom of the hole. The strain cell is then lowered, until the end of the cylinder bursts the bag and penetrates into the cement. The cement is thus displaced, to flow and polymerize as an annulus between the strain cell and the borehole wall. The cell is centered in this annulus by strips of plastic 1 mm thick, and its orientation in azimuth is defined by the setting rods. After allowing a sufficient time for the cement to polymerize and hold the cell firmly to the rock wall, the borehole containing the cell is overcored in the usual way, and the relief-strains measured. The core is then removed from the borehole, to check the quality of the cement/rock bond, which must be strong and intact if the measurements are to be accepted. Rocha and Silverio give details of the theoretical relationships between the measured strains and the regional stress components, for which a computer solution is obtained.

Low-modulus, solid inclusions

Basically, the LNEC multi-component gauge is a low modulus, solid inclusion, a theoretical examination of which is presented by Argawal and Boshcov. The general expression for the radial deformation of such an inclusion, on overcoring, in a plane stress field is

$$u_r = \frac{-\epsilon(1-v_i)}{\frac{E_i}{E_r}(1+v_r)+(1-v_i)} - \frac{as_1(1-v_i)}{E[(1+v_r)\frac{E_i}{E_r}+(1-v_i)]}$$

$$+ \frac{2a}{E_r}S_1 \frac{(3+v_i)\cos 2\phi}{[(5-v_r)\frac{E_i}{E_r}+(3+v_i)]}$$

where u_r is the radial deformation of the hole along the radius a, inclined at angle ϕ to the reference direction. $(a + \epsilon)$ is the initial radius of the soft inclusion, which is assumed to be stretched so that it has radius a when inserted into the hole. If the stretch is released the inclusion comes into contact with, and presses against, the wall; a is the hole radius before overcoring, s is the uniaxial stress applied in the direction $\phi = 0°$; E_i, v_i are Young's modulus and Poisson's ratio, respectively, of the inclusion; E_r, v_r are Young's modulus and Poisson's ratio, respectively, of the rock.

u_r is relatively insensitive to variations of Poisson's ratio for the rock and for the inclusion, and terms in the equation containing E_i/E_r may be dropped with negligible error when E_i/E_r approaches or becomes less than 1/100. In practical terms the error may be considered to be acceptable for values of E_i/E_r approaching 1/10 (see Table 10.1).

Table 10.1. *Influence of Ratio of Modulus of Inclusion (E_i) to Modulus of Rock (E_r) on the Radial Deformation (U_r) of a Circular Borehole Radius "a" subjected to a Uniaxial Stress S_2 (Argawal and Boshcov)*

$\dfrac{E_i}{E_r}$	U_r	Remarks
1/10	$-0.857A + 1.745B$	
1/100	$-0.984A + 1.970B$	
1/1000	$-0.998A + 1.997B$	$a = 1$
0	$-1.00A + 2.00B$	$E_i = 0$ (open hole)

$$A = \frac{a}{E_r} S_2$$

$$B = \frac{a}{E_r} S_2 \cos 2$$

The USGS solid-inclusion borehole probe

Nichols, Abel, and Lee describe the development and testing of a solid inclusion borehole probe which is intended for the determination of the three-dimensional stress state in a rock

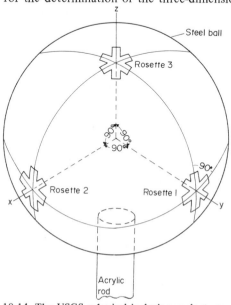

Fig. 10.14. The USGS spherical inclusion rock stress gauge.

mass. The probe consists of a 25 mm diameter chrome alloy steel ball, on which are mounted three electrical resistance strain-gauge rosettes, in the configuration shown in Fig. 10.14. The electrical connections for the gauges are carried along an acrylic rod support, cemented on to the ball to serve as a handle, and the complete assembly is encapsulated in a cast epoxy-resin cylinder 38 mm diameter by 41 mm long. The theoretical basis of the instrument is that of the high-modulus elastic inclusion in an elastic host material. For a spherical inclusion the strain in a given direction at the center of the sphere is equal to the strain measured in the same direction at a point on the surface of the sphere.

The probe is cemented in place using an epoxy-resin grout combined with carborundum filler, and it is not prestressed. It is therefore primarily suited for measurement of relative stress increase, after the time of its insertion into the host rock, and it can respond to a tensional field, or to the relief of stress if overcored, only to the extent that the cement bond to the rock will hold. The instrument is reported to give a satisfactory response, corresponding to that which is predicted to approximate from two-dimensional elastic theory, when tested in elastic models under hydrostatic and triaxial stress fields. No complete three-dimensional elastic analysis yet exists that will correspond to such a probe in rock. The evidence obtained from field tests with the instrument is still more difficult to assess and very little has, as yet, been reported. In comparative tests applied to field stress measurements at the Rangely anticline, Colorado, readings obtained by the use of the USGS probe gave figures that appeared to be lower than those obtained by other devices, as reported by De La Cruz and Raleigh.

Determination of Rock Stress Using Hydraulic Pressure Cells

The Flat-jack Technique

The flat-jack technique of rock stress measurement consists of attaching extensometers, or reference measurement points, in a suitable arrangement over the rock face, and then cutting a slot into the rock with the object of relieving the rock of the ambient stress. Some investigators attempt to measure stresses in a biaxial field by placing two slots at right angles. The length of the slot is commonly from $33 - 48$ cm long, and of similar depth, about 4 cm thick, to receive flat-jacks $30 - 45$ cm^2.

Construction of the slot produces local stress relief and the resulting rock deformation is measured, over a period of 3 to 4 days, applying the extensometer over various pin combinations. The flat-jack is then inserted and grouted into the slot, with its edge flush with the rock surface and left another 3 to 4 days for the grout to harden. Hydraulic pressure is then applied to the flat-jack and the pressure increased in stages, measurements being taken across the various pin combinations until these are seen to be at their original preslot values. Pressure is then released and increased in from two to four cycles over a period of several days and the mean cancellation pressure determined. The complete operation takes from 2 to 3 weeks.

Alexander gives formulae based on elastic theory, assuming an elliptical slot and plane stress, from which it is deduced, if Poisson's ratio = 0.2:

$$S = aP + bQ$$

where S = rock stress normal to the jack,

Q = rock stress parallel to the jack,

P = mean cancellation pressure,

a and b are constants depending upon the dimensions of the flat-jack and the geometry of the gauge positions relative to the flat-jack.

When the measurement points are on a line through the central axis of the slot and located at a distance $L/3$ from the slot, where L is the length of the slot, it is assumed that the cancellation pressure equals the stress in the rock.

In Alexander's theoretical treatment the cancellation pressure depends upon the dimensions of the slot and flat-jack, the biaxial stress field, and Poisson's ratio. It is independent of the

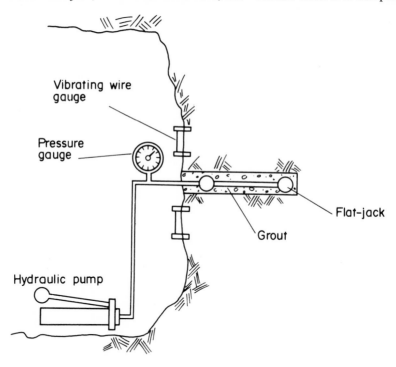

Fig. 10.15. The flat-jack technique of measuring rock stress.

modulus of elasticity of the rock. In practice, although there is no dependence on linearity of the stress-strain relationship, the success of the technique depends upon the existence of the same deformation characteristics on relief as on reloading to cancellation pressure. This may not always occur. However, the major objections to the flat-jack technique are that the measurements must be made at the edge of the excavation in a zone which is subject to unknown and irregular stress distribution, and which may itself be destressed.

Borehole Hydraulic Pressure Cells

The borehole jacks and pressure cells, which have already been described in connection with the determination of *in situ* rock deformability, can also be applied for the determination of *in situ* stress. Sellers describes the theory for such measurements, using either cylindrical borehole pressure cells or borehole flat-jacks. For the cylindrical pressure cell it is shown that, for the plane strain condition,

$$\Delta S + \Delta T = \frac{\Delta PNE}{2\pi La^2 (1 - v^2)}$$

where ΔS and ΔT are the rock stress changes within the rock mass, at some distance from the borehole; ΔP is the change in hydraulic pressure within the cell; a is the radius of the borehole; r and θ are the polar coordinates describing the position at which the radial stress σ_y and the tangential stress σ_θ are measured. L is the length of the pressure cell. N is the pressure/volume characteristic of the cell, including tubing and pressure gauge (see Fig. 10.16).

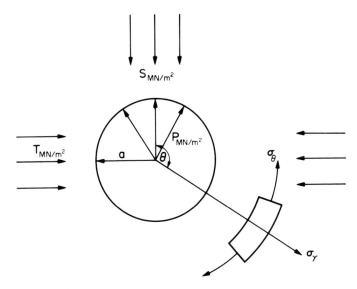

Fig. 10.16. Stresses around a hydraulic borehole cell which exerts a uniform radial pressure on the borehole wall (Sellers).

Therefore, if N, E, and v are known, ΔS and ΔT can be calculated from the observed fluid pressure changes ΔP.

The borehole flat-jack type of hydraulic cell consists of a copper envelope, 25 mm wide by 25 cm long, constructed of flattened tubing, into which is sealed a length of thin hydraulic tubing and a pressure gauge. The system is filled with hydraulic fluid and the flat-jack is encapsulated into a cylindrical plug of cement (epoxy, micro-concrete, or "hydrostone"). This encapsulated flat-jack is then inserted into the measurement borehole (see Fig. 10.17). The pressure/volume characteristic of the cell is determined experimentally, using a screw pump, and for plane stress conditions

$$\Delta S = \frac{\Delta PNE}{2\pi La^2}$$

while the corresponding relationship for plane strain is

$$\Delta S = \frac{\Delta PNE}{2\pi L a^2 \, (1 - v^2)} \, .$$

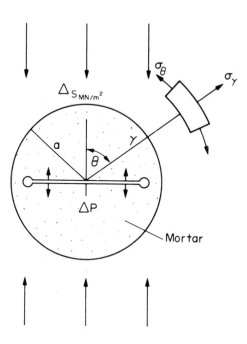

Fig. 10.17. Stresses on the borehole flat-jack type of hydraulic cell (Sellers).

The response of these pressure cells is improved by reducing the volume of contained hydraulic fluid, and also by increasing the rigidity of the system. The substitution of a photoelastic pressure gauge for the Bourdon-type gauge achieves this purpose. Hydraulic pressure cells may also be used to determine absolute *in situ* stress by the stress-relief technique, if the cell is first calibrated at specific values of internal hydraulic pressure. The calibration may be performed *in situ*, using large flat-jacks to impose a controlled and measured load, or be performed on representative rock specimens in the laboratory. The observed cell-pressure change, due to expansion of the rock and cell on relief, is then related to the absolute *in situ* stress level.

Determination of in situ Rock Stress at Depth

Fairhurst has reviewed the theory and practice of *in situ* rock stress determination, with particular reference to the problem of measurement in rock masses at great depth from the surface. One of the major practical difficulties associated with techniques involving deformation meters, borehole strain cells and inclusion stressmeters is the limitation on the depth at which they may be used. This is seldom as much as 30 m and sometimes less than 5 m. Many of the techniques are limited also to observations in holes inclined above the horizontal. Exploration of the state of stress at depth in the Earth's crust is important in relation to better understanding of earthquake phenomena and the possibility of earthquake control. It is also of great importance in petroleum and natural gas reservoir engineering, and in relation to problems associated with the possible disposal of radioactive fluid waste materials. As a result of his critical study of the problem, Fairhurst suggests that hydraulic fracturing is the most readily adaptable method of exploring the regional stresses at depth in the rocks of the Earth's crust.

A prototype instrument operating on this principle has recently been described. The Deep Stress Probe (DSP) consists of a hydraulically operated tool containing a combined straddle-packer and impression-packer, to determine the orientation and azimuth of induced fractures at depth in a borehole.

Determination of the in situ State of Stress in Non-elastic Earth Materials

Strictly speaking, the stress-relief strain-measurement techniques are only valid if applied to homogeneous, isotropic, strong rocks that display elastic, or near-elastic, deformation characteristics. They are of little value in soft, porous sediments, or in materials having marked anisotropy. They are of no value whatsoever in materials that display predominantly time-dependent deformation properties under load, such as ice, permafrost, rock salt, and potash. Neither, in these materials, should a soft inclusion be used, because it is unable to differentiate between elastic and creep deformation. However, if an effective "average" deformation modulus can be established for the material concerned, over the range of stress involved, then a high modulus inclusion stressmeter, if calibrated at an appropriate "pre-stress" level, might yield an approximate value for the absolute regional stress component, in the direction of measurement.

The determination of relative increase or decrease of stress above or below the prestress level is a much easier undertaking, and one that may be made accurately, without reference to the deformation modulus of the host rock, provided that the inclusion stressmeter is sufficiently rigid. High modulus inclusions that are applied without prestress, such as the USGS borehole probe and the photoelastic glass insertion stressmeter, should be regarded as being suited only for the observation of stress increase, because any decrease of stress upon them, after the time of their insertion, places tension on the cement bond. On a rigid inclusion, this should not be relied upon. Since the response of a high modulus inclusion to stress is independent of change in the effective modulus of a relatively soft host material the inclusion may be calibrated to read directly in terms of stress, and that calibration will hold good on a short-term basis, even though the host material is subject to creep and flow. The photoelastic glass insertion stressmeter acts as such an inclusion in all materials of uncertain or varying modulus, up to an effective E value of 34475 MN/m^2.

Skilton reports that the meter maintained its calibration figure to within 5%, in rock salt, even though the salt underwent considerable creep during the period of observation, the change in lateral strain, at constant stress, being as depicted in Fig. 10.18. When inserted at a zero con-

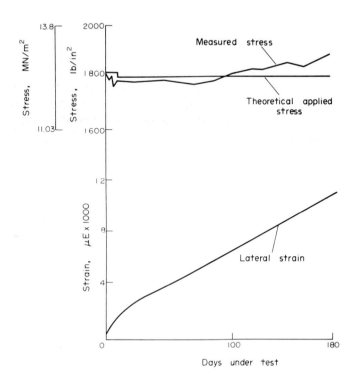

Fig. 10.18. Stressmeter readings and measured lateral strain under constant load (Skilton).

dition of stress and the rock salt subsequently placed under increasing load, the calibration figure was again maintained. Hawkes also reports stress observations made in frozen sand, using a glass inclusion, in which measurements were effective up to the point when the host material began to disintegrate around the inclusion, due to the extent of creep deformation involved.

A rigid inclusion that is inserted into a plastic or visco-elastic material which is already under load, will, in the course of time on a long-term basis, "pick up" load from that material, as a result of the creep around the inclusion, until, ultimately, the inclusion becomes strained to an extent that is determined by the stress-strain characteristics of the inclusion and the magnitude of the total ambient stress in the host. At the same time, it will respond immediately to any relative stress increase in the host. This is illustrated by Skilton's experiment, illustrated in Fig. 10.19, in which a glass stressmeter was inserted into rock salt at a datum stress level of 12.55 MN/m^2, at which time the meter read zero. After 165 days had elapsed the meter registered 72% of the theoretical absolute stress in the host. During this period the load on the salt was maintained constant for 100 days, and then was increased, in one increment, to 13.65 MN/m^2. This increment was instantaneously registered on the stressmeter.

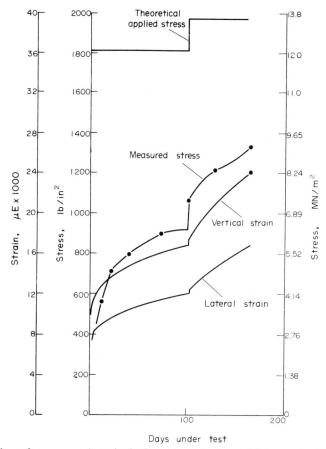

Fig. 10.19. Build-up of stress on a photoelastic glass stressmeter inserted into a rock salt specimen already under load (Skilton).

Pressure Cells

Hydraulic pressure cells may also be of considerable assistance when studying the stress distribution in non-elastic earth materials. In this case it is appropriate to apply the cell by the "active principle, in which the ambient stress, or strata pressure, is balanced by an equal and opposite fluid pressure in the cell, and this is measured. The principle is applied to concrete structures in the Carlson stressmeter, and in soft rocks and soils by the Gloetzl cell. The Gloetzl cell consists of a pressure-sensing pad and a hydraulic bypass valve assembly. When in use the cell functions as a pressure-actuated bypass valve in an individual hydraulic circuit. The cell is maintained in a "closed" condition by the action of pressure on the sensing pad. To measure the magnitude of this pressure the hydraulic pressure in the cell delivery line is slowly increased, at a constant rate. When the cell delivery pressure equals the pressure on the sensing pad the valve system opens, so bypassing hydraulic fluid to the cell return line. The pressure at this point is indicated by a precise manometer on the delivery line.

Geotechnology

Fluid Inclusions in Rock Salt

Dreyer describes a hydraulic measurement system for the determination of *in situ* stress in rock salt. The arrangement is illustrated in Fig. 10.20. Hydraulic oil is pumped into a sealed borehole, 38 mm diameter, which is drilled into the rock wall with a slight downward inclination.

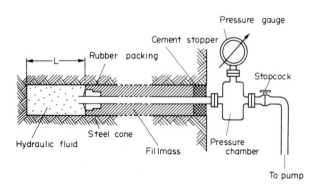

Fig. 10.20. Arrangement of fluid inclusion in rock salt (Dreyer).

Rubber packing-seals are installed in the hole immediately after it has been drilled, so that oil can be pumped behind the seal. The length of hole that is thus pressurized is about 30 cm, and the pressure is maintained at the limit that the seals can withstand, while the remainder of the hole is filled with an impervious plug of cement or epoxy resin. The purpose of this initial pressurization, without delay after drilling the hole, is to minimize the extent of creep that can take place in the material near the borehole wall, before the measurement process begins. After sealing-off and pressurizing the fluid inclusion, creep of the surrounding borehole wall takes place until the fluid pressure attains a value which totally resists any further deformation of the rock. This "equilibrium pressure" is equivalent to the maximum shear stress in the rock $(\sigma_1 - \sigma_2)/2$ where σ_1 and σ_2 are the principal stresses. At shear stresses below the equilibrium pressure the rock behaves elastically and creep does not occur.

For potash rock, Baar estimates that the maximum value for the difference between the maximum and minimum local stress components is about 1379 kN/m², and hence the initial equilibrium pressure in the borehole should be around 689.5 kN/m². The value of the maximum principal stress component is then determined by pumping in more fluid, to further pressurize the inclusion to a value that is estimated to be higher than σ_1. The hydraulic pressure line is then closed. If the pressure were taken sufficiently high in a brittle material, then the rock should fracture, as in the "Hydrofrac" technique, in a direction perpendicular to σ_2 and the fluid pressure, once the fracture had been initiated, would fall to that required to propagate the fracture, that is, to a new equilibrium value σ_2. In a plastic material, however, the effect of pressurizing the fluid inclusion above the yield strength of the containing rock material is to cause load transfer from the rock to the fluid inclusion, because creep takes place in the host rock until that load transfer is complete and balance is restored. There will therefore be a new equilibrium value, to which the inclusion pressure will fall, for any setting of inclusion fluid pressure above the initial equilibrium level. The maximum equilibrium pressure in the inclusion will be equivalent to σ_1.

Indirect Measurement of Rock Stress

Geophysical Techniques

Various attempts have been made throughout the years to make use of the change of sonic velocity with stress in rock as a means of stress measurement. In general, however, the results have been disappointing as also have been attempts to make practical correlations between stress and resistivity in the field. More success has attended the use of microseismic-acoustic techniques for the observation of rate of stress increase in rocks under pressure.

Qualitative Observations of High-stress Zones

Qualitative information on the build-up of stress in rock and the location of highly stressed zones is possible by observation of the cores from rotary diamond drills, or the cuttings from auger or scrolled rotary drills. In highly stressed hard rock the drill cores break into thin discs and the size and number of fragments can be treated statistically to give useful information concerning the relative strength of the rock mass in relation to strata pressure. In softer rock, where auger-type drills are used, the size of the drill cuttings becomes very much less in zones subject to stress concentrations.

Measurement of Stress by Hydraulic Fracture

The "Hydrofrac" technique gives an indirect measurement of the magnitude of the minimum compressive principal stress, in the wall of a borehole drilled into a rock mass. In a homogeneous elastic material the fracture should propagate in a plane perpendicular to that stress, if it is assumed that one of the principal stress directions coincides with the axis of the borehole.

Choice of in situ Stress Measurement Technique

Apart from practical problems of application in the field, the weaknesses in rock stress-measurement techniques lie in problems of calibration and in the assumption that the observed measurement is actually the physical quantity desired. In the determination of absolute stress, borehole deformation meters and borehole strain cell techniques depend entirely upon assumptions of elasticity, and they measure only the components of elastic rebound that may occur on overcoring. Subsequent delayed elastic and time-dependent deformations are not taken into account. The effects of heterogeneity and anisotropy in the rock mass are ignored. The methods are therefore of limited application. Even in materials that demonstrate reasonably approximate elastic behavior, the problem of determining the appropriate deformation modulus remains. Some investigators have avoided this problem by adopting indirect compensation techniques, in which the overcored rock specimen, containing its deformation meter or strain cell, is placed under load in a biaxial pressure jacket, which is then pressurized to restore the meter to its initial reading. The pressure thus applied is then equated to the *in situ* rock pressure. Such methods have obvious attractions but they do not remove the fundamental difficulty in cyclic compensation methods, which is the requirement of identical stress-strain characteristics on relief as on reloading.

There is also great practical difficulty in producing controlled biaxial stress fields by means of which to compensate the observed stress relief, and there must be considerable uncertainty as to the stress distribution that is produced by external platens or jacks, over the cross-section of rock cores, sometimes 150 mm or less in diameter. Experience with photoelastic gauges, in which the observed signal is a direct indication of the state of strain in the gauge itself, shows how dangerous can be the assumption that the strains at a point of measurement in a rock specimen are related simply by the dimensions of the specimen and the magnitude and direction of forces externally applied to the rock.

Instruments which depend on the bonding of strain gauges and rosettes to rock must be used with great care. Photoelastic gauges "fail safe" in this respect. If the bond is not good their signal will be unreadable. But electrical strain gauges will give out a signal irrespective of the condition of the bond, and this may be read as rock strain. The condition of the bond should be directly investigated before accepting the readings given. To improve the "measurement success" ratio the drill holes should be inclined upwards at an angle of not less than 5° from the horizontal, so that water can drain out. The hole should then be effectively cleaned and dried before attempting to attach the Leeman multi-component gauge, or any of the biaxial strain gauges or strain cells. The deformation meters are less critical in this respect, and can be used in down holes. New cements are continually being introduced, and some of these will set under water. The strength of the bond in such cases is somewhat uncertain. They are more likely to be satisfactory with soft inclusions, such as the LNEC multi-component probe, and less so with rigid inclusions. Photoelastic glass inclusions have been known to display a bond strength sufficient to withstand 2 fringe order (600 psi (4.137 MN/m^2) on a 1.5-in. (38 mm) length stressmeter) in tension, before separating at the bond. In compression the same meter would be used up to 4 fringe order, and new meters set successively at similar increments of stress increase, to extend the range as far as necessary. The limit of range, for a single photoelastic glass stressmeter in compression, is primarily determined by the surface finish when drilling the glass sheet from which the stress-plug is fabricated. Well-drilled glass stressmeters, with carefully ground edges, can withstand nine or more fringe orders, but the optical pattern becomes progressively more difficult to read, and the chances of glass fracture become increasingly great, after the fifth fringe order.

Comparative assessments of the various *in situ* stress measurement techniques are few in number, and the results generally inconclusive. Factual, objective, comparisons are impossible to make, for it is impossible to establish any absolute standard against which the various measurements may be compared. The criterion of consistency is not valid, for it is possible for a technique to produce consistent results that are quite erroneous, and the observer usually has no means of detecting whether or not this is so, except for the fact that it is most unlikely that *in situ* stress measurements taken at different points in time or in space should coincide. Experimental comparisons of stress measured in elastic materials under known and controlled load are another matter, but they are largely irrelevant to the problem. Neither can theoretical models reproduce the real situation. There may be conditions at the field site that will restrict the possible choice of technique; for example, the measurement may be required at remote depth, in a down hole, or the rock may be non-elastic, such as potash or rock salt. In these cases the methods that are particularly suited to these respective conditions should be applied, as described earlier in this text.

The investigator has a wider range of selection when choosing a technique to use in elastic or near-elastic rocks, at close range. The "best" technique to use is very largely a matter of opinion and circumstance. Once having chosen to experiment with a technique, and after reaching a stage of proficiency in it that produces an acceptable "measurement success" ratio for him, an investigator will tend to prefer his technique to others, since he becomes practised in it and learns to understand its limitations. In the end it is the interpretation and application of the result that matters. If one is aiming to establish a precise measured value for the regional *in situ* stress it is logical to prefer a technique that is least objectionable on theoretical grounds. The investigator might therefore prefer to use borehole deformation meters or the Leeman multi-component gauge, which measure at some intermediate depth within the borehole, rather than the borehole strain cells or biaxial gauges which are attached to the back of the hole. But there would be less point in aiming at precision if the measurement were being made around an active engineering excavation, where the stresses must vary in time and space. When one also tries to take into account all those other factors associated with the rock mass, such as non-linearity of stress/ strain, anisotropy, heterogeneity, etc., it becomes obvious that the result, no matter how it is obtained, can only be, at best, an approximation. How crude an approximation no one can tell, but a number of such observations made in a geotechnical context should permit of some broad generalized conclusions of a qualitative nature, which is what Hast has done, with the aid of his prestressed inclusion meter, in Scandinavia.

Similarly, when involved in a practical strata control or support problem in geological, civil, or mining engineering, instead of seeking to establish a precise value for the *in situ* stress, it is more useful for the engineer to seek answers to such questions as "Are the stresses relatively high or relatively low, in this locality, compared with others?, Are the stresses increasing, or decreasing, at this point, in relation to time, and if so, at what rate?, and, In which directions are the stress components acting?" When looked at from this point of view, the borehole strain cells and biaxial gauges prove themselves to be very useful and practical engineering tools, in spite of the fact that they may have a less-satisfactory theoretical basis than the deformation meters, and they also have a propensity to indicate anomalously high tensile components of stress. The question of availability of the equipment is also important. The deformation meters and prestressed inclusions are not all generally available but the equipment required for strain cell and biaxial gauge techniques is available commercially, or can be constructed with little difficulty.

Hall and Hoskins have reported on a survey of *in situ* stress measurement methods, as applied by operators from various parts of the world. Besides containing an extensive bibliography on the subject this report contains many useful comments on practical aspects of the techniques reported on. Table 10.2 summarizes the results obtained by De La Cruz and Rayleigh, when comparing measurements made at the Rangely anticline, Colorado, using five different methods. It shows the photoelastic plastic biaxial gauge producing average readings much closer to the

Table 10.2. *Comparison of Average Values Obtained at One Site, Using Five Different Methods in situ Stress Measurement (De la Cruz and Rayleigh)*

Method of measurement	Azimuth reckoned north (deg)	Principal stresses (MN/m²)		Maximum shear stress (MN/m²)
USBM Borehole deformation gauge	63	0.54	0.27	0.13
Direct strain gauge technique	71	1.92	0.93	0.46
CSIR "Doorstopper" strain gauge technique	88	2.22	−2.40	2.31
USGS Spherical inclusion gauge	41	0.17	−0.08	0.12
Photoelastic plastic biaxial gauge	77	1.50	−0.74	1.12
Average values	68	1.27	−0.36	0.83

combined mean than did any other of the techniques used. The true significance of this is a moot point, but it does indicate that the photoelastic plastic biaxial strain gauge may be worth trying first, as a reconnaissance tool, before deciding which of the more sophisticated tools might be best to use, or indeed, whether any of them is likely to be any more useful. It may well tell the investigator all that he may hope to learn from attempts to measure the absolute *in situ* stress, at a fraction of the time, cost, effort, and degree of elaboration involved when using either deformation meters or multi-component gauges. Its major disadvantage is the difficulty of reading biaxial strains at less than half a fringe order, hence it is relatively insensitive to low stress magnitudes, but it is very sensitive to stress directions, which can be readily identified at very low stress levels. And if the investigator is interested only in observing any build-up of stress in a rock mass, or in a concrete structure, at close range, then there is no question as to which is the most suitable technique to use. The photoelastic glass-insert stressmeter can identify all the characteristics of a biaxial stress field, under increasing load, with a degree of sensitivity, precision, cheapness, and ease of application, that no other practical field technique in geotechnology can rival. So much, in fact, that, to the uninitiated, and to those who have been conditioned to believe that no modern scientific investigation can be conducted without a massive display of electronics, coupled to a computer, the photoelastic techniques are often looked upon as being somewhat mysterious, akin to a conjuring trick, or at best (see Hall and Hoskins) "a bit of an art".

The ideal stressmeter would be an inclusion of high rigidity, so that it could function independent of the deformation modulus, or change of modulus, of the host material. It would be inserted coincident with excavation by the drill, so that no deformation of the rock occurs around the hole, prior to its insertion, and it would function as an "active" gauge. The pressure imposed by the surrounding rock would be counterbalanced by the measured pressure within the gauge. As yet no one technique has all these ideal properties, so that, when choosing a technique for a given field situation, and in his assessment and interpretation of the results obtained, the engineer should use theoretical concepts as a guide, and ally these to his experience, intuition, and common sense.

Selected References for Further Reading

ARGAWAL, R. K. and BOSHCOV, S. Theory of the "soft inclusion" as a deformation gauge in boreholes. *Int. J. Rock Mech. Min. Sci.,* vol. 3, pp. 319-323 (1966).

BAAR, C. A. *Determination of Stresses in Salt and Potash Mines,* Saskatchewan Res. Council, Engineering Division, Report E 72-8 (1972).

DE LA CRUZ, R. V. and RALEIGH, C. B. Absolute stress measurements at the Rangely anticline, Colorado. *Int. J. Rock Mech. Min. Sci.,* vol. 9, pp. 625-633 (1972).

FAIRHURST, C. *Methods of Determination in situ Rock Stresses at Great Depths,* U. S. Army, Corps of Engineers, Missouri River Division, Technical Report No. 1-68, 1967.

HALL, C. J. and HOSKINS, J. R. *A Comparative Study of Selected Rock Stress Instruments,* Tech. Rept. No. V1-BMR-2, Advanced Projects Research Agency, Washington DC, 1971.

HAST, H. The state of stress in the upper part of the earth's crust. *Tectonphysics,* vol. 8, pp. 189-211 (1969).

HAWKES, I. Theory of the photoelastic biaxial strain gauge and its application in rock stress measurement. U. S. Army CRREL, Hanover, N. H., Technical note (March 1967).

HAWKES, I. and MOXON, S. The measurement of in situ rock stress using the photoelastic biaxial gauge with the core-relief technique. *Int. J. Rock Mech. Min. Sci.,* vol. 2, pp. 405-419 (1965).

HERGET, G. First experience with the CSIR triaxial strain cell for stress determinations. *Int. J. Rock Mech. Min. Sci.,* vol. 9 (1972).

HERGET, G. Variation of rock stress with depth at a Canadian iron mine. *Int. J. Rock Mech. Min. Sci.,* vol. 10, pp. 37-51 (1973).

LEEMAN, E. R. The measurement of stress in rock. *South African Inst. Min. Met.,* Sept. and Nov. 1964.

LEEMAN, E. R. The determination of the complete state of stress in rock in a single borehole. *Int. J. Rock Mech. Min. Sci.,* vol. 5, pp. 31-56 (1966).

MERRIL, R. H. *Three-component Borehole-deformation Gauge for Determining Stress in Rock,* U. S. Bur. Mines Rept., Invest. No. 7015, 1971.

MERRIL, R. H. and PETERSON, J. R. *Deformation of a Borehole in Rock,* U. S. Bur. Mines Rept., Invest. No. 5881, 1961.

NICHOLS, T. C., ABEL, J. F. and LEE, F. T. A solid inclusion borehole probe to determine three-dimensional stress changes at a point in a rock mass. *U. S. Geol. Survey Bulletin* 1258-C (1968).

OKA, Y. and BAIN. A means of determining the complete state of stress in a single borehole. *Int. J. Rock Mech. Min. Sci.,* vol. 7, pp. 503-516 (1970).

ROBERTS, A. The measurement of stress and strain in rock masses. In *Rock Mechanics in Engineering Practice,* Stagg and Zienkiewicz (Eds.), pp. 156-201, Wiley, London, 1968.

ROBERTS, A. *et al.* The photoelastic stressmeter. *Int. J. Rock Mech. Min. Sci.,* vol. 1, pp. 441-457 (1964) and vol. 2, pp. 93-103 (1965).

ROCHA, M. and SILVERIO, H. A new method for the complete determination of the state of stress in rock masses. *Geotechnique,* vol. 19, pp. 116-132 (1969).

ROEGIERS, J. C., FAIRHURST, C. and ROSENE, R. B. The DSP – a new instrument for estimation of the *in situ* stress at depth. *Soc. Petroleum Engrs of AIME, 6th Conf. on Drilling and Rock Mechs., Austin, Texas,* 1973. Paper in SPE 4246.

SEAGAR. Pre-mining lateral pressures. *Int. J. Rock Mech. Min. Sci.,* vol. 1, pp. 413-419 (1964).

SELLERS, J. B. The measurement of rock stress changes using hydraulic borehole gauges. *Int. J. Rock Mech. Min. Sci.,* vol. 7, pp. 423-436 (1970).

SKILTON, D. Behavior of rigid inclusion stressmeters in viscoelastic rock. *Int. J. Rock Mech. Min. Sci.,* vol. 8, pp. 283-289 (1971).

VAN HEERDEN. Potential fracture zones around boreholes with flat and spherical holes. *Int. J. Rock Mech. Min. Sci.,* vol. 6, pp. 453-463 (1969).

CHAPTER 11

Strata Pressures and Support Loads

The determination of strata pressure is of interest in geotechnology, not only in relation to the absolute or relative state of stress in earth and rock masses, but also in relation to the loads imposed by the soil or rock on structures erected within, against, or upon, the earth material. In soil mechanics and foundation engineering the correlation between calculated earth pressure and soil-structure interaction is a perennial problem. In rock mechanics, while considerable advances are being made, in theory and in practice, to solve the problem of determining *in situ* field or regional stress, and while the observation of direct loads imposed upon simple struts in tension or compression is commonplace, very little is known of the nature of the contact loads between rock masses and excavation support structures involving more complex and unknown stress distributions.

Earth Pressure

Retaining walls

Calculated estimates of vertical earth pressure underneath point loads or distributed loads may be made using the Boussinesq formula as a basis. Estimates of horizontal pressure on retaining walls, or on sheet piling are commonly made using the Rankine formula in which it is assumed that a cohesionless material exerts on the retaining wall a horizontal pressure which increases from zero at the top to a maximum at the base, and which is directly proportional to depth at any intermediate point. This triangular, straight line, stress distribution determines the resultant horizontal pressure, P_a,

and $$P_a = K_a \zeta H^2$$

where K_a is the coefficient of active pressure,

ζ is the soil density,

H is the height of the retaining wall (see Fig. 11.1).

P_a is termed the resultant thrust or active earth pressure. It is assumed to act horizontally at a height of $h/3$ from the base.

$$K_a = \frac{1 - \sin \varphi}{1 + \sin \varphi} \quad \text{or} \quad \tan^2 (45° - \varphi/2)$$

where φ is the angle of internal friction for the soil.

For the earth pressure to be "active" the retaining wall must yield slightly to the resultant thrust. If the retaining wall is so rigid as to resist the thrust without yielding to it, or if the wall is constrained so that it cannot yield, then the pressure developed upon the wall is termed "passive" pressure, P_p,

where $$P_p = K_p \zeta h$$

and K_p is the coefficient of passive earth pressure.

$$K_p = \frac{1 + \sin \varphi}{1 - \sin \varphi} \quad \text{or} \quad \tan^2 (45° + \varphi/2).$$

The cohesionless condition may be assumed to exist in materials such as broken rock, dry sand, or gravel. Other granular materials that may be considered to be cohesive will be partially self-supporting, and they will exert less pressure on the retaining wall than would cohesionless materials of the same density and internal friction characteristic.

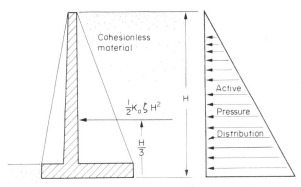

(a) Active pressure on an earth-retaining wall

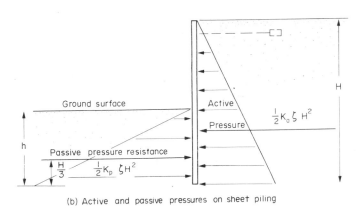

(b) Active and passive pressures on sheet piling

Fig. 11.1. Active and passive pressures on earth-retaining walls.

The horizontal pressure at a depth h below the horizontal surface of a cohesive soil is given by Bell's development of the Rankine formulae

$$P_a = \zeta h \tan^2 (45° - \varphi/2) - 2c \tan (45° - \varphi/2)$$

for the active pressure, and

$$P_p = \zeta h \tan^2 (45° + \varphi/2) + 2c \tan (45° + \varphi/2)$$

for the passive pressure.

By extending the depth of foundation of a masonry or concrete retaining wall, the passive pressure may be utilized to help the effect of base friction in resisting any tendency of the wall to fail by lateral sliding (Fig. 11.2(a)). If the retaining wall is constructed of sheet piling, then the passive resistance pressure generated in the foundations must withstand the active thrust from the loaded side. An insufficient depth of burial might allow the toe of the retaining wall to be pushed out and result in failure (Fig. 11.2(b)).

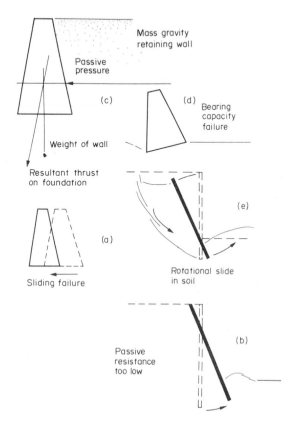

Fig. 11.2. Possible modes of failure of retaining walls.

Masonry and concrete retaining walls are designed to resist being overturned and they maintain their stability by virtue of their mass. Their design is usually such that the resultant of the gravity load imposed by the weight of the structure and the horizontal thrust imposed by the retained earth material acts within the middle third of the base (Fig. 11.2(c)). In this case the passive pressure resistance that is afforded by the depth of wall on the unloaded side is relatively insignificant. The stability of such a retaining wall is also dependent upon the bearing capacity of the ground underneath the foundation. If this is insufficient it could lead to failure of the wall by overturning (Fig. 11.2(d)). A safety factor of 3 is commonly recommended, i.e. the resultant pressure should not exceed one-third the bearing capacity of the foundation material.

The ultimate bearing capacity of cohesive clay soils is largely dependent on the shear strength of the foundation material. Failure in such a case may take the form of shear rotation along a circular slip surface (Fig. 11.2(e)). The safety factor in such conditions is seldom as high as 3 and may be less than 2.

The pressure distribution upon the face of a retaining wall is determined not only by the friction, cohesion, and density characteristics of the retained earth material, but also by the manner in which the wall responds to the load that is imposed upon it. If the wall is so rigidly constructed, or is constrained by neighboring structures to the extent that it cannot yield at all, then the pressure distribution takes the triangular pattern, increasing proportionately with depth at the passive value. If the wall yields appreciably by rotation about the loaded end of its base, or by sliding, then the pressure distribution is again triangular and proportionate to depth, but this time at the active value. But if the yield is only small (say around 0.1% of its height), then a pressure distribution results which takes a curvilinear form. The upper part of the curve approximately follows the passive pressure line, but since the total thrust remains equivalent to the active value the pressure distribution curves away from the passive line at depth below the surface of the retained material. The maximum pressure intensity and the center of pressure lie somewhere around mid-height above the base of the wall.

Trench supports

The sides of trenches cut through soils or in incompetent rock require support, which is commonly provided by vertical sheeting retained by cross-struts (Fig. 11.3(a)). Theoretically, the lateral pressure imposed by a cohesionless material on these supports, which are not rigid and yield slightly under load, will conform with the triangular active pressure distribution

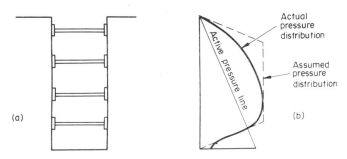

Fig. 11.3. Pressure distribution on trench supports.

pictured by Rankine. In practice, however, the real stress distribution over the flexible shuttering will be variable, and it takes a distribution pattern depending upon the circumstances. The pressure distribution in any particular case can be observed experimentally but in the general case a parabolic distribution of pressure may be assumed, similar to that indicated in Fig. 11.3(b). For the purpose of design this is equated to a trapezoidal distribution of area 1.44 times that of the active pressure triangle. The total thrust is then $1.44P_a$ and this is assumed to act at mid-depth in the trench.

"Arching" in unconsolidated earth materials

The curvilinear pressure distribution over the surface of trench siding, and in some retaining walls, is due to the development of friction forces within the retained earth material. These forces are generated as a result of the tendency of the earth material to yield along shear slip surfaces, to follow the yield of the retaining wall or the trench support. The net result is that those parts of the retaining support that are immediately in contact with the yielding earth material are partially relieved of that pressure, while pressure is intensified at other areas due to the transfer of load to more rigid parts of the system.

A similar load-transfer process is responsible for "arching" in unconsolidated and granular materials. Terzaghi described this effect and developed a theoretical analysis for it in 1943, and the subject has been explained extensively by subsequent investigators. It is of considerable importance to many aspects of geotechnology, such as the construction and support of tunnels in soil and loose rock, the design and operation of storage bins and transfer-chutes for granular materials, the operation of systems of mining by caving and fracture of large rock masses, together with their transfer and transport as granular-blocky materials.

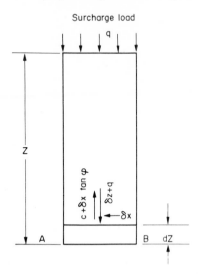

Fig. 11.4(a) Yield above a tunnel in soil (after Terzaghi).

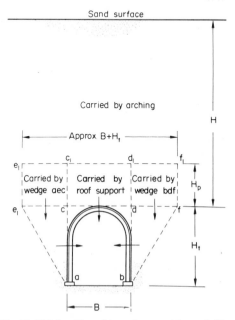

Fig. 11.4(b) Loading of tunnel support in sand (Terzaghi).

Terzaghi's analysis of arching in soil is illustrated in Fig. 11.4(a). This represents a cross-section of a prismatic mass of soil, of height Z. Consider an element at the base of the strip, of height dZ which is excavated on the floor AB. The soil immediately above the excavation yields and tends to move vertically downwards. This yield is resisted by friction forces on slip surfaces within the soil, so that the vertical pressure on AB is reduced by an amount equal to the vertical component of the developed shearing forces.

If it is assumed that the sliding surfaces are vertical then the normal force on the plane of sliding $= \sigma_x$ and $\sigma_x = K_{az}$ where $K = \sigma_x/\sigma_z$.

The shear force on the plane of sliding $= \tau_{dz}$

and $\tau_{dz} = C + \sigma_x \tan \varphi$

$\qquad = C + K_{az} \tan \varphi.$

Terzaghi developed an expression for σ_z, and showed that the vertical yield at AB has virtually no effect on the state of stress in the soil at distances $5B$ or more above the yielding strip. In the soil above $5B$ distance from AB no shear stresses are developed, this material acts simply as an overburden pressure or surcharge supported by the arch.

The arching concept in loose rock and sand can be used as a means of estimating the roof loads imposed on the linings of tunnels in these materials. Terzaghi imagined the arching to produce a transfer of load by friction onto the material at the sides of the tunnel (H_t in Fig. 11.4 (b). The remainder of the yielding material, to an additional height H_p, is carried by the tunnel supports, and that above H_p acts as the surcharge, carried by the arching.

The characteristic arching in the roof above excavations in granular and broken materials suggests that, in these circumstances, the load on a tunnel lining is the weight of that mass of rock, or soil, which tends to fall out into the tunnel. It also suggests that if the tunnel is not lined then its roof should be arched up to the zone of a zero pressure surface, because below this surface the earth material is in tension. A number of investigators have proposed rock load theories of this nature. Couter states that the height of loose rock that should be provided for, in this way, is determined by the width or the excavation and the spacing of discontinuities in the rock mass. Using his classification of "Blocky" when the joint spacing is more than 30 cm and "Broken" if they are closer than 30 cm then the height of loose rock to be expected above an excavation of width b is about $2b$ for "broken" rock and less than $0.7b$ for "blocky" material.

The rock load concept, as a design criterion, has been extensively used, since Terzaghi first described it. Terzaghi classified rock and soil on the basis of the rock pressure concept, and gave recommended design data, shown in Table 11.1. Several other investigators have also produced rock load data and classifications, including Stini, Bierbaumer, Lauffer, and Rabcewicz. These are tabulated in the report by Deere *et al*. Szechy's text, which is a standard reference on tunneling, discusses other rock load theories, notably those of Kommerell, Haroscy. Forchleimer, Ritter, Protodiakonov, and Engesser, in a comprehensive review, including original references.

The height of the expected upbreak, corresponding to H_p in Fig. 11.4(b), and hence the shape and mass distribution of the rock loading upon the tunnel lining, were determined more exactly by Szechy on the basis of a static model. His basic assumption is that it is the tensile strength of the rock itself that will terminate the upbreak above a rectangular cavity. With the justification of Dischinger's theory he assumed an inner stress distribution, indicating a field of compression stresses above and one of tensile stresses below the neutral axis, considering that in a homogeneous solid rock mass no change is to be expected either in the strength properties or in loading. The only factor which leads to a gradual decrease of inner stresses is the decrease of effective span due to progressive arching. Denoting this factor by α he quotes the height of the upbreak zone (h) as

$$h = [1.13\,(1 + \beta/_2)^2 - \alpha]\,b_0$$

and gave the following values for α and β:

α = 0.5 if the disc-like beam under the arch is fixed,
α = 0.39 if the disc-like beam is simply-supported,
β = $\frac{1}{2} \sim \frac{1}{3}$ for plastic material,
β = 2 $\sim$ 3 for elastic material (in the absence of more precise information it may be assumed that $\beta = 1$ in solid rock and $\beta = 2.5$ in plastic rock and soils).

TABLE 11.1. *Rock load Hp in feet of Rock on Roof of Support in Tunnel with Width B (ft) and height H_t (ft) at depth of more than 1.5 (B + H)* (Terzaghi)*

Rock condition	Rock Load Hp in feet	Remarks
1. Hard and intact	zero	Light lining, required only if spalling or popping occurs.
2. Hard stratified or schistose. §	0 to 0.5 B	Light support
3. Massive, moderately jointed	0 to 0.25 B	Load may change erratically from point to point
4. Moderately blocky and seamy	0.25 B to 0.35 $(B + H_t)$N	No side pressure
5. Very blocky and seamy	(0.35 to 1.10) $(B + H_t)$	Little or no side pressure
6. Completely crushed but chemically intact	1.10 $(B + H_t)$	Considerable side pressure. Softening effect of seepage towards bottom of tunnel requires either continuous support for lower ends of ribs or circular ribs
7. Squeezing rock, moderate depth	(1.10 to 2.10) $(B + H_t)$	Heavy side pressure, invert struts required. Circular ribs are recommended
8. Squeezing rock, greath depth	(2.10 to 4.50) $(B + H_t)$	
9. Swelling rock	Up to 250 ft irrespective of value of $(B + H_t)$	Circular ribs required. In extreme cases use yielding support

* The roof of the tunnel is assumed to be located below the water table. If it is located permanently above the water table, the values given for types 4 to 6 can be reduced by 50%.
§ Some of the most common rock formations contain layers of shale. In an unweathered state, real shales are no worse than other stratified rocks. However, the term shale is often applied to firmly compacted clay sediments which have not yet acquired the properties of rock. Such so-called shale may behave in the tunnel like squeezing or even swelling rock.
 If a rock formation consists of a sequence of horizontal layers of sandstone or limestone and of immature shale, the excavation of the tunnel is commonly associated with a gradual compression of the rock on both sides of the tunnel, involving a downward movement of the roof. Furthermore the relatively low resistance against slippage at the boundaries between the so-called shale and rock is likely to reduce very considerably the capacity of the rock located above the roof to bridge. Hence, in such rock formations, the roof pressure may be as heavy as in a very blocky and seamy rock.

 In all these applications of the rock load concept the extent of the protective arch zone only is considered, and the depth of cover, or overburden, is disregarded. However, in pseudoelastic and plastic rocks stressed beyond the yield point the effect of depth is a relevant factor in determining not only the magnitude of rock pressure (cf.Horvath) but also the time required for the development of a protective arch zone. Using this standpoint other investigators have developed theories for the rock load, for example, Bierbaumer's theory applied to cavities excavated at depth in materials displaying high shear strength. According to Bierbaumer the tunnel is loaded by a rock mass bounded by a parabola of height $h = \alpha H$, and the rock pressure is given by $P = \alpha_1 H \rho$

where $\alpha_1 = 1 - \dfrac{\tan \phi / \tan^2(45° - \phi/2)/H}{b + 2m \tan(45° - \phi/2)}$

ρ = specific weight of the rock,
ϕ = internal friction angle,
m = height of the tunnel.

Relative yield

Smith modified the arching concept, to allow for the effects of varying degrees of rigidity in the tunnel lining. Using Smith's approach of "relative yield" Lane developed equations for the rock load, considering large finite blocks bounded by vertical planes through the tunnel springlines, to simplify the mathematics.

There have been few attempts to correlate the design characteristics of tunnel linings and behavior of the various installations with measured loads upon the linings. It is impossible therefore to estimate the factor of safety or the degree of overdesign that is commonly provided by the design procedures using the rock load concept. The increased attention that is now given towards systematic site investigation, in the attempt to assess the engineering qualities of the rocks encountered in underground excavations, suggests that in the near future it should be possible to do this. For example, Coon has put forward a qualitative correlation between Deere *et al's* Rock Quality Designation (RQD) and the support requirements for tunnels through the rock. If this were combined with records of active load measurements in the field, for a given type of lining it should be possible to design supports to resist a load that is a function of the rock quality, the size of opening, and the construction technique involved.

Stress Distribution Around Excavations at Depth in Rock

Theoretical stress distributions around excavations at depth in rock may be calculated on the basis that (a) the rock around the opening is a fractured and discontinuous granular material, (b) the rock behaves as an ideal elastic mass, and (c) the rock displays both elastic and plastic deformation characteristics. For case (a) the soil mechanics approach of Terzaghi may be used, but for rock masses at depth either case (b) or case (c) is more applicable.

The effect of making an opening in stressed rock is to produce a variable stress distribution in the rock around the opening. The ratio between the stress at a point in the new stress field and the original virgin stress component, in the same direction, is termed the stress concentration. The magnitudes and distribution of the stress concentrations around an excavation will depend upon many factors. These factors include the shape of the opening, the magnitude and directions of the imposed loads, and the ratio between major and minor principal stresses in the regional stress field. They are independent of the elastic constants of the material and the size of the excavation. Although they vary with distance into the rock sides, this variation is in terms of a dimensionless ratio, which for a circular tunnel is r/a where a is the tunnel

section radius and *r* the radial distance from the tunnel axis.

In general the stress concentrations are larger where the perimeter of the excavation changes its alignment such as around the periphery of a circular tunnel, or at the corners of a rectangular excavation. They are smaller in the middle of straight-sided boundaries.

Detailed and comprehensive elastic stress analyses for single and multiple rock excavations of various shape and circumstances are described in the publications of Obert, Duvall, and Merrill, to which the reader is referred. The classical study of stress distribution around a circular hole in an infinite elastic plate is illustrated in Fig. 11.5(a) and (b).

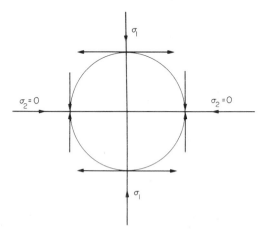

Fig. 11.5(a) Tangential stress concentrations around the perimeter of a circular hole in an infinite uniaxial stress field.

In a uniaxial stress field where $\sigma_2 = 0$ the boundary tangential stress concentrations vary around the periphery of the hole from a maximum of 3 on the minor stress axis to -1 on the major stress axis. This means that a compressional regional stress field generates a tension tangential to the boundary of the excavation at each end of the major stress axis, its magnitude being equal but opposite in sign to the imposed stress.

At each end of the minor stress diameter a tangential stress three times the regional stress is developed, in the same direction as the regional stress.

In a biaxial, hydrostatic, stress field where $\sigma_1 = \sigma_2$ the tangential stress concentration around the boundary of a circular hole in an infinite elastic plate is uniform throughout at 2.

The tangential stress concentration diminishes rapidly with increase of depth into the side walls, and both for the uniaxial and hydrostatic conditions approaches 1 on the minor principal stress axis when the ratio $r/a = 4$.

On the major principal stress axis the tangential stress concentration also diminishes with increase of depth, to attain zero when $r/a = 2$ in a uniaxial field, and approaches 1 when $r/a = 4$ in the hydrostatic field.

The radial stress concentration is zero at the immediate wall boundary but increases rapidly to reach about 0.5 when $r/a = 0.5$. In a hydrostatic stress field it continues to increase and to approach a value of 1 when $r/a > 4$. In a uniaxial field the radial stress concentration reaches a maximum about $r/a = 0.5$ and then diminishes slowly with further depth into the rock walls.

The stress analysis of excavations in rock which displays visco-elastic and plastic deformation characteristics is yet more complex than the elastic analysis. The problem is of major importance in connection with the design of tunnel lining and support systems in which the function of the support is to impose added rigidity upon the rock walls, and so increase their resistance to shear.

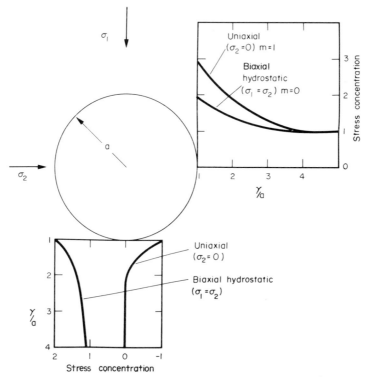

Fig. 11.5(b) Variation of axial tangential stress concentrations with depth into the sidewalls of a circular hole in an infinite elastic stress field.

On the basis that a rock will behave elastically provided a critical yield shear stress in the walls is not exceeded, but will display time-dependent creep if the shear stress exceeds that value, then, if the shear strength of the rock walls can be reinforced by a mechanical support system to a higher value than the critical *in situ* shear stress, then an elastic regime can be maintained in the rock material, and elastic design procedures may then be justified.

Design of Tunnel Linings

The rational design of reinforcement support lining systems for tunnels driven through earth materials is not possible without certain fundamental prerequisites. These are:

1. Understanding of the physical strength characteristics of the earth materials, including detailed knowledge of their mode of yield, deformation, and approach to failure, in response to the increase of imposed loads.

2. Determination of failure criteria for the material, in terms of which maximum permissible stresses, strains, and loads may be specified.

3. Knowledge of the stress distribution within the soil or rock mass through which the

tunnel must be constructed. This involves the determination of three-dimensional stress components, in magnitude and direction, and the determination of the magnitudes and directions of stress concentrations around the periphery of the excavation.

4. Knowledge of the contact conditions between the earth material and the tunnel support system. This involves the interaction between the earth material and the lining, which determines the load distribution that is imposed upon the lining from without, i.e. from the solid.

5. Specification of a structural design for the tunnel support system, using materials whose engineering strength properties are known, incorporated into a structure capable of carrying the imposed loads with a predetermined factor of safety.

6. In the case of pressure tunnels, which are also loaded by fluid pressure from within, the design procedure must follow in the reverse direction, and the resultant load imposed upon the lining, fluid pressure from within, and rock pressure from without, will determine the stability of the system.

From what has already been described in this text, it will be apparent to the reader that, as far as items 1 to 4 are concerned, our knowledge is far from being sufficiently adequate for us to pursue such a design procedure with absolute confidence. But in structural engineering generally, absolute confidence in a design specification is not always attained. If it were otherwise there would be no need for codes of practice to include "factors of safety". In other words, the engineer attempts to guarantee the stability of his structure by making it stronger than it need be – or by "overdesign". But engineering is, or should be, also a matter of economics, and a structure that is overdesigned is more expensive than it need be. At the same time a structure that is underdesigned may collapse and fail, and may have to be repaired or rebuilt. It may cause loss of life and property. In any event, it is likely to cost much more in the end than would the overdesigned structure. It should therefore be the aim of the engineer to achieve a sound design, one that will stand without failure, but at the same time not too far overdesigned, and it is one of the prime functions of continuing research in structural engineering and materials science to reduce the margin of overdesign that is required to guarantee stability. It is also one of the ironies of those aspects of structural engineering, such as geotechnology, in which a large measure of ignorance still lies, that the fact that the engineer's structure stands is no guarantee that his design was right, but when his structure fails then he knows that either his design or its execution is likely to have been at fault.

This, of course, assumes that the engineer intended that his structure should stand permanently. There are some aspects of geotechnology, particularly in mining engineering, when the structure need only stand for a limited period of time, and where the excavations must be structurally supported to withstand time-dependent yield of the rocks concerned. In such circumstances a sound design would be one that provided for the excavations to remain open, with a minimum of maintenance cost, only over the period of time that they are required to be in use. Anything beyond that would represent uneconomic overdesign. On the other hand, there are other aspects of geotechnology, such as the construction of foundations for structures to withstand earthquakes, or the foundations, tunnels, and earthworks, associated with dams, hydro-electric, water supply and sewerage schemes, when the humanitarian and economic consequences of failure cannot be contemplated and accepted by the designer. In these circumstances substantial overdesign is justified.

The design and support of excavations in soil and rock is therefore far from being an exact science. Until such time as the gaps in our knowledge are filled, the wise geotechnologist, while being alert to the accelerating advances that are being made in analytical design methods, will draw much from case studies of what is going on and what has already been achieved in similar circumstances elsewhere.

Flexible buried cylinders

The construction of tunnels at shallow depth by the "cut and fill" method is becoming increasingly popular amongst engineers, since it is more economical to do this by modern earth-moving equipment than it is to make tunnels by underground boring. The tunnel is constructed by cutting a trench, then placing a cylinder in the trench, and subsequently covering this with earth-fill. Where high overpressures must be provided for, the analysis of Duns and Butterfield is relevant. The cylinder geometry and nomenclature is shown in Figs. 11.6(a) and (b).

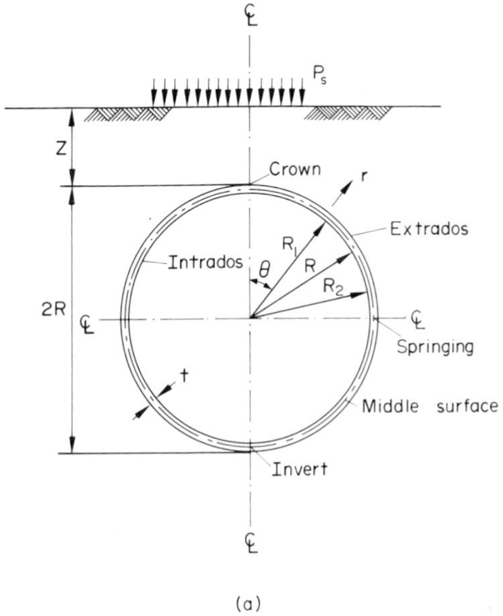

(a)

Fig. 11.6(a). Flexible buried cylinder (Duns and Butterfield). Geometry and nomenclature.

On the basis of a theoretical approach using two approximate methods and a finite-element analysis it is shown that the resultant stresses in the cylinder, under static load are

$$\text{Axial thrust } T = P_0 R - \left[\frac{P_2 R}{3} + \frac{2S_2 R}{3}\right]\cos 2\theta$$

and

$$\text{Bending moment } M = \left[\frac{P_2 R^2}{3} + \frac{S_2 R^2}{3}\right]\cos 2\theta$$

The interaction stresses on the shell-earth medium interface are

$$\sigma_r = -P_0 - P_2 \cos 2\theta,$$
$$\sigma_\theta = 0,$$
$$\tau_{r\theta} = S_2 \sin 2\theta$$

where P_0 = dilatational interaction pressure,
P_2 = radial interaction pressure,
S = tangential interaction pressure,
R = radius of cylinder.

217

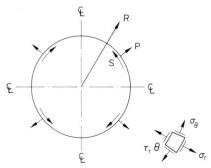

Loads on elastic medium

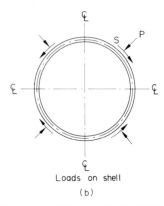

Loads on shell

(b)

Fig. 11.6(b) Flexible buried cylinder (Duns and Butterfield). Loads on shell/medium interface.

The results obtained by the analytical methods of Duns and Butterfield, as compared with experimental observations, under static loading conditions are shown in Figs. 11.7(a) and (b).

No significant deviation in the shape of the thrust and moment diagrams was observed as the result of time-dependent loading, and the results of a dynamic finite-element analysis suggested that a dynamic load factor over the static condition in an idealized soil medium should not exceed 1.2, provided the depth of burial is more than about $0.5R$. A dynamic load factor of 1.20 should be adequate for design purposes. A further experimental and theoretical study of the buckling behavior of brass and steel tubes buried at shallow depth in medium-dense sand provided the basis from which was deduced:

Critical ground overpressure

$$P_0 = \frac{(2 - \nu_S)}{3(1 - \nu_S)} \cdot \frac{F_k E_S}{4n}$$

and the characteristic buckling node number

$$n = \sqrt[3]{\frac{3 F_k E_S}{2 E_c} \cdot \frac{R}{t}}$$

where ν_S = Poisson's ratio for the soil,

F_k = Foundation modulus factor.

$$F_k = \frac{1 - (R/b)^2}{[1 + (R/b)^2 \ (1 - 2\nu_S)]}$$

where $b = R + Z$ (see Fig. 11.7)
(for $Z > 4R$ F_k is approx. 1.0),
E_s = Young's modulus for the soil (plane strain),
E_c = Young's modulus for the cylinder (plane strain),
t = cylinder thickness,

$$E \text{ (plane strain)} = \frac{\text{Young's modulus}}{(1 - v^2)}.$$

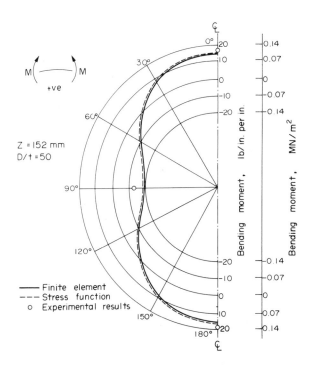

Fig. 11.7(a) Bending moment diagram around a flexible buried cylinder under static loading (Duns and Butterfield).

Buried tunnels at shallow depth

In a comprehensive review of the design problems of tunnel liners and support systems, Deere *et al.* describe the factors that are operative around a tunnel constructed underground in soil. The question of stability is considered in two localities: (1) at the advancing face of the tunnel and (2) around the lining.

The stability of the exposed material at the face of the tunnel is determined by several factors. These include the shear strength and the deformation characteristics of the earth material, the overburden pressure, the overall geometry (depth/width ratio) and the engineering construction procedures. The short-term stability of an opening in clay can be predicted if its undrained shear strength is known. From an examination of data collected from various sources, Deere *et al.* conclude that if the tunnel is located at a depth three to four times its height then the exposed face should be stable if the total overburden pressure or surcharge is less than about six times the undrained shear strength. For shallow depths the critical value is stated to be less than 6.

219

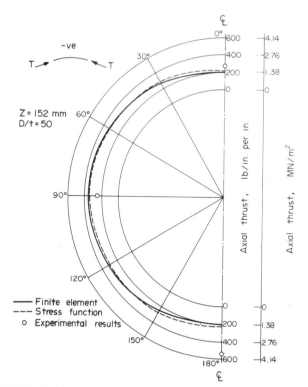

Fig. 11.7(b) Axial thrust diagrams around a flexible buried cylinder under static loading (Duns and Butterfield).

Deere *et al.* propose a basis for determining the ring stress for which the lining could be designed, using the diagram, Fig. 11.8 which illustrates the interaction of the soil and tunnel lining, for which the average radial stress is plotted as a function of average radial displacement. If the tunnel wall were perfectly rigid, and inserted concurrently with excavation there would be no radial deformation at all. In these circumstances the average radial stress in the lining could be the mean of the vertical and horizontal earth pressures that existed before construction of the tunnel, measured on the central axis of the tunnel. This is represented by the point A on Fig. 11.8. If the diameter of the tunnel were progressively reduced the total ring load would decrease proportionately in a manner depending on the ground reaction relationship. This relationship must be determined experimentally for the particular soil concerned. If the soil were elastic then the ground reaction relationship would follow the straight line AB, but in practice it is more likely to be curvilinear, say along AC.

During construction of the tunnel the soil moves radially inwards, and before the lining can be inserted, some radial deflection must already have occurred. At the instant when the circumferential lining is placed in contact with the soil suppose the radial deflection u_1 had occurred. Then the minimum average radial stress required in that lining, to prevent all subsequent inward radial displacement of the tunnel wall, is represented by the value D.

Suppose that before the support ring could be inserted a total wall deflection of u_2 occurs, then a perfectly rigid support ring should be designed for a radial stress E. But, in practice the ring will probably not be so rigid, and it will yield to the soil pressure by an increment u_3.

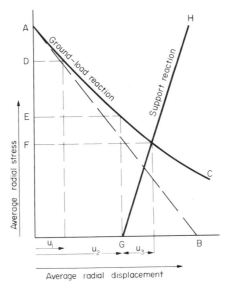

Fig. 11.8. Interaction of soil and tunnel lining (Deere *et al.*).

The final radial stress in the support ring will then be *F*. The line GH represents the relationship between the radial load on the lining and the corresponding radial yield of the lining structure. It is termed the support reaction curve. The final load on the lining is given by the intersection of the ground reaction and the support reaction curves.

Effects of Construction Procedure

The design and construction procedures for the lining of tunnels at shallow depth may be conducted on the basis that the lining is required to be (a) flexible or (b) rigid.

Flexible lining

Assuming that the earth material is perfectly elastic and that the biaxial components of field stress over a cross-section of the tunnel are σ_1 vertical and σ_2 horizontal. Suppose that it is possible to insert a flexible lining before excavating the ground, the lining upon its insertion being perfectly circular, capable of carrying ring stress but not capable of carrying bending stress. If the ground inside the lining were then removed the equilibrium of the system could only be maintained without imposing bending stresses in the lining if the lining deformed from its circular shape to an elliptical shape. The shape of the ellipse would be exactly matched to the magnitudes and directions of σ_1 and σ_2, so that radial stresses in all directions in the deformed lining were equal and no bending stresses existed.

In practice the lining, inserted after excavation, will be strong enough to take a limited bending stress and the earth material is not elastic. The final shape of the excavation will be determined by the deformation characteristics and shear strength of the soil as well as by the field stress characteristics, the material strength, and the dimensions of the tunnel lining. Soil strength and lining properties thus combined together will determine the stability of the system.

221

Following this design principle the strength properties of the earth material itself can be utilized to the full, and a very economical tunnel support system may be devised.

The success of the method greatly depends on the yield characteristics of the earth material and the wall contact conditions. Localized stress concentrations and marked anisotropy should not exist, so as to allow the circular lining to deform to a shape which will resist the field strata pressures without buckling.

Rigid lining

If marked anisotropy or heterogeneity exists in the earth material, or in the anticipated imposed stress distribution, or if the earth material is fractured, fissured, or blocky, then a more rigid lining might be preferable since this could carry an appreciable bending stress as well as ring stress. The cross-sectional shape of the lining may take any one of many forms, of which two may be taken as representatives: (a) circular and (b) rectangular.

A common method of construction for a circular lining is to make this in segments of eight or sixteen to a ring, depending upon the diameter of the tunnel. The segments, usually of steel or reinforced concrete, are assembled and connected together, building the rings in two halves, separated along the horizontal axis of the tunnel at mid-height or at the spring line. This horizontal split provides the abutment surfaces for vertical jacks, which, when pressurized, force the two half-rings into contact with the earth material and impose the desired ring stress into the support. The two half-rings are then connected at the horizontal split to take the load, and the jacks are removed.

A circular shape is able to take non-isotropic imposed stresses with equal ability, no matter what the characteristics of the directional stress field may be, in the surrounding medium. It will distort from its original circular shape to a form in which bending stress and ring stress combine to produce a resultant radial stress in the lining, which must be uniform in all directions if the lining is to be in equilibrium.

The lining of a rectangular cross-section is less able to cope with variations of directional field stress. The choice of this section is usually decided by other matters than the structural stability of the lining, and primarily by the purpose for which the tunnel is to be used. For example, it is often to be seen in vehicular tunnels and in underground excavations containing machinery or other installations that require a solid base. Having chosen this shape then the lining must be installed to keep it stable. In tunnels at shallow depth, unless there are particular circumstances that dictate otherwise, the biaxial stress components in the section are assumed to be vertical and horizontal. The imposition of bending stresses in the lining, from these stress components, is inevitable. It is common, but not always justifiable, to assume that the major stress component is vertical, in which case the top of the tunnel may be constructed semicircular in shape like "barrel vaulting". The vertical sides of the lining may then incorporate jacking systems during construction to generate a ring stress in the roof lining and thus reinforce its resistance to bending.

Tunnels in Homogeneous, Massive, Elastic Rock

The design of linings to shafts and tunnels in hard, massive, rock may sometimes be approached by the simple assumption that the lining is a hollow cylinder of elastic material. The cylinder is subject to an internal pressure (p) which counterbalances the imposed strata pressure from outside the lining. The strata pressure is assumed to be uniform and hydrostatic.

If the thickness (t) of the cylinder is small, relative to its diameter (d), it may be assumed that the ring stress (σ) in the cylinder wall is uniform, and

$$\sigma = \frac{p\,d}{2t}\ .$$

However, in the majority of cases, in practice, the thin-walled assumption will not be valid, for the thickness of the lining is appreciable and the stress within it changes with depth into the lining.

Three possible cases must be considered:

(a) The lining is subject to uniform pressures both from within the tunnel and from the outside walls.
(b) The elastic properties of the lining are exactly matched to the elastic properties of the rock and there is a perfect welded lining-to-rock contact.
(c) The lining is loaded from without by the field strata pressure through a continuous contact, and there is no pressure within the tunnel.

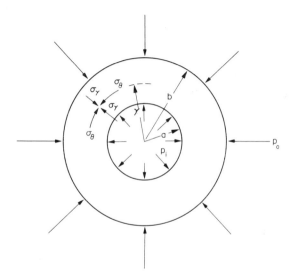

Fig. 11.9. Biaxial stresses in an elastic thick-walled cylinder, loaded externally and externally.

For case (a) illustrated in Fig. 11.9 the radial and tangential stresses in the lining are given by the relationships

$$\text{radial stress } \sigma_r = \frac{a^2 p_i - b^2 p_o}{b^2 - a^2} - \frac{a^2 b^2 \ (p_i - p_o)}{b^2 - a^2} \cdot \frac{1}{r^2} \ ,$$

$$\text{tangential stress } \sigma_\theta = \frac{a^2 p_i - b^2 p_o}{b^2 - a^2} + \frac{a^2 b^2 \ (p_i - p_o)}{b^2 - a^2} \cdot \frac{1}{r^2} \ .$$

In case (b) the lining forms an integral part of the elastic host material. The stresses around the walls of this excavation correspond to those around a circular hole in an infinite elastic plate with the stress concentrations that exist in such a case.

For case (c), with no internal pressure on the lining,

$$\text{radial stress } \sigma_r = \frac{b^2 p_o}{b^2 - a^2} [1 - \frac{a^2}{r^2}],$$

223

$$\text{tangential stress } \sigma_\theta = \frac{b^2 p_O}{b^2 - a^2} \left[1 + \frac{a^2}{r^2}\right]$$

where a = internal radius,
    b = external radius,
    p_i = internal pressure,
    p_O = external pressure,
    r = radius at measurement point.

Excavations in Massive Elasto-plastic Earth Materials

The support of excavations at shallow depth in clay, or at greater depth in weak rock, is affected by plastic deformation of the earth material. It is assumed that at depth in the material, where the material is constrained, the mass behaves elastically, but the construction of the excavation and relief of confinement in its vicinity weakens the wall strata, and if the stresses there exceed the yield strength of the material a plastic zone is created around the periphery of the excavation.

If the excavation is not lined then yield occurs when the *in situ* stress level exceeds the unconfined compressive strength of the material. In the case of a plastic clay with cohesion (c) but no frictional resistance this would be when the strata pressure at the limit of the plastic zone (p_2) = 2c. The insertion of a stressed lining in this excavation would impose a confining pressure on the rock walls and thus restrict the development and extent of the plastic zone. Deere *et al.* show the relationship between the radius of the plastic zone (R), the unconfined compressive strength (2c), the strata pressure at the limit of the plastic zone (p_z) and the pressure imposed by the lining (p_i) for a frictionless soil.

If the earth material has frictional strength as well as cohesion then it may be considered to yield according to the Mohr-Coulomb failure criterion, in which case the unconfined compressive strength will be

$$\frac{2c \cos \varphi}{1 - \sin \varphi}$$

where φ is the angle of internal friction. The component of frictional resistance must now be combined with the strata pressure and the radial pressure in the lining, to determine the extent of the plastic zone.

A similar approach may be used to study excavation design and support problems, in rock materials at depths where elastoplastic transitions may occur. Horvath describes one such approach. It is postulated that, assuming uniform overburden strata, for a given rock material there will be a critical depth below the surface. At depths less than the critical value the rock will behave elastically.

An excavation in the rock under such conditions, assuming that the rock is competent and not loaded beyond its strength, will be supported by the rock itself. If a support lining is inserted, that lining will only be subject to the stresses imposed during the process of its construction and those subsequently imposed by fragments detached from the rock mass behind the lining. The lining will not be subject to stresses imposed from the rock mass as a whole.

But if the excavation is at a depth beyond the critical value, then the rock walls will deform plastically. A lining inserted into the excavation would then be subject to stress generated by plastic yield of the rock mass. To maintain equilibrium the force system imposed from within the excavation, by the lining, must balance the distortion potential of the *in situ* stress field at the wall surface.

Horvath pictures the stress condition in the surrounding walls of a circular shaft driven vertically into the rock mass, as shown in Fig. 11.10 and quotes the deformation potential on the wall surfaces as

$$VT_\theta = \frac{1+\nu}{E} \left[\frac{H\rho}{1-\nu}\right]^2 \cdot \frac{7\nu^2 - 4\nu + 1}{3}$$

where ν = Poisson's ratio,
E = Young's modulus,
H = depth,
ρ = rock density.

The critical depth

$$H_{crit} = \frac{\sigma_F}{\rho} \cdot \frac{1-\nu}{(1 - 4\nu + 7\nu^2)},$$

$$\sigma_F = H\rho \left(1 - \frac{\nu}{1-\nu}\right).$$

and the force system which must be applied by the lining on the excavation wall is equal to the rock pressure p, where

$$p = \frac{\nu}{1-\nu} \cdot H\rho \pm \sqrt{\left(\frac{H\rho}{1-\nu}\right)^2 - \frac{(1-2\nu)^2}{3} + \frac{\sigma_F^2}{3}}$$

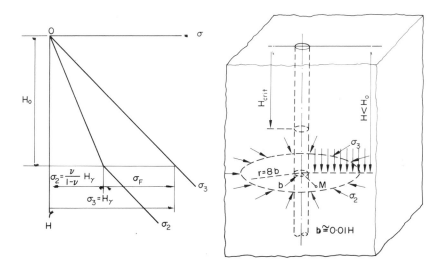

Fig. 11.10 Stress conditions around a shaft of circular section (Horvath).

Interaction between Earth Material and Excavation Support Structures

There are two inherent difficulties associated with the use of load cells and stress gauges in earth materials: (1) the gauge does not behave like the soil or rock it displaces, and (2) insertion and placement of the gauge disturbs the stress field that is to be measured. Chapter 10 dealt with devices and techniques intended for the determination of free field stress. The problems to be solved are no less difficult, indeed, in some respects are even more difficult, when we attempt to investigate the contact characteristics of load-transfer between earth materials and excavation support structures.

Earth pressure cells

Earth pressure cells of various design have been used in soil mechanics and foundation engineering, to investigate soil-structure interaction, for many years. In general they consist of hydraulic pressure cells, or mechanical/elastic load cells, in which the deflection of a diaphragm or a piston is measured and calibrated in terms of pressure.

There are two possible locations for the contact-pressure measurement devices: (a) in the soil or rock, abutting on to the structure, (b) in the structure abutting on to the earth material. A complete study of the contact interaction requires that instruments be set in both positions, on each side of the contact being studied. This has seldom, if ever, been done. Ideally, the instruments used in either situation should perfectly match the stress-strain characteristics of the material in which they are placed. Consider a gauge that is inserted with its measurement surface flush with a wall and in contact with soil in a foundation, and suppose that the soil is not disturbed. If the gauge is more rigid than the wall, there will be a concentration of stress upon it and it will read in response to a pressure which is higher than the soil pressure on the wall. Conversely, if the gauge is less rigid than the wall it will give too low a reading.

But the result will also be affected by any disturbance of the soil in contact with the gauge. If the soil in contact with the gauge has not been properly compacted, so that it is softer than the soil in the rest of the foundation, or if the contact is not good, then the sensing element of the gauge will give too low a reading. This may cause a rigid gauge to read too low, although normally it would read too high.

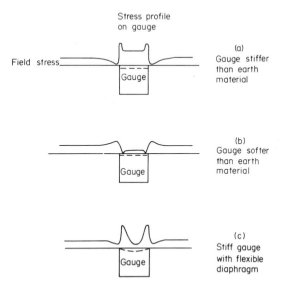

Fig. 11.11. Stress distribution around earth-pressure cells.

The redistribution of stress on an earth pressure cell is illustrated in Fig. 11.11. When the gauge is uniformly more rigid than the earth material the stress level over the gauge is increased to a value above that of the field earth pressure, with high stress concentrations adjacent to the peripheral walls of the gauge. At the same time, the earth pressure in the outer vicinity of the gauge is partially relieved, so that it is below the field earth pressure. The converse situation exists when the gauge is uniformly less rigid than the earth material. In this case the field earth pressure around the outside perimeter of the gauge is increased, while the pressure over the gauge itself is considerably reduced. A gauge with non-uniform rigidity, such as one which embodies a flexible diaphragm carried in a rigid case, would combine both these effects, to an extend depending upon its design. Figure 11.11 shows such a gauge producing high stress concentrations over the case, and pressure relief in the earth outside the perimeter as well as over the central portion of the diaphragm.

These variable pressure distributions are reduced in magnitude if the gauge is thin, compared with its diameter, when the pressure to be measured is normal to the face of the gauge. It is generally accepted that the ratio T/D should be as small as possible if the effects of rigidity mismatch are to be minimized. In a review of the problem of soil-structure interaction, Selig quotes the relationship

$$\frac{P_c - p}{p} = \frac{P_e}{p} = C_a \frac{T}{D}$$

where P_c is the average pressure on the gauge,
$\quad p$ is the true earth pressure,
$\quad P_e$ is the gauge error,
$\quad T$ is the gauge thickness,
$\quad D$ is the gauge diameter,
$\quad C_a$ is the cell action factor which depends on gauge/soil stiffness ratio, soil properties, and gauge T/D ratio.

The cell action factor increases with increase of the gauge/soil stiffness ratio, and approaches a limiting value, being approximately constant when

$$\frac{E \text{ gauge}}{E \text{ soil}} > 10,$$

other things being constant.

Loads on Excavation Support Structures.

Pillar Loads

In the evolution of underground mining technology the "room and pillar" technique figures largely. The essence of the technique is that the mining excavations must be supported by pillars of unmined rock on either side. It is a method of working that is applied, with many variations as to detail, in a wide range of sedimentary and massive mineral deposits. In vein-type deposits also, pillars of rock and unmined ore are frequently left to support certain sections of the mine. The maintenance of ground control and stability around the excavations depends upon whether or not the pillars are able to carry the loads that are imposed upon them, and this will be determined by their size and strength, in relation to the ambient stress distribution. During the years an enormous amount of documented experience has been gained which enables engineers to lay out pillar-support systems, using empirical rules and intuitive design parameters. Nevertheless, it not infrequently happens that complete strata control is not achieved and hence the stability of the excavations is threatened.

It is at this point that the strata-control engineers and the rock mechanics specialists may be called in for advice and assistance. Theoretical concepts of load and stress distribution, however, can be of little avail without factual information as to what is really happening. Much can be learned from a detailed visual and instrumented study of modes and rate of deformation, and directions of movement in the strata surrounding the excavations, but added information as to stress magnitudes and distribution is also required. This is particularly important when the threatened instability takes the form of rock bursts, or gas and rock outbursts, which are liable to occur suddenly in material that, to all outward appearances, may appear to be rigid, strong, and stable.

Location of abutment zones

On the analogy that the redistribution of strata pressures around excavations in rock causes "pressure arching" above the excavations, the regions of high pressure that are produced by the deflection of load cause stress concentrations termed "abutment zones" in the solid rock. While, in theory, the maximum stress concentrations around an excavation in an elastic material occur at the periphery of the excavation, in practice, owing to yield and fracture of the immediate wall rock, they are deflected to an unknown depth within the walls. The location of these abutment zones may be approximately deduced by qualitative techniques, such as observation of core "discing" and rock fragmentation in rotary borehole drilling, or they may be precisely located by applying the techniques of *in situ* stress measurement already described.

When applied to exploration of the stresses around the walls of a tunnel, results similar to those illustrated in Fig. 11.12 are likely to be obtained.

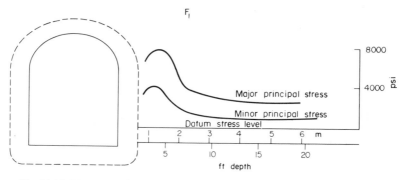

Fig. 11.12. Plane stress data measured at various depths in a tunnel wall (Royea).

From a zero plane stress condition at the wall, and after passing through an indeterminate stress zone, the extent of which depends upon the degree of yield and fracture of the strata around the periphery of the tunnel, the measured stresses will increase to a maximum value in the abutment zone beyond which they decrease again, and (in the absence of interrelated effects from neighboring excavations) ultimately attain the original virgin stress level. The depth at which constancy of stress level is attained commonly ranges from about 2 to 6 m. It may be less in very strong rock, if the walls are sound, and more if the rock has been seriously weakened, as for example by blasting. Royea quotes the results depicted in Fig. 11.12, while Stephenson and Murray quote evidence of interference with the virgin stress pattern at a depth of four excavation diameters within the rock walls.

Abel and Lee observed stress changes occurring at 2 to 7 diameters ahead of model tunnels drilled in various materials, with compressive stress concentrations 1 to 6 diameters ahead of

228

the face. No such stress peaks are to be predicted by elastic theory.

The abutment zones associated with wide excavations, such as stoping areas and longwall extraction faces, may extend to greater depths into the solid. The conditions here differ greatly from those which exist around excavations supported on each side by solid rock. The excavations now may extend over a width of from 50 to 200 m, and sometimes over a length of 1000 m or more. However, only certain parts of these excavations are required to remain open, to provide access roadways along the "ribsides" in an "advancing" system, or through the solid in a "retreating" system. The working area which must be supported, consists of a strip, usually some 4.5 to 6 m wide, extending along the working "face" (see Fig. 11.13).

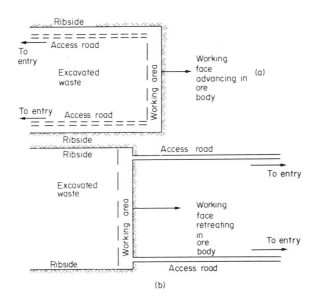

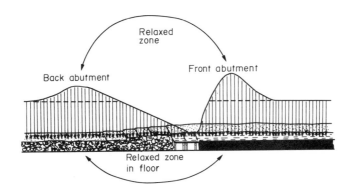

Fig. 11.13. Wide workings in mineral extraction. (a) Plan view, advancing. (b) Plan view, retreating. (c) Section showing approximate distribution of pressure across a face or a ribside. (N.B. In some instances the back abutment is much more diffuse than is shown here.) (After Sprath).

TABLE 11.2. Location and Magnitude of Front Abutment Pressure in Longwall Coal Mining as Reported by Different Investigators

Place of investigation	Jacobi and Everling (1)	Creuels and Hermes (2)		Suzuki (3)	Borecki and Salustowicz (4)		Leigh (5)	Banerjee (6)
	Givondele Seam	Wilhelmina	Hendrik Seam VIII and IX	—	Miechowice Seam 413	Miechowice Seam 504	Easington Low Main Seam	Waterloo Main Middleton Little Seam
	Germany	Germany	Germany	Japan	Poland	Poland	U.K.	U.K.
Instrument or technique of measurement	Borehole diametral strain	Hydraulic pressure capsules		Metal capsule with strain gauges	Dynamo-meter	Dynamo-meter	Pott's Stress-meter	Encapsulated hydraulic cells
Depth of seam from surface (m)	799	479	640	475	549	731	518	85.3
Thickness of seam (m)	1.2	1.2	2.4	2.3	1.8	—	1.7	1.5
Approximate position of peak pressure from face into solid (m)	—	3.7	7.6	10.1	15.2	—	2.4	2.4
Maximum increase of pressure in terms of cover load	450%	425%	120%	116%	6%	28%	70%	96%
Approximate limit distance from face at which effect of front abutment can be felt (m)	213	42.7	51.8	16.1	79.2	97.5	100.6	18.2

(1) Jacobi, O. The pressure on seam and goaf. Everling, G. Calculation of stress from measurements in boreholes. Int. Strata Control Congress, Essen, 1956.

(2) Creuels, F. H. and Hermes, J. M. Measurements of changes in rock pressure in the vicinity of a working face. Int. Strata Control Congress, Essen, 1956.

(3) Suzuki, K. Investigations of rock pressure variations around underground longwall workings. Int. Strata Control Conference, Paris, 1960.

(4) Borecki, M. and Salustowicz, A. Experimental work in the field of rock mechanics. Int. Strata Control Conference, Paris, 1960.

(5) Leigh, R. D. Strata pressure and rock mass movements. Ph.D. Thesis, Univ. of Durham, 1962.

(6) Banerjee, S. F. The Stress in the Front Abutment Zone. Res. Rept., Postgraduate School of Mining, Sheffield, Feb. 1968.

The superincumbent strata over such wide workings are carried partly by the solid rock ahead of the face and over the ribsides, and partly by the support given by waste rock which "caves" or collapses into the excavated area, or which is "stowed" or "backfilled" as part of the mining process.

The location of the abutment zones over the solid has a marked influence on the strata control conditions along the working face, and on the support afforded by the "ribsides". The magnitude of the stress concentration in the abutment zone, and the lateral extent of that zone, also affect the support conditions that must be maintained along the access roadways of a retreating system, since these roadways have to pass through the abutment zone.

The recorded observations of several investigators, working in various localities, are reported in Table 11.2 from which it is apparent that considerable variation in the observed data exists, from which no general rules can be drawn as to where the abutment zone lies, or what its magnitude is likely to be. Measured values of the stress concentration range from 6% to 450% at distances ranging from 2.5 to 15 m into the walls.

Determination of the extent of the fractured zone.

It will be apparent that the pressure abutment zone must lie beyond the zone of yielding and fractured rock around the periphery of the excavation. Geophysical techniques may be applied to determine the extent and degree of fracture in the wall rock. In general there are three methods. These involve the measurement of sonic velocities and the determination of either "time delay along the surface" or "time delay in depth". In both these procedures the sound pulse is produced by a hammer blow and the pulse is detected by a transducer in contact with the rock some distance away. It is assumed that the fractured rock is broken uniformly to the extent that the velocity of dilatational sound waves V_1 within this layer is isotropic, constant, and less than the dilatational sonic velocity V_2 in the solid rock. From which, referring to Fig. 11.14, the depth of the fractured zone h is given by

$$h = \frac{X_0}{2} \cdot \frac{V_2 - V_1}{V_2 + V_1} ,$$

X_0 is determined by an experimental plot of time delay t against distance x along the face. Two straight-line plots of slopes V_1 and V_2 are obtained, and their point of intersection locates X_0 (Fig. 11.14(b)).

In the "time delay in depth" method a number of holes are drilled perpendicularly into the rock face. The sound pulse is generated by a hammer near the mouth of the hole and received inside the hole by a transducer fixed to the end of a probe. The measured time delay at various depths is measured and plotted as a function of the depth of the probe, and the gradient of the plot at depths beyond the fractured zone should correspond to the velocity of sound in solid rock. At depths less than that of the fractured zone the measured sonic velocities are less than that for solid rock. A block diagram representing the equipment used in both time delay methods is shown in Fig. 11.15.

The third method uses ultrasonic pulse reflections. Ultrasonic pulses are generated by a barium titanate transducer, vibrating at a characteristic frequency, say 500 kilocycles per second. This is placed in a borehold probe to explore a zone of rock around the borehole (Fig. 11.16). The ultrasonic pulses are reflected from discontinuities, and the reflections are observed on the screen of a suitable recording apparatus, typical records such as those of Fig. 11.16(b) being obtained.

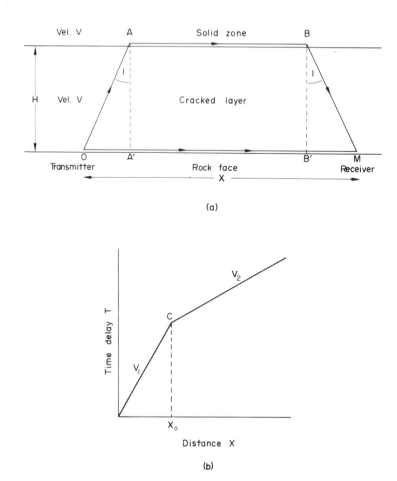

Fig. 11.14(a) Sound paths from transmitter to receiver in the "time delay along the surface" method (Lutsch and Szendrei). (b) Plot of results, "time delay along the surface" method.

Loads in Concrete Tunnel Linings

The loads on concrete tunnel linings may be deduced from measurements of stress in lining segments. The measurement transducers may take the form either of vibrating wires or photo-elastic stressmeters. The vibrating wire gauges are incorporated into bricks or tiles which in turn are embedded in the main structure. Photoelastic stressmeters may be inserted into segments of the concrete wall, each segment with its contained stressmeter being calibrated in a loading press so that the stressmeter fringe order gives a direct read-out in terms of load when it is subsequently incorporated into the tunnel lining.

When the lining is preloaded to provide a ring stress on insertion, compression load cells may be fixed between the segments to monitor the ring stress after the setting-jacks have been withdrawn.

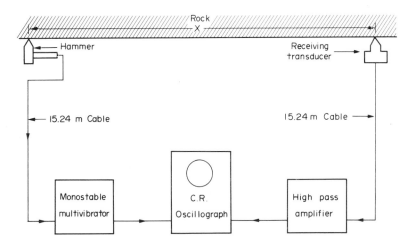

Fig. 11.15. Block diagram of equipment used in time delay methods of fracture-zone exploration.

Prop and Strut Loads

In underground excavations, while the main superincumbent loads are distributed by the strata spanning the abutment zones, a zone of yielding and fractured strata within the "pressure-relief" zone which immediately surrounds the excavation, may require to be supported from within the abutments and possibly from within the excavation itself. In the case of narrow excavations in "competent" rock the tendency for the walls to move into the excavation sets up a local "arching" action which is capable of carrying the loads imposed by the forces within the "pressure-relief" zone (Fig. 11.17). In "incompetent" ground some additional support will be required. This extra support may be provided in excavations by strata reinforcement such as rock bolts, and by steel or concrete linings. Wide excavations in incompetent or moving ground are supported by props or struts which, in a horizontal or near horizontal excavation, are set vertically between roof and "floor". In a steeply inclined excavation the struts are usually set between "hanging" and "foot" walls at an angle approaching the perpendicular to the walls, so that any tendency for the hanging wall to move gravitationally acts to tighten the struts. In former years these struts were very commonly simple timber "props" but today steel supports are widely used in preference to wood. The steel props may be rigid lengths of tube or H-section girders, but telescopic props, embodying friction grip, screw jack, or hydraulic jack principles, are commonly used. The telescopic props have an advantage over rigid non-telescopic lengths in that they can be preloaded to exert a known, predetermined, resistance to strata convergence when they are set. Friction-grip and hydraulic-jack props may also be designed to yield at a required value of loading, while still exerting some resistance to convergence.

Modern underground mining technology makes increasing use of hydraulic supports because these can be readily incorporated into mechanized systems, in which the support units are combined with excavating machinery, which in some instances is capable of automation and remote control. However, the capital cost of such installations is very high, so that considerable care must be exercised in the design and selection of equipment that will be capable of

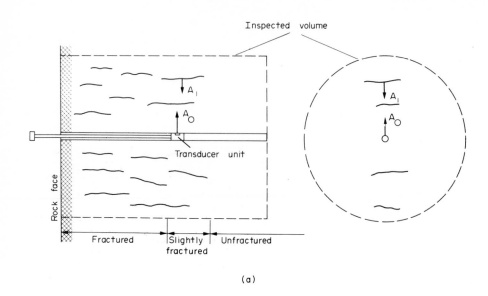

(a)

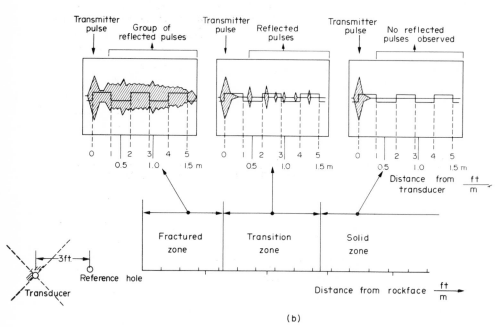

(b)

Fig. 11.16(a) Ultrasonic pulse reflection method of fracture-zone exploration. (b) Typical records obtained by ultrasonic pulse method (Lutsch and Szendrei).

operating satisfactorily under the prevailing conditions. This requires that study and research be pursued, including the systematic measurement of prop loads at various stages of the mining process. A simple method of prop-load measurement on hydraulic systems is the hydraulic dynamometer prop, in which the fluid pressures are measured by a Bourdon element, sometimes coupled to a clockwork recorder.

Some support units consist of several hydraulic jacks operating together in sequence, some being extended to give support while others are being retreated to permit lateral movement of associated, attached, machinery. The placement of hydraulic dynamometers of conventional design in the various fluid circuits of these units presents practical problems, mainly owing to the bulk and vulnerability of the recorders. A practicable alternative is to use photoelastic glass pressure gauges. Saltuklaroglu describes an investigation of hydraulic chock loads, conducted by this means. Each supporting "chock" embodied five hydraulic jacks and these were distributed along the working face. The distribution of load on the supporting legs of chocks at various positions along the face was measured. From such records the "mean support load density", an index which combines the magnitude of support resistance and the distance between the supports, required for effective roof control, may be determined. The method of computation of this index is described by Shepherd.

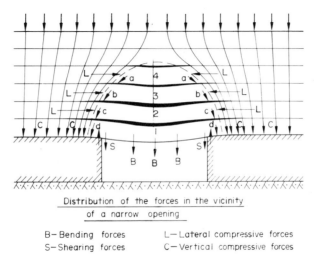

Distribution of the forces in the vicinity
of a narrow opening

B— Bending forces   L— Lateral compressive forces
S— Shearing forces   C— Vertical compressive forces

Fig. 11.17. Distribution of forces in the vicinity of a narrow opening in laminar sediments
(Walker, Alder, and Potts).

The contact loads between prop and horizontal support element, or between prop and roof or floor stratum, may be measured by conventional strain-gauge-type load cells (Fig. 11.18) or by photoelastic load cells (Fig. 11.19). When using load cells of any type, it is often overlooked that the calibration characteristics of the device are greatly dependent upon the system of loading, particularly the surface contact conditions and the maintenance of axiality in the direction of loading. Underground site conditions in geotechnology, in mining and civil engineering, are liable to impose very variable and arduous loading conditions which differ greatly from laboratory calibration conditions. It is therefore important that the load-measurement device should maintain its calibration characteristic over a wide range of possible end-contact conditions and be able to do this as well as accommodate considerable eccentricity in the direction of loading (Fig. 11.20).

The errors in measurement due to variation in response to load under various end-contact conditions can be obviated by measuring strain at some point on or within the prop, and calibrating this directly in terms of load. On cylindrical tube-type props this involves setting at least three strain gauges 120° apart around the circumference of the prop, preferably at its mid-height. On H-section girder props strain gauges are set on both sides of the central web and a mean reading observed. Alternatively, linear extensometers may be used in a similar alignment, but over a longer gauge length (Fig. 11.21). Another alternative is to use small photoelastic load cells set into ring gauges. These are placed in holes on the web, the holes being taper-reamed to fit the gauge when the prop is under zero load (Fig. 11.23).

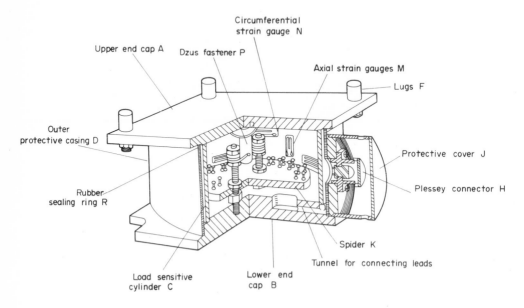

Fig. 11.18. MRE mine support load cell.

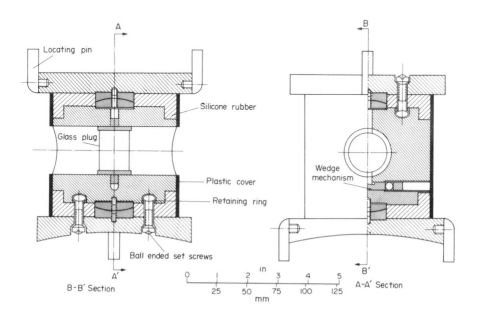

Fig. 11.19. Photoelastic pit-prop load cell.

A typical application of this load cell, when inserted to determine the loads on the supporting legs of tunnel arches, is shown in Fig. 11.23. The legs of each arch were supported on telescopic "stilts", the function of which is to allow the arch to yield under a high incident roof load. The resistance of the support to convergence between roof and fllor thus builds up as the stilt compresses, with increasing distance from the advancing face.

Fig. 11.20. Photoelastic pit-prop load cell under eccentric loading.

Tension in rock bolts and cables

Strata reinforcement, such as rock bolts and steel cables, is only effective so long as the reinforcement is in tension. This compresses the rock over the length extending between the rock bolt anchor and the bearing plate (Fig. 11.24). In such strata reinforcement systems the bolts are tensioned at a predetermined setting load, measured by direct pull or less satisfactorily, by a torque wrench. Subsequent observation of bolt or cable tension is possible if a measurement device is incorporated into the face plate and bolt connection. Precise measurements require load cells, either of strain gauge dynamometer or photoelastic type.

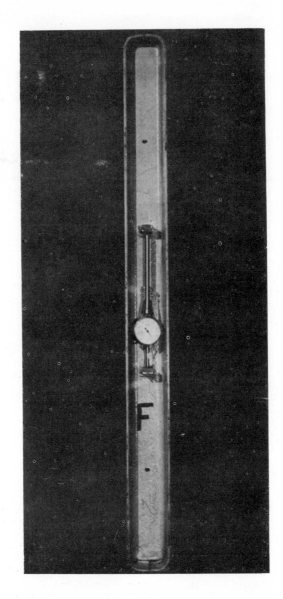

Fig. 11.21. The Winder gauge mounted on a rigid mine support-prop.

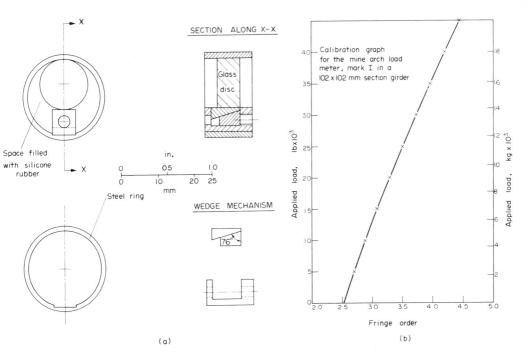

Fig. 11.22. Mine arch load cell. (a) Mode of construction. (b) Calibration characteristic.

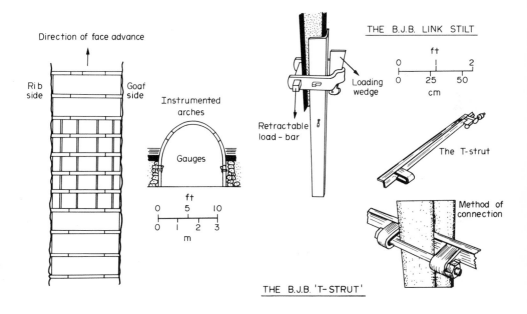

Fig. 11.23. Arrangement of steel arches on stilts, to provide a yielding support in a mine roadway.

G—I

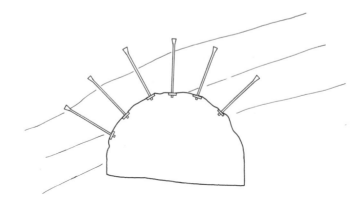

Fig. 11.24. Arrangement of rock bolts for tunnel reinforcement.

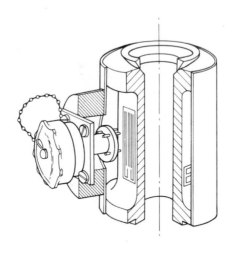

Fig. 11.25. MRE roof-bolt dynamometer.

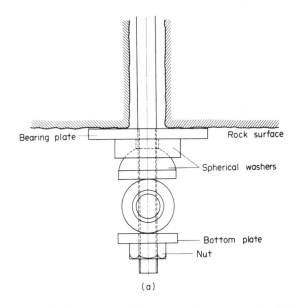

Bearing plate

Rock surface

Spherical washers

Bottom plate

Nut

(a)

(b)

Fig. 11.26. Photoelastic rock-bolt dynamometer. (a) Mode of construction. (b) Dynamometer in service.

241

Selected References for Further Reading

ABEL, J. F. and LEE, F. T. Stress changes ahead of an advancing tunnel. *Int. J. Rock Mech. Min. Sci.,* vol. 10, pp. 673–698 (1973).

BIERBAUMER, A. *Die Dimensionierung des Tunnelnauerwerks,* Leipzig, Engelmann, 1913.

CHELEPATI, C. V. Arching in soil due to yielding of a rigid horizontal strip. *Proc. Sym. Soil-structure Interaction, Univ. of Ariz.,* 1964.

COON, R. F. Correlation of engineering behavior with the classification of *in-situ* rocks. Ph.D. Thesis, Univ. of Illinois, Urbana, 1968.

DANIEL, I. M. and ROWLANDS, R. E. Study of lined and unlined cavities in biaxially loaded rock. *Proc. of 12th Sym. on R. Mech.,* 1971, p. 877.

DEERE, J. V., PECK, R. B., MONSEES, J. E. and SCHMIDT, B. *Design of Tunnel Liners and Support Systems,* P.B. 183.799, National Technical Information Service, Washington DC, 1969.

DIXON, J. D. *Analysis of Tunnel Support Structure with Consideration of Support-rock Interaction,* U.S. Bur. Mines Rept., Invest.No. 7526, 1971.

DUNS, C. S. and BUTTERFIELD, R. Flexible buried cylinders. *Int. J. Rock Mech. Min. Sci.,* vol. 8, pp. 577–627 (1971).

FISECKI, M. Y. The determination of strata pressures and support loads in mines. M. Eng. Thesis, University of Sheffield, 1965.

HOOPER, J. A. The theory and design of photoelastic load gauges incorporating glass element transducers. *Int. J. Rock Mech. Min. Sci.,* vol. 9, pp. 363–401 (1972).

HORVATH, J. Calculation of rock pressure in shafts and roadways of circular section. *Int. J. Rock Mech. Min. Sci.,* vol. 8, pp. 239–276 (1971).

LANE, K. S. Effect of lining stiffness on tunnel linings. *Proc. 4th Int. Conf. on Soil Mech. & Foundation Engineering,* vol. 2, p. 223 (1957).

LANE, K. S. Garrison dam—evaluation of results from tunnel tests section. *J. Soil Mech. Found. Div., Proc. of ASCE,* vol. 83, SM4 (1959).

LAUFFER, H. Gebirgsklassifizierung für den Stollenbau. *Geologie und Bauwesen,* vol. 24, H. 1 (1958).

LUSCHER, U. The interaction between a structural tube and a surrounding cylinder of soil. Ph.D. thesis, MIT, Cambridge, Mass., Aug. 1963.

LUSCHER, U. and HOEG, K. The beneficial action of the surrounding soil on the load-carrying capacity of buried tube. *Proc. Sym. Soil-structure Interaction, Univ. of Ariz.,* 1964.

MARSTON, A. *The Theory of External Loads on Closed Conduits in the Light of the Latest Experiments,* Bull. 96, Engineering Experimental Station, Iowa State College, 1930.

MARSTON, A. and ANDERSON, A. O. *The Theory of Loads on Pipes in Ditches and Tests on Cement and Clay Drain Tile & Sewer Pipe,* Bull, 31, Iowa Engineering Experimental Station, Ames, Iowa, 1913.

MEYERHOF, G. G. and BAIKE, L. D. Strength of steel culvert sheets bearing against compacted sand backfill. *Highway Research Record,* No. 30 (1963).

MINDLIN, R. D. Stress distribution around a tunnel. *Trans. ASCE,* vol. 105, p. 1117 (1940).

NEWMARK, N. M. and HALTIWANGER, J. D. *Air Force Design Manual,* Report AFSWC-TDR-62-138, Air Force Special Weapons Center, 1962.

OBERT, L. and DUVALL, W. I. *Rock Mechanics and the Design of Structures in Rock* (Chapters 16 and 17). Wiley, New York, 1967.

ROYEA, M. J. Rock stress measurement at the Sullivan Mine. *Proc. 5th Canadian Rock Mechs. Symposium, Toronto,* pp. 59–74, Dec. 1968.

SALTAKLAROGLU, M. The determination of loads on mine support systems. Ph.D. Thesis, University of Sheffield, 1968.

SHEPHERD, R. Study of strata control on mechanized coal faces. *4th Int. Conf. on Strata Control and Rock Mechanics, New York,* May 1964.

SHEPHERD, R. and WILSON, A. H. The measurement of strain in concrete shaft and roadway linings. *Trans. Inst. Min. Engng.,* vol. 119, pp. 561–577 (1959).

SMITH, G. H. *Elements of Soil Mechanics for Civil & Mining Engineers* (Chapter 6) Crosby Lockwood, London, 1968.

STEPHENSON, B. R. and MURRAY, K. J. Application of the strain rosette relief method to measure principal stresses throughout a mine. *Int. J. Rock Mech. Min. Sci.,* vol. 7, pp. 1–22 (1970).

SZECHY, K. Approximate determination of rock pressure on the basis of a statical analogy. *1st Int. Symp. on Rock Mechs., Lisbon,* vol. 2, pp. 385–388 (1966).

SZECHY, K. *The Art of Tunneling,* Akademizi Kiado, Budapest, 1966.

TERZAGHI, K. *Theoretical Soil Mechanics,* Wiley, New York, 1943.

CHAPTER 12

The Observation of Mass Deformations in Geotechnology

Methods of observing and recording the deformations of earth and rock masses under load use a wide variety of instruments and techniques, and provide limits of precision in measurement that may range, for example, from the ultra-precise needs for determination of earth-structural strains in earthquake-prone areas, to the more general aspects of long-range reconnaissance provided by aerial photography.

Photographic Techniques

Photogrammetry

The science by means of which photographs may be used for the purpose of measurement is photogrammetry. In terrestrial photogrammetry the camera, termed a phototheodolite, is set firmly at a base station, at which its position may be precisely located by photographing other base stations in the field. At least two photographs, taken from different positions, are required to locate the object viewed. The technique of stereophotogrammetry may then be applied, using mechanical and optical devices to reconstruct the field of view, to scale.

There are many useful applications of photogrammetry in geotechnology. One of the most important is in geological site investigations concerned with engineering foundation projects such as dams and bridges, and in relation to the stability of large rock excavations and open-pits. The strength and stability of such rock masses depend greatly upon the relationships between the predominant structure of geological discontinuities, such as cleavage and jointing, and the ambient stress situation. Photography provides a very convenient means of permanently recording rock structural features for subsequent detailed examination and analysis, to an extent which is not possible on the field site itself. The information so obtained is analyzed statistically, and may be incorporated into optical data-processing systems, to bring out the relevant information quickly and automatically. Pincus describes optical data-processing of vectorial rock fabric data, which is applicable to studies of rock mass structure and rock petrofabric structure.

The camera is particularly useful for observing and recording structural data on large vertical rock cuts, direct access to which may not be possible, or to rock cuts exposed in engineering work and in open-pit mining, where access may be impeded or dangerous, due to the operations concerned. Time-lapse photographic techniques may be used to observe and record the extent and rate of deformation of earth and rock structures under load in large-scale site-investigation tests, and to observe the development of deformation leading to potential failure and collapse of earth and rock masses in open pits and in landslide areas.

They are also useful to observe the extent and rate of ground movements in areas where the underground extraction of minerals or fluids results in subsidence of the ground surface. The method has been used too, with great affect, to study glacial ice movements, and avalanche phenomena.

Aerial photogrammetry is primarily a reconnaissance tool. Not infrequently, after a major landslide or earthquake, aircraft may provide the only means of access to and over the site concerned. In a more general context the method has been used very effectively to trace fault zones in geologically active areas, and to map particular areas of interest. The limit of accuracy in fixing a point of measurement on a scale of 1/500 would be equivalent to about 5 cm horizontally and 10 cm vertically.

Direct Surveying Techniques

Conventional surveying techniques have developed through the years, depending upon optical telescopes for the maintenance of alignment. These are coupled to circular protractors for angular measurement, and to levels and plumb-lines for the establishment of horizontal and perpendicular ordinates. The limits of precision in measurement using these techniques depend upon the instrument sensitivity and the accuracy of graduation in the various scales employed, together with the skill and practice of the surveyor. Measurements to an orientation accuracy of 5 sec standard deviation are equivalent to a positional error of 2.5 cm over a length of 1 km, while a standard deviation of 1 minute of arc would mean a positional error of 30 cm over the same distance. A precise geodetic leveling could result in a probable error in elevation to around $0.5 \sqrt{K}$ mm, where K is the distance (km).

Conventional surveying techniques attained a standard of proficiency to the limits of accuracy attainable with optical instruments, that has remained more or less static for several decades. In geodetic surveying distances are measured by a system of triangulation, consisting of measured angles from a reference base line. Recent years, however, have seen some major developments in the introduction of instruments operating on revolutionary new principles. These include gyro-theodolites, in which the reference direction, from which angular measurements are made, is maintained by gyroscope.

Apparatus for refined distance measurement has also been introduced, making use of lasers, and a variety of electromagnetic devices such as microwave beams. The instruments are capable of measuring distances to an accuracy of ± 1 part per million over a daylight measuring range of 30,000 m (Laser Geodolite). A more general-purpose Tellurometer has an accuracy of 2 parts per million over a range of 2000 m. They are especially useful in engineering surveys since they enable distances to be measured from one instrument setting, in a matter of seconds. They all operate on similar principles in that the distance to be measured is deduced from the time interval which elapses during the transmission and reception of a signal bounced between the instrument and its target. The interval is measured in terms of a phase shift, and in some of the more recent introductions this is displayed as a direct digital read-out.

The use of precise leveling, by conventional surveying techniques, forms an essential part of geotechnical observations in connection with subsidence of the ground surface over mine workings. The absolute movement of the boundaries of excavations, on the surface and underground, may be determined only by reference to fixed base lines outside the zone of influence of the disturbed ground. In underground investigations, a change in the vertical direction is easiest to determine with accuracy. Ortlepp and Cook describe the use of precise leveling of reference "benchmarks" established for the measurement and analysis of the deformation around deep hard-rock excavations. The benchmarks consisted of stainless-steel links attached to rock bolts anchored at a depth of 200 mm in the roof stratum of the underground roadways. Measurements made on invar staffs, suspended from these links, were accurate to ± 1.0 mm.

Earth-strain Measurements

Measurement of the elastic deformation of the earth's surface, in response to large-scale tectonic and other natural forces, is possible, using geodetic surveying methods or by direct strain measurement. Geodetic measurements are made over base lines, which range from about 1 km to some tens of kilometers in length. The maximum strain sensitivity of geodetic survey measurements is about 1 microstrain (1 part per million), so that only comparatively large tectonic strains may be detected by this means, and these may take considerable time to develop. Consequently geodetic survey earth-strain measurement is a long-term project, observations being made only at about 6 monthly intervals.

Direct Measurement of Earth Strain

Two types of instrument are currently in use for the direct measurement of earth strain. One uses a laser beam, combined with an optical interferometer to detect small changes in length. The other, more common, instrument uses a quartz tube 60 m long, constructed of several sections each about 3 m long, supported so that each section abuts against its neighbor and all being pushed against a concrete block anchored firmly to bedrock (Fig. 12.1). The position of the free end is measured precisely, relative to the earth, by means of a capacitance strain

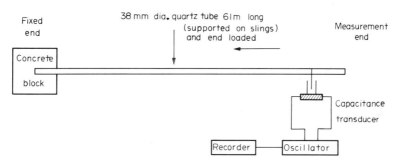

Fig. 12.1. Arrangement of quartz-rod extensometer for determination of seismic strains.

transducer, the position of which is continuously monitored and recorded. The quartz strain meter system incorporates equipment for periodic in-place calibration while measurements are being recorded. These instruments are extremely sensitive and will detect and record to within 1/100 microstrain on a long-term basis, and better than this over a short-term. Being so sensitive, the instruments must be very carefully housed behind airtight doors, in a temperature-controlled atmosphere. They are used to measure earth-tidal strains, secular or long-term changes, and sudden strain changes associated with seismic events.

Mass-deformation Extensometers

Observation of mass deformations is possible using a variety of instruments, incorporating wires, tapes, rods, or tubes, to any desired length. All incorporate certain basic features. (1) At

Geotechnology

one end, a firm anchor or reference stop. (2) At the other end, a reference index and scale, or linear measurement transducer. (3) If temperatures are not stable or controlled, correction for temperature variation may be required. (4) Where flexible wires or tapes are used, a standard tension must be applied. The transducer system for linear measurement may be mechanical, electrical, or photoelastic, identical with those used in laboratory measurements.

Convergence Recorders

Extensible rods, in screwed sections, or telescopic tubes, are convenient to use when measuring deformation over lengths up to 3 or 4 m, or, when applied vertically, up to about 6 m length. It is important that the rod be perfectly rigid, without sag between the reference marks, which are usually metal pins let into the rock. The sensitivity and range of the measurement head may vary from the direct linear trace formed by a recording pen on a rotating drum, to more refined measurements over a shorter range.

The convergence recorders that have long been used to strata-control observations underground are often of the rotating-drum type (Fig. 12.2). Typical records provided by such instruments set in one of the access roads to an extraction face are shown in Fig. 12.3. The record

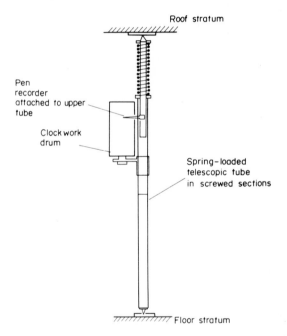

Fig. 12.2. SMRE-type convergence recorder.

shows the rate and extent of convergence between the roof and floor strata as the face advances from a position 140 m in advance of the recorder, to a point 90 m behind it. In horizontal strata the magnitude of such movements is generally proportional to the seam thickness, assuming that the excavated area is allowed to cave, and is not "backfilled". The type of roof rock is important too. A strong, inflexible sandstone rock roof would be likely to produce measurable convergence at a greater distance from the approaching face than would be caused by a more flexible shale. The reason for this probably lies in the fact that the overhanging roof weight in

246

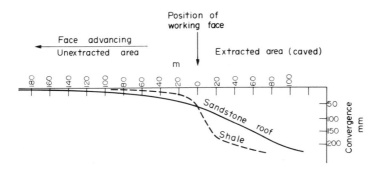

Fig. 12.3. Convergence record in advance of and behind a moving excavation wall (Hoffman).

the case of the sandstone would be much higher than that for the shale. The shale breaks off more easily behind the face supports, in the process of caving. This relieves the weight over the working face. The S-bend in the convergence record for the shale is much steeper than that for the sandstone, showing that the build-up of support from the caved rock, in the extracted area, is also much quicker under the shale roof than it is under the sandstone roof.

Another useful function of convergence recorders is to aid observation of the load-yield characteristics of excavation-support structures.

Stope Stability Meter

In hard-rock mines the movements producing relative displacement of roof and floor strata, or between hanging-wall and foot wall, may be quite small. But at the same time such small movements at depth may represent the release of considerable amounts of strain energy in the surrounding rock mass. Observation of the rate of release of strain energy, as demonstrated by the convergence of the excavation walls, is of prime importance in mines subject to rockburst.

Measurement of "Ride"

A necessary ancillary to the observation of convergence or closure of the rock walls in an excavation is to determine the relative lateral movement across the axis of closure. This is because "convergence" is not always a simple process of a reduction in one linear dimension. It is often accompanied by displacements in other directions also. Such displacements may be observed most easily when closure is measured along a vertical ordinate, in which case a plumb-weight suspended from a pin in the roof, or hanging-wall, will indicate any lateral movement with reference to a location peg which, at the start of the investigation, is set in the floor stratum vertically below the pin.

Precise measurements may be made by fixing a flat metal plate on the floor under the plummet. The plate is fitted with lateral micrometers which locate the tip of the plummet on a rectangular coordinate system. Another similar instrument is the MRE "Romometer' (Fig. 12.4). This consists of a spring-loaded telescopic tube with a ball at the top to locate in a roof plate, and a pivoted gimbal at its foot, to locate in a floor plate. The foot of the romometer carries a divided circle and an index plate, mounted on a vertical center. An adjustable spirit level, coupled to a tangent micrometer, and carrying another spirit bubble at right angles to

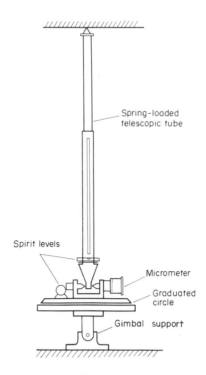

Spring-loaded telescopic tube

Spirit levels

Micrometer

Graduated circle

Gimbal support

Fig. 12.4. The MRE "Romometer"

itself, serves as a means of detecting and measuring any linear separation of the initial reference indices, relative to a selected datum direction.

Rock-stability Alarms

Rod and wire extensometers may incorporate alarm systems to give warning when any abnormal rate of strata movement occurs, or when the extent of deformation has reached a prescribed limit, at which the alarm is set.

Such devices are particularly useful for setting at inaccessible points on an exposed rock face, say at some point overlooking a transport route or a working face, or at any location where a potential earth or rock fall hazard exists. Similarly they may be set to give warning of untoward strata movement in roof strata above a large tunnel or a mine stope.

Reed's stope convergence warning light is intended for such a purpose (Fig. 12.5). It is attached to coupled sections of tubing, to the appropriate length, and arranged so that convergence beyond the limit set by an adjustable micrometer operates a micro-switch, to close the electrical circuit and light up the alarm signal. In open-pit installations or in hazardous rock-slide areas wire extensometers coupled to audible signals may be used, sometimes in combination with remote sensing and monitoring equipment, operated from a central pit control and observation station.

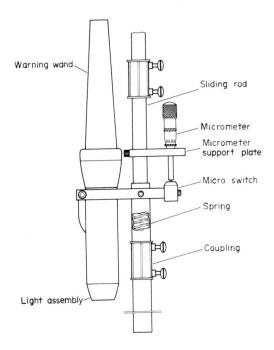

Warning wand

Sliding rod

Micrometer

Micrometer support plate

Micro switch

Spring

Coupling

Light assembly

Fig. 12.5. Reed's roof stability alarm (Soiltest)

Tape Extensometers

Precision steel tapes make convenient general-purpose extensometers for lengths from 3 m to 60 m. When combined with a fine-reading attachment, such as a micrometer or dial gauge, an accuracy of measurement to ± 0.125 mm can be achieved. The extensometer tape is hooked on to anchor points, usually composed of rock-bolt expansion shells set into drillholes in the rock. If the extensometer is anchored at both ends of the measurement base, it must include a spring-loaded connection, the tension of which is adjusted to a standard value before taking the reading. Alternatively the tape may be anchored at one end of the line only, and the other end passed over a pulley, to be loaded by deadweight, at the reference station.

Borehole Extensometers

Wire extensometers are frequently used, anchored at depth in boreholes drilled into the rock walls. Multiple-wire extensometers, embodying up to eight different wires, anchored at different points along the borehole, enable differential strata movements to be detected and measured along the length of the hole. The total depth so observed ranges up to 200 m, in holes 50 to 76 mm diameter. Each wire element must include its own tension member, which, in a simple instrument may consist of deadweights suspended vertically outside the mouth of the borehole, at the reference station. In the more sophisticated arrangements the wires may embody spring standard tension, each wire being connected to a sensing rod. The position of each sensing rod is then precisely measured at a sensing head placed at the mouth of the hole. Measurement may

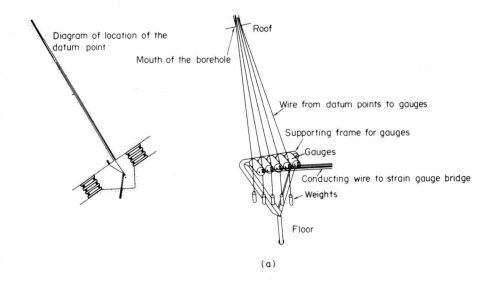

Diagram of location of the
datum point

Mouth of the borehole

Roof

Wire from datum points to gauges

Supporting frame for gauges

Gauges

Conducting wire to strain gauge bridge

Weights

Floor

(a)

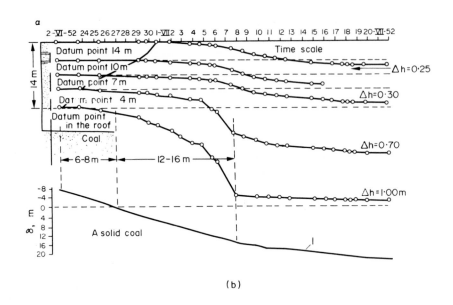

(b)

Fig. 12.6 (a). Multiple-wire borehole extensometer (Avershin), (b) Bed separation measurements in the roof of a coal mine (Avershin)

be manual, using dial gauges and micrometers, or electrical, using either resistance or inductive strain gauges, sometimes with remote monitoring. Figure 12.7 shows an eight-point wire extensometer, in which the sensing head sits in a 152 mm diameter socket at the mouth of the borehole. Measurement is effected by a removable dial indicator, which is screwed on to the

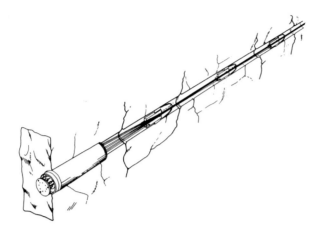

Fig. 12.7. Multiple-wire borehole extensometer – Type LT. (Terrametrics).

reference plate when required. The gauge reads the travel of each sensing rod relative to the reference plate, over a total range of 90 mm at one setting. A similar instrument, modified to provide a higher accuracy of reading (to ± 0.025 mm on a borehole length up to 60 m), is shown in Fig. 12.8. Each sensor movement is detected by a stainless-steel cantilever and electrical strain transducer, coupled to an indicating bridge box for either local or remote read-out.

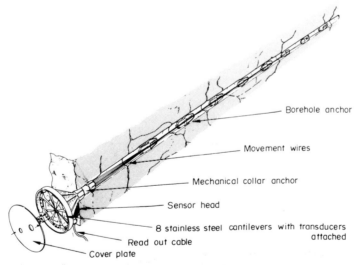

Fig. 12.8. Multiple-wire borehole extensometer – Type F. (Terrametrics).

The data so obtained is reduced manually or by computer, to yield: (1) The displacement of each fixed anchor relative to the sensor head. (2) The displacement of each anchor relative to its adjacent anchors. (3) The time rate of displacement of each anchor relative to its adjacent anchors, per unit distance between anchors, that is, the strain rate, and (4) the time rate of change of the strain rate. These strain gradients are usually plotted graphically, so that the observer may deduce the location of strata zones that are actively being compressed, or perphaps undergoing tensional displacement. Active fracture zones and planes of movement may be recognized and located.

Typical applications of these instruments are shown in Fig. 12.9. Figure 12.9(a) is a rock slope, in which the relative axial extension of the various stratification layers is being monitored. Figure 12.9(b) shows an investigation to measure the strain gradients in the strata around the tunnel, in addition to changes in dimensions of the tunnel and deformation of the tunnel lining.

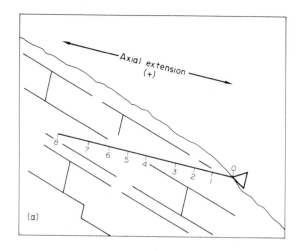

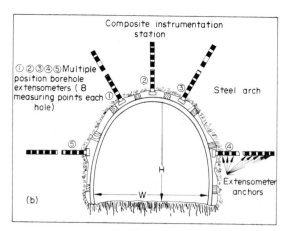

Fig. 12.9. Applications of multiple extensometers. (a) In a rock slope. (Terrametrics). (b) Measuring strata movements around an excavation in rock. (Terrametrics).

Any tendency to creep in the rock walls will be reflected in differential movement between two wires anchored at the same depth but subjected to different values of tension. In rocks subject to appreciable creep it is important that this should be measured, so that the appropriate corrections may be made for anchor creep to the extensometer readings. Multiple-point extensometer readings are also liable to be impeded in fracture zones, or where appreciable non-axial movement occurs relative to different anchor locations. The observer must be constantly alert to the possibility of fouling between the several wires in the hole, or between the wires and the hole perimeter. It is important that no localized high frictional contacts should develop, unbeknown to the observer. This means that he should periodically check each separate wire, by incremental loading whenever possible, to determine its load/extension characteristic, which should remain unchanged during the course of the investigation.

Water-level Apparatus

Potts describes a water-level apparatus for measuring the relative vertical displacement between two rock surfaces. This equipment is of interest in that it was a principle used in the ancient world when surveying and levelling. A reference level is established in two containers connected by flexible tubing. Any movement of one wall relative to the other separates the measurement pointer from this reference surface and adjustment of the pointer back again to the reference surface serves to measure the amount of displacement on a scale and vernier.

Borehole Sonde Techniques

Idel Sonde

The Idel Sonde (Terrametrics) is a probe device to measure vertical and horizontal displacements. It is used for the observation of settlement and deformation in earth embankments, and rockfill dams. The equipment consists of three major components; the probe, a number of metal plates, and the transmitter/recorder. The plates are normally 400 mm diameter, with a 90 mm central hole. They are placed in the earth or rock fill, at regular intervals, as the embankment is constructed, and a 1 m length of plastic pipe 75 mm diameter is fitted through the central hole of each plate, leaving the connection with sufficient flexibility to allow the plate to move, relative to the pipe, as settlement and deformation occur. These pipes are joined together to form a continuous tube, running through the embankment.

The amount of settlement and displacement is observed by measuring the movement of the metal plates. This is observed by lowering the probe in vertical holes, or drawing the probe along horizontal and inclined holes, using a metered cable and line. The signal, transmitted through the probe, reflected from the plates, and recorded at the surface, locates the plates to within 0.8 mm (see Fig. 12.10).

Slope Meter

Another probe for observing the stability and movement of earth embankments, tailings dams, earth and rock slopes, and open-pit walls is the C-350 Slope Meter (Soiltest) (Fig. 12.11). This measures angular deviations from the vertical, from which any lateral displacement may be deduced. Basically it consists of a stainless-steel probe which embodies a suspended pendulum weight. Any movement of the weight is detected by a strain gauge sensor. The probe is lowered into a 44 mm square steel casing inserted in a borehole. The probe fits across one diagonal of the casing in which it rides on two wheels, bridging the angular corners. The depth of the probe

Fig. 12.10. Idel Sonde. (Terrametrics).

is observed, together with resistance values on the strain gauge indicator, on a digital read-out. Numerical readings, calibrated in units of slope, are obtained. These are checked by a second run of the probe, aligned across the opposite diagonal. Changes in slope are plotted graphically against distance along the casing, to give records of any progressive earth movements.

Fracture Plane Location

Detection and measurement of the location and magnitude of shear movements at depth in a rock mass may be affected by installing shear strips, grouted into boreholes drilled into the rock. The shear strips are essentially printed-circuit resistors, made in 76 cm sections, which are clipped together to the desired overall length. The strips are coupled by cable to a monitoring read-out system to identify and locate any shear movement.

Choice of Strain Transducer for Underground Measurements

The choice of instrumentation system for a geotechnical investigation, either surface or underground, will be greatly influenced by the physical conditions under which the investigation

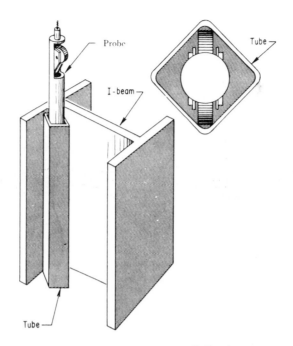

Fig. 12.11. C-350 slope meter (Soiltest).

must be made. Sometimes it is possible to provide a suitable environment in which delicate instruments may be installed and protected, so that refined measurements may be made. When the site is an active engineering operation, either in civil or mining engineering, the investigation often has to be conducted in such a way as to present a minimum of interference with the operations concerned. As a general rule it must be expected that the operations will be noisy, dusty, or wet, possibly hot and humid or sub-zero freezing, accompanied by mechanical vibrations and the proximity of heavy moving machinery. Some interference with the engineering operations is inevitable, particularly when the instruments are being installed, for they must be carefully housed and protected so that they will continue to function usefully and remain accessible for periodic examination. During the periods of installation and subsequent examination the safety of the investigating personnel must be ensured.

In these circumstances it is essential that the instrumentation system be robust and serviceable. It is seldom the case that equipment, designed and intended for laboratory use, will also be suitable for a field investigation, without modification.

For many purposes there will be a choice of basic operating principle, electrical, mechanical, or photoelastic. Electrical and electronic systems have the advantages of versatility and flexibility, and are easily adapted for continuous recording, monitoring, and remote control. They have the disadvantage of being highly susceptible to the adverse environment. The instruments themselves are often delicate in construction. Perfect insulation cannot always be guaranteed. The investigator may take the display of ammeters, voltmeters, and digital read-outs at their face value and they may not be measuring the desired parameter in accordance with laboratory calibration conditions.

255

Mechanical (including hydraulic) systems have the advantages of robust construction and simplicity of design. If continuous recording and remote monitoring are not required they may be preferred to electrical systems. Photoelastic field instruments are, as yet, relatively new and untried, but they have the robustness of mechanical systems, with the capability for wide adaptation to a variety of field problems. Direct access for viewing the transducers is usually a necessary requirement, but to a limited extent remote viewing periscopes, telescopes, and optical monitoring accessories may sometimes be feasible. On the score of accuracy and ease of application in strain measurement, for example, Schmidt and Frenk report that for large deformations, in excess of 200 microstrains, the three systems are, for all practical purposes, equivalent. For strains in excess of 500 microstrains the range of error for all the three procedures, on comparable tests, was found to be below 10%. While the mechanical and photoelastic systems showed marginal errors below 10% down to 100 microstrains (mechanical) and 180 microstrains (photoelastic), the electrical devices exceeded 20% error at 250 microstrains (Fig. 12.12).

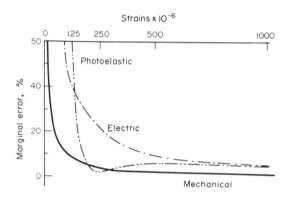

Fig. 12.12. Percentage variation in readings of electrical, mechanical and photoelastic strain indicators, used underground, (Schmidt and Frenk).

Interpretation of Measurements

The essence of a geotechnical field investigation lies in the interpretation of the measurements made, and in their application towards the satisfactory conduct of the engineering operations concerned. Sometimes the issues may be capable of detailed definition and examination, or they may have been explored to an extent that permits codes of practice to be formulated. These may include acceptable factors of safety in approved design procedures, as in many aspects of soil mechanics. At other times the issues are less clear, and this is so in many aspects of rock mechanics. Here the objectives to be gained by measurement of load and deformation, and the deduction of stresses, are sometimes described in somewhat general terms. For example, Merrill and Johnson, reporting on an investigation involving measurement of strain and displacement caused by undercutting in block caving, stated:

" . . . the stresses in the rock and concrete have been determined both before and after the undercut is mined. Also the strain measurements have been reduced to principal strains and some evaluations are in progress to relate the change in radial and tangential strains to the applied stresses. The objective of the study is to develop qualitative and quantitative design concepts and formulas for the design and control of openings used for block caving."

Sometimes the results of an investigation can permit definite conclusions to be drawn. For example, the work reported by Ortlepp and Cook, in which extensive measurements of underground strata movements were compared with theoretical elastic displacements in two South African gold mines, with similar geology although geographically widely separated, showed a degree of agreement which provides a rational basis for the structural design of excavations in those mines according to established principles of elasticity. Again, King, describing the motivation for surface subsidence measurements, stated that if accurate measurements were made at the surface they should relate to the underground mining process:

> "If you can relate strain and subsidence and can predict subsidence, then you can also predict strain. Knowledge of this strain will enable buildings to be designed to resist this strain."

The objective of providing structural design data for engineering in rock is more likely to be attained when dealing with strong massive elastic and pseudo-elastic materials, than with non-elastic and discontinuous earth and rock masses. In the general case the engineer has to attempt an assessment of measurements, often incomplete in themselves, sometimes too few in number to permit of rational statistical treatment, and influenced by factors, some unknown, and others determined by the environment. Nevertheless useful results may ensue if the investigation is conducted in such a manner as to bring out information that is of critical importance to the project concerned. When measuring the deformability of a rock mass, for example, there are various methods from which to choose — *in situ* compression tests, plate-bearing tests, radial-jack and pressure-chamber tests, and borehole jacks. The tests vary in scale and character and it is most unlikely that all would produce the same result at any one site. The engineer must decide which of the test methods is likely to produce information that is most relevant to his structural design problem. In a discussion of these matters Heuze points out:

> "The manner in which the test is conducted, and the point of view from which the data are analyzed are just as critical as the test size. Expressedly, one must distinguish between behavior under loading or unloading, under fast or slow loads, and under static or cyclical effects. The response of a dam abutment, a lock structure, a rotating tower foundation, or a wave protection structure, will be predicted from tests designed or analyzed in different manners."

As an example of the interpretation of deformability measurements in a discontinuous rock mass, Heuze quotes the results of a mathematical model analysis by Duncan and Goodman, from which it is deduced

$$E = \frac{1}{1/E_r + 1/K_n S}$$

where　E　is the overall deformability modulus of the rock mass,
　　　　E_r　is the elastic modulus of the rock material,
　　　　S　is the spacing between discontinuities,
　　　　K　is the joint stiffness.

E_r can be determined by laboratory tests, the other factors are observed in the field. This relationship is shown in graphical form in Fig. 12.13 correlated with Deeres RQD (Rock Quality Designation). The reader should compare this figure with Fig. 11.8 as an illustration of

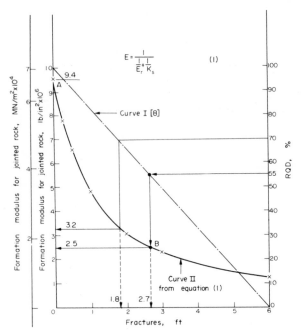

Fig. 12.13. Deformation modulus of a tri-orthogonally jointed rock mass as related to fracture spacing, joint stiffness, and RQD (Heuze).

the application of the technique by Deere *et al* to the design of tunnel linings and excavation support systems.

Measurements of mass deformation, material strain, and deductions of stress, in geotechnology, are often obscured by residual effects. For example, Fig. 12.3 shows a roof/floor convergence curve observed at a point initially in advance of a moving excavation wall. It is one such as would be observed at position A (Fig. 12.14). Excavation of the narrow access road to the measurement site must inevitably produce some deformation of the strata before the convergence recorder could be set. The recorder thus measures only part, residual in the sense of "remnant", of the total convergence. The difference between residual and total convergence deformation at position A, Fig. 12.14 would probably be relatively small, and might, if the rock is strong, be insignificant, but this would almost certainly not be the case if the recorder were installed at position B, behind the advancing excavation wall. In such a situation, no matter how close the instrument station was placed, near the wall, or how soon it was installed after the excavation were made, some appreciable deformation would have occurred that could not be measured. Heuze and Goodman report that, in an investigation to measure strata convergence in a limestone mine, the measurement record (Fig. 12.15), was observed. This shows the convergence measured during a period when an excavation face, 9m wide, was advanced 60m beyond the measurement station, then halted, and then widened in strips of 3m width to a total span of 21m. On the basis of a projected mathematical finite-element analysis to deduce absolute movement from the residual readings, Heuze and Goodman estimate that from 40% to 50% of the deformation was not recorded (Fig. 12.16).

Residual stresses in the sense of "locked-in" stresses are also important. When observing the

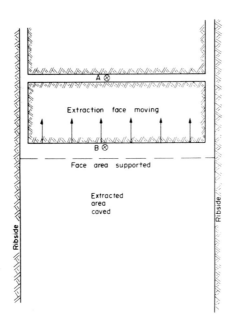

Fig. 12.14. Convergence measurement stations in advance of and behind a moving excavation wall.

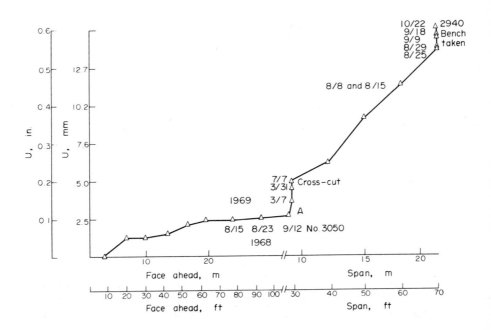

Fig. 12.15. Data from closure tape station (roof to floor) 8m — high room in marble (Heuze).

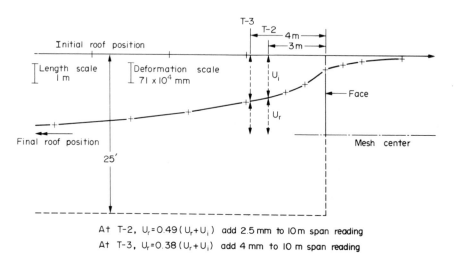

At T-2, $U_r = 0.49(U_r + U_i)$ add 2.5 mm to 10 m span reading
At T-3, $U_r = 0.38(U_r + U_i)$ add 4 mm to 10 m span reading

Fig. 12.16. Initial (U_i) and residual (U_r) room closure. Interpretation of data from tape stations nos. 2 and 3, using finite element analysis (Heuze).

response of structural support members to load it is often forgotten that the material may already be in a condition of stress, as a result of the processes involved in its manufacture, and these stresses may be of considerable magnitude, even though the support member is carrying no external load. Sometimes the residual stresses may be measured, for example the effect of residual stresses in metals can be observed by coating the surface with a sheet of photoelastic material through which a hole is drilled into the metal. The residual stress may then be deduced by observing the birefringence which is displayed in the coating owing to elastic stress-redistribution caused by the hole in the metal.

Again, in multi-component earth materials, such as rocks and concrete, considerable residual strain and inherent stress may be contained, as a result of dehydration and shrinkage within the material, and the different responses of the material mineral components, cement, aggregate and matrix, to any change in the ambient environment.

The matter can be explored in detail by constructing a square slab of concrete 25 to 40 mm thick, in which each piece of aggregate consists of a shaped element of plate glass, the aggregate being held in a matrix of cement.

If this slab is placed under load on a compression testing machine each piece of aggregate becomes, in effect, a stressmeter, and on being observed by transmitted polarized light isochromatic fringe patterns are seen which display the pattern of stress distribution within the slab.

The manner in which load is taken by, and transmitted through, the aggregate is clearly visible, in such an experiment. If the same slab had been coated with photoelastic plastic, and then subjected to load, a strain pattern of isochromatics would be visible under reflected polarized light. At low loads comparatively high strains would occur over the cement matrix, while the coating over the aggregate would display comparatively low strains (Fig. 12.17(a) and (b)).

We would thus have a material under external load and a display showing that this load is taken mainly by the aggregate which takes the stress, while the cement matrix experiences the major component of deformation. Clearly, the deduction of stresses from measured linear strains, on such a material, can lead to wide errors. The results obtained from such a procedure

Fig. 12.17.(a) Birefringence in a photoelastic coating cemented over a concrete slab under load.

can only be fortuitous if electrical or mechanical strain gauges or extensometers are used on the external surface of such a material, depending upon the position of the linear measurement in relation to the spatial distribution of matrix and aggregate in the material.

If the strain gauge is of the vibrating-wire type, inserted into the concrete, another complication arises. The wires in such gauges are commonly stretched between steel anchor points which, being of high modulus relative to the concrete, must form points of high stress concentration around which high stress gradients will exist. The strains thereby measured are likely to be very different from those which an average stress in the area of influence of the gauge would produce.

In the case of the photoelastic stressplug, used in rocks and concrete, provided the plug has a diameter not less than that of the aggregate, or hard mineral rock constituents in its vicinity, the birefringence in the glass is calibrated in terms of the average stress within the material, in the area of influence of the plug (5 diameters) (Fig. 12.17(b)).

Stresses are defined in terms of forces and areas and, considering non-homogeneous materials, it is possible to think in terms of point stresses and average stresses, the average stresses being defined in terms of areas determined by the degree of heterogeneity. Basically, engineers are interested in the average stresses which can be equated with the areas over which they act to give a resultant total force. This approach is more or less straightforward when dealing with stable materials, but in the case of concrete and rocks the problem is rendered more difficult because stresses may be generated by internal changes of dimension associated with differential shrinkage and creep.

In many structures the stresses generated by internal changes are important, for example in

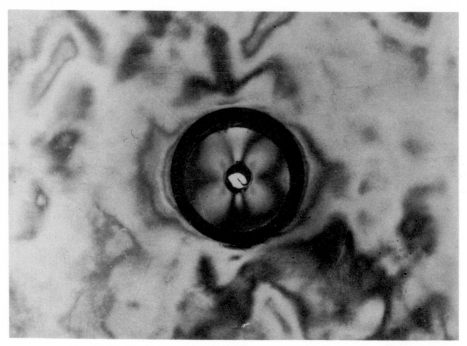

Fig. 12.17.(b) Stressmeter set in center of slab measures average stress in concrete.

causing shrinkage cracks, and they must be taken into account when assessing the safety and stability of the structure. Stresses in concrete and rocks may be classified as follows:

(a) Stresses produced by externally applied forces. The average stress level, as earlier defined, can in this case be equated to the applied loads.

(b) Stresses produced by differential changes in volume, caused either by humidity or thermal gradients. These stresses affect the average stress level locally, but do not produce an external resultant force.

(c) Local macroscopic stresses around individual pieces of aggregate, produced by shrinkage of the cementing matrix around the aggregate. If these stresses are random they do not produce a resultant force to influence the average stress level. The material may then be considered to be isotropic. But if the material is not isotropic the internal local stresses will generate directional internal stresses. The material will also demonstrate "preferred direction" of strain and strain-relaxation in the material, when an external load or a confining pressure is removed. This has obvious significance in connection with stress determination by the "stress-relief" technique.

The effects of humidity, thermal gradients, and shrinkage can be very different in mass concrete from those demonstrated by reinforced concrete. For example, in a mass concrete column the shrinkage stresses build up to a maximum in compression after the concrete is poured and then diminish again to zero as the concrete dries out and cures, but in a reinforced-concrete column the stresses in the interior are tensional and increase exponentially to approach a maximum residual value.

The explanation for this is that in a mass concrete column loss of moisture occurs progressively from the surface towards the interior, and the surface layers dry out first and shrink, thereby putting tensional longitudinal and hoop stresses round the column. The interior is thus held in compression so long as a moisture gradient exists, with the interior at a higher humidity than the exterior which is in tension. Ultimately, when the concrete has completely cured and dried, the material attains a zero state of stress throughout (see Fig. 12.18(a)).

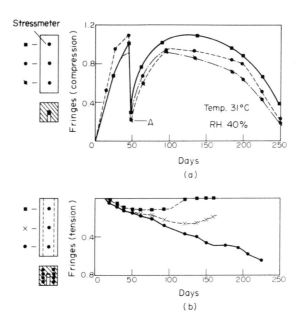

Fig. 12.18(a) Shrinkage stresses in a mass concrete column. (b) Shrinkage stresses in a reinforced concrete column (Dhir and Hawkes).

On the other hand, in a reinforced-concrete column the shrinkage of the concrete is restrained by the reinforcement, so long as the concrete remains bonded to the reinforcement. The concrete is thus placed in a tensional stress state and this tension increases in magnitude as the concrete dries out. It thus remains as a residual stress in the column provided that the concrete maintains its bond with the reinforcement and does not creep or crack (see Fig. 12.18(b)).

Such a reinforced-concrete column in which stressmeters are inserted will display a stress-strain characteristic in which the effects of residual shrinkage stress counteract the effects of compression due to an externally applied load. Only after the tensional shrinkage component has been counterbalanced will the concrete go into compression, and the stressmeter shows this. During the whole of this time, however, the deformation of the column under the externally applied load, as measured by strain gauges or extensometers applied to the external surface of the column, would give a strain-time characteristic like that shown in Fig. 12.19(a).

The conventional strain-stress elasticity conversion from superficial strain measurements thus may indicate an apparently compressional stress when, in fact, the concrete is in a state of tension.

A common example of a situation where incorrect deductions may be made from what appears to be obvious evidence arises in connection with the measurement of support loads or pillar loads. When the support of an underground excavation is being shared between a number

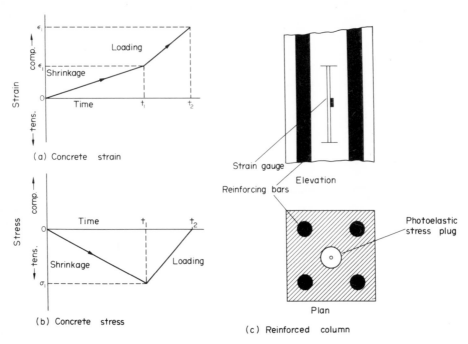

Fig. 12.19. Stress and strain in a reinforced concrete column (Dhir and Hawkes).

of similar columns it is commonly assumed that the total load will be shared equally between them. In that event the measurement of progressive deformation on one of the columns will enable the load taken by that column to be estimated assuming that the material is behaving plastically and its stress/strain/time characteristics are known. Alternatively, if the support material is elastic then the measured deformation might be taken to indicate that the support is being subjected to an increasing load, and subjected to stress above its yield strength but below its ultimate or "peak" strength.

But the evidence might also mean that the instrumented support, due to some inherent weakness or structural failure within it, is not taking an increasing load or its share of the total load, but is, in fact, deforming under a decreasing applied load, so far as itself is concerned, because it is shedding its load on to the neighboring supports. Therefore a correct interpretation of the evidence can only be made if the neighboring supports are also instrumented and observed. The deduction would be assisted if the supports were rock pillars, by the fact that the deformation moduli for unloading and loading would be different, and therefore the deformations of the pillars would be correspondingly different, relative to one another.

Somewhat similar considerations must also be applied to observations of yield on hydraulic supports that are designed to yield at a specific load, which is determined by placing an appropriate relief valve in the hydraulic-jack circuit. Where, as in many contemporary designs of support, several separate jacks are required to operate together it is important that all should function in exactly the same way. Otherwise they exert unequal and unbalanced resistance to the strata that they are required to control. If the relief valve on any one jack should malfunction and not open then that leg will be overloaded. On the other hand, if the relief valve opens too easily or does not close properly, then the leg will shed load onto its neighbors and they will take more than their designed load. It is not possible to tell, by observation of deformation alone, which, if any, of these malfunctions exists. To all outward appearances the supports may appear to function perfectly. Only the indications of pressure gauges in each separate hydraulic circuit can show what is really happening.

In the same way, when an investigation is in progress to determine the mean support-load density required to give effective roof support and control of an excavation in rock, it is not sufficient to instrument only a representative sample of separate supports, distributed over the area concerned. Each measurement making up the sample should include a family of supports, distributed radially around a central support unit. Only then will it be possible to guarantee that the load measured on the central unit of each family is truly representative.

Selected References for Further Reading

AVERSHIN, S. G. Some new methods and results of experimental investigations into manifestations of rock pressure around the face area during the extraction of coal seams. *Int. Conference on Strata Control, Paris,* 1960.

BENIOFF, H. A linear strain seismograph. *Bull. Seism. Soc. Am.,* vol. 25, pp. 283-309 (1935).

BURGER, J. and LORBERG, R. H. Earth strain measurements with a laser interferometer. *Science,* vol. 170 (3955), p. 296 (1970).

CHRZANOWSKI, A. New techniques in mine orientation surveys. *Proc. 1st Canadian Symposium on Mine Surveying and Rock Deformation Measurements,* Univ. New Brunswick, 1966, pp. 23-46.

DHIR, R. K. and HAWKES, I. The measurement of stress change in concrete using the biaxial photoelastic stressmeter. *Conference on Experimental Methods of Investigating Stress & Strain in Structures,* Prague, 1965.

DORRER, E. Applications of photogrammetry in mining surveying. *Proc. 1st Canadian Symposium on Mine Surveying and Rock Deformation Measurements,* Univ. New Brunswick, 1966, pp. 76-85.

DUNCAN, J. M. and GOODMAN, R. E. *Finite Element Analysis of Slopes in Jointed Rock,* U. S. Army Waterways Exptl. Station, Vicksburg, Miss., Rept. No. S-68-3 (1968).

HEUZE, F. E. Sources of error in rock mechanics field measurements, and related solutions. *Int. J. Rock Mech. Min. Sci.,* vol. 8, pp. 297-310 (1970).

HODGES, D. J. Electro-optical distance measuring instruments. *Proc. 1st Canadian Symposium on Mine Surveying and Rock Deformation Measurements,* Univ. New Brunswick, 1966, pp. 47-65.

HOFFMAN, H. Results of measurements carried out in a seam subject to rockbursts (in German), *Gtuckauf.,* vol. 87, p. 101, Essen (1957).

KING, H. J. and SMITH, H. G. Surface movements due to mining. *Leeds Univ. Mining Soc. J.,* pp. 127-142 (1955).

MAJOR, M. W., SUTTEN, G., OLIVER, J. and METSGER, R. On elastic strain of periods 5 secs. to 100 hrs. *Bull. Seism. Soc. Am.,* vol. 54 (1), pp. 295-346 (1964).

MALONE, S. Earth strain measurements in Nevada and possible effects on seismicity due to the solid earth tides. Ph.D. Thesis, Univ. of Nevada, 1972.

MERRILL, R. H. and JOHNSON, G. H. Changes in strain and displacement caused by block caving. *4th Int. Conf. on Strata Control and Rock Mechs.,* Columbia Univ., New York, 1964.

NISHIDA, M. and TAKABAYASHI, H. Thickness effects in "Hole Method" and applications of the method to residual stress measurement. *Scientific Papers of Inst. of Physical & Chemical Research,* vol. 59, no. 2, pp. 78-86, Tokyo (1965).

ORTLEPP, W. D. and COOK, N. G. W. The measurement and analysis of the deformations around deep hard-rock excavations. *4th Int. Conf. on Strata Control and Rock Mechs.,* Columbia Univ., New York, 1964.

PINCUS, H. J. Optical data processing of vectorial rock fabric data. *Proc. 1st Congress Int. Soc. for Rock Mechs.,* vol. 1, pp. 173-177, Lisbon, 1966.

PINCUS, H. J. Sensitivity of optical data processing to changes in fabric. *Int. J. Rock Mech. Min. Sci.,* vol. 6, pp. 259-276 (1969).

POTTS, E. L. J. Underground instrumentation. *Q. Jl. Colo. Sch. Mines,* vol. 52, no. 3, pp. 135-182 (1957).

POTTS, E. L. J. *Rockburst Research Unit, Kolar Gold Mining Undertaking,* Report No. 1, Govt. of Mysore, 1958.

SCHMIDT, C. M. and FRENK, B. W. Accuracy of various techniques used to measure strain in sub-surface operations. *Int. J. Rock Mech. Min. Sci.,* vol. 9, pp. 1-6 (1972).

YOUNG, C. Lasers – a review. *Proc. 1st Canadian Symposium on Mine Surveying and Rock Deformation Measurements,* Univ. New Brunswick, 1966, pp. 193-209.

CHAPTER 13

The Response of Earth Materials to Dynamic Loads

Dynamic Phenomena in Geotechnology

Many problems in geotechnology involve dynamic factors. These include seismic phenomena, such as earthquakes and rockbursts, shock loading on rock materials by mechanical impact and blasting, the study of energy transmission in rock-breaking tools and earth-excavation machinery, and the control of vibrations and noise generated by mining and civil engineering operations. Interest in the dynamic properties of earth materials also arises in rock testing and in some model studies, when gravitational body forces are imposed by rotating the specimen in a centrifuge — the method of barodynamic testing.

Our major concern, however, usually lies with the response of rocks and earth materials to dynamic loads, either generated by the forces of nature, as in an earthquake or rockburst, or in the course of engineering operations such as blasting, pile-driving, and rock-cutting. In all these processes, energy is generated and released at the source of impact and thereafter passes into the surrounding material in the form of pulses or vibrations. Vibrations of different modes of origin have fundamental similarities, but differ in their amplitude, wavelength, and frequency. For example, a minor earthquake may yield more than 10 times the energy released from a 15-megaton nuclear explosion. But the nuclear blast generates energy several orders of magnitude more than a normal quarry blast, which, in its turn, is of a higher order of energy than that generated by mechanical impact. Similarly, the seismic vibrations at some distance from an earthquake epicenter may have a wavelength of several miles, whereas those from a quarry blast will be measured in terms of feet or so, while the energy pulses transmitted along a mechanical component, such as a drill rod, have a much higher frequency and a correspondingly shorter wavelength, measurable in fractions of an inch.

Shock Waves

If the energy source at the point of impact is sufficiently high, the impacting pulse may destroy the coherency of the surrounding earth material, so that it no longer behaves as a solid. Some of the energy is consumed in this process and the remainder passes into the surrounding material as a compressive shock wave, which travels at a velocity depending upon the pressure in advance of, and behind, the shock front, but greater than sonic velocity in the material. This shock wave front is stable, so long as the stress generated by the shock wave exceeds a critical value, termed the Hugoniot elastic limit. If the shock stresses decline to the uniaxial yield strength of the material, referred to the bulk and shear strength moduli, the shock front becomes unstable, and it breaks up into a two-front wave, the first front travelling at elastic velocity. Under certain conditions the second front may catch up with the first, elastic, front, to re-establish a stable shock wave, but if this does not happen, then, as the

266

disturbance propagates further from its origin, it passes into a transitory plastic wave and then settles into a stable elastic wave. In brittle materials such as quartz and quartzitic rocks the plastic wave is relatively unimportant.

The Rankine-Hugoniot equation, derived from considerations of the conservation of mass, momentum, and energy in the material, theoretically describes the conditions whereby shock waves are transmitted through the material.

$$\text{Mass} \frac{v_1}{v_0} = 1 - \frac{u_1 - u_0}{U_1 - u_0},$$

$$\text{Momentum } \sigma_1 - \sigma_0 = \rho_0 (U_1 - u_0)(u_1 - u_0),$$

$$\text{Energy } En_1 - En_0 = \frac{\sigma_1 - \sigma_0}{2} (v_0 - v_1)$$

where v is the specific volume (reciprocal of density), i.e. $v = 1/\rho$,
     u is the mass velocity or particle velocity,
     U is the shock velocity,
     σ is the stress normal to the front,
     En is the specific internal energy.

The subscripts 0 and 1 refer to the conditions ahead of the shock-wave front and immediately behind the front, respectively.
The shock velocity

$$U = v_0 \sqrt{\frac{\sigma_1 - \sigma_0}{v_0 - v_1}}.$$

Elastic Waves

Two basic types of elastic waves are generated by an explosive blast or seismic shock — *body waves* which travel through the interior of the rock and *surface waves* which can only travel along the surface of a discontinuity in the material.

Body Waves

There are two modes of body wave — compression or primary (P) waves and shear or secondary (S) waves. P-waves induce longitudinal oscillatory particle motions and when they impinge on a free boundary or discontinuity which crosses their path in any direction other than that at right angles, displacements occur which give rise to S-waves. In S-waves the particles oscillate in a transverse direction, without compressing the material. P-waves travel in any material which resists compression, but S-waves can only be generated in solid materials that can offer resistance to change of shape.

The passage of body waves through a material is dependent upon the density and elastic properties of the material, as defined by the characteristic equations:

$$V_p = \left[\frac{K + 4/3G}{\rho}\right]^{\frac{1}{2}}$$

and

$$V_s = \left[\frac{G}{\rho}\right]^{\frac{1}{2}}$$

where *K* is the bulk modulus, that is, the ratio between the applied pressure and the fractional change in volume when the material is subjected to a uniform hydrostatic compression,

    G is the shear or rigidity modulus, that is, the ratio between shear stress and shear strain,

    ρ is the mass density of the material,

    V_p is the longitudinal wave velocity; it is sometimes termed the dilatational wave velocity, since the passage of longitudinal waves involves changes in the volume of the material through which they pass,

    V_s is the shear wave, or secondary wave, velocity. This wave is sometimes termed the distortion wave because, although it does not alter the volume of the material through which it passes, it does cause the material to rotate or shear in a direction perpendicular to the wave motion.

When the primary and secondary waves from a deepseated seismic disturbance reach the upper region of the Earth's crust and mantle they impinge upon discontinuities along which the surface waves are then propagated. These surface waves are of a complex character, but, since their wavelength is generally appreciably greater than those of the primary and secondary waves, they are collectively termed the L or long wave. The separate arrival of the P-, S-, and L-waves emanating from a seismic disturbance can be identified on seismograph records, and since the P- and S- waves travel at characteristic velocities, which theoretically are determined only by the density and elastic properties of the materials through which they pass, observation of their separate times of arrival, relative to the time at which the initial shock occurred, provides a means of exploring the physical properties of these materials indirectly. Also, seismograph records, observed at two or more separate localities on the Earth's surface, provide a means of determining the relative magnitude of earthquakes, and also serve to locate their origin, focus, and epicenter.

Surface waves

In the case of shallow disturbances the greater part of the energy is carried by surface waves, of which four modes are known, termed, in decreasing order of velocity, Coupled (C), Hydrodynamic (H), Love (Q), and Rayleigh (R) waves. Each is defined by a characteristic particle motion, Q- and R-waves being the best known and most easily detected.

In Rayleigh waves the particle describes an ellipse described by

$$u = 0.42\, a \sin \eta, \qquad w = 0.62\, a \cos \eta,$$

where *u* is the displacement in the direction of transmission,

    w is the displacement in the vertical direction,

    η is a parameter which decreases with increase in time.

The ellipse is retrograde and the maximum displacement in the direction of transmission is about two-thirds of that in the vertical direction. The Rayleigh wave is plane, there being no lateral displacement at right angles to the direction of transmission. The velocity of the wave is $0.92 V_s$ for a medium with Poisson's ratio of 0.25 but decreases with increase of Poisson's ratio. Rayleigh motion is calculated to diminish to zero at a depth of 0.192 of a wavelength, reversing sign below this nodal point. Experimental confirmation for small blasts was obtained by Dobrin *et al.*, who found the depth of crossover to be 12.16 m, equivalent to 0.136 of a wavelength.

In Love (Q) waves the particle motion is transverse to the direction of transmission and there is no displacement in a vertical plane. The conditions for the existence of Love waves require the transmitting medium to be bounded above by a reflecting surface, such as a rock-air discontinuity or a medium of lower body shear wave velocity, and bounded below by a medium in which the velocity of the body shear wave (V_S) exceeds that in the transmitting medium. For short wavelengths the Q-wave velocity is equal to V_S for the upper boundary stratum and for longer wavelengths it is equal to V_S for the substratum.

Hydrodynamic (H) waves move particles in elliptical orbits in a progressive sense, as does the path of a surface wave on water. The Coupled (C) wave has combined motion in longitudinal and transverse directions, the particles describing an ellipse, which can be either retrograde or progressive, inclined at some angle less than $90°$ to the direction of transmission.

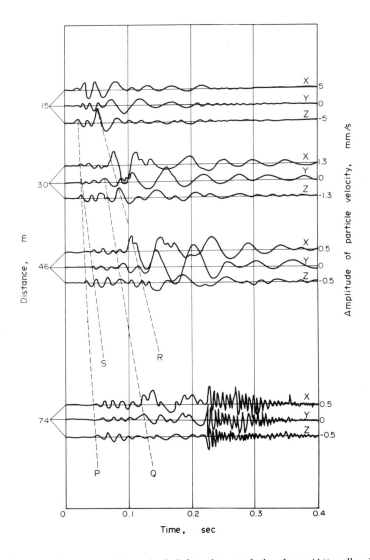

Fig. 13.1. Wave profiles generated in sandy shale by a 4-oz. explosive charge (Attewell and Farmer).

269

H-waves and C-waves are, as yet, imperfectly understood. They are identified best on records of vibrations in unconsolidated surface materials, and less easily in other materials, where it is difficult to identify the separate arrivals of C-, H-, Q-, and R-waves over short distances of travel.

An example of wave profiles obtained at various distances from small explosive charges in sandy shale is given in Fig. 13.1 (Attewell and Farmer), in which the separate arrival of P-, S-, Q-, and R-waves is identified.

Stress Waves in Rocks

The dynamic properties of rock materials may conveniently be studied using rods and bars, for this considerably simplifies the theoretical treatment, and also lends itself readily to experimental investigation.

The arrangement is also directly relevant to problems of analysis and design in percussive rock drills, pile drivers, and other impact devices. The experimental set-up may consist of a vertical drop-hammer impacting on the rock specimen or the firing of small explosive charges impacting upon the end of horizontal cylinders of rock suspended in a ballistic pendulum.

Although the wave configuration at the impacted end of the specimen is complicated by multiple reflections, this effect diminishes with distance, and after a longitudinal distance of approximately three diameters from the source the wave may be considered to be propagated with a plane front.

Wave Velocity and Particle Velocity

An impacting load on the end of the rod or cylinder is transmitted through the material with a finite propagation velocity or wave velocity.

The velocity of movement of the solid particles vibrating in the wave path is termed the "particle velocity".

The wave velocity and the particle velocity may be measured by various techniques. The particle velocity may be obtained as u, where

$$u = \frac{\sigma}{\rho V_p} \quad \text{or} \quad u = \frac{1}{\sigma(\rho E)},$$

but it is often more convenient to measure the "free surface velocity", or the velocity of movement of a free rock face when the stress pulse arrives at, and is reflected by, that face.

Laboratory measurement of velocities.

It can be shown quite easily that the speed at which the rock-air discontinuity jumps forward when the stress pulse reaches it is twice the particle velocity. This movement may be observed and recorded by high-speed cameras, or measured by means of capacitance gauges, inductance transducers or photoelectric and photomultiplier circuits. The latter method is particularly suitable for the laboratory measurement of fast transients. (Fig. 13.2) shows a typical arrangement. The flat end of the cylindrical rock specimen acts as a shutter across a small gap which is machined in the photomultiplier housing. When the stress wave generated by the detonated explosive charge reaches the end of the specimen, light is obscured from the photocell. The associated change in voltage, over the small time interval involved, is amplified and recorded on the oscilloscope. The circuit thus records the free surface velocity, the delay time for the passage of the stress wave along the specimen, and hence the wave velocity.

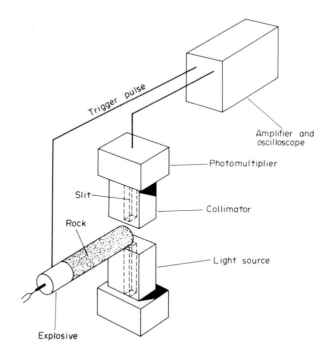

Fig. 13.2. Photomultiplier arrangement for measurement of free surface velocity (Attewell).

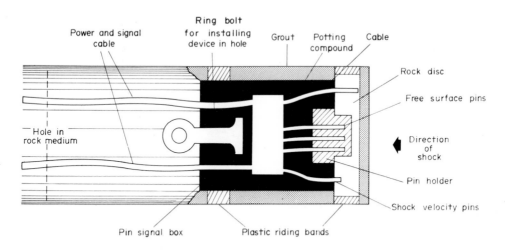

Fig. 13.3. Borehole pin-contactor instrument for measurement of *in situ* wave and particle velocities generated in rock by blasting (Lombard),

Field measurement of velocities

   A borehole device for the field measurement of shock and particle velocity in close proximity to major explosive blasts, is described by Lombard. Fig. 13.3 shows a longitudinal section of the device in position. It is a pin-contactor instrument in which the passage of the surface of measurement between the ends of the pins is timed to provide a record of velocity. The device is comprised of a 127 mm diameter by 25 mm thick disc, machined from a piece of rock cored from the strata in which the field observation is to be made. A circular recess is machined into the center of the disc and a plastic pinholder supports several pins at accurately located distances from the flat bottom of the recess. These pins measure the free surface velocity and hence the particle velocity. Other pins, inserted within the rock, record the passage of the shock wave and hence measure the shock velocity.

Measurement of ground vibrations generated by conventional mining or small-scale explosive charges, or natural earthquake vibrations, may be made by means of a vibration transducer or seismograph. The basic component of such an instrument may be a weight, supported on springs or some flexible support system, so that, when the body of the instrument is moved by an incoming ground wave, the inertia of the suspended or supported weight acts so that the weight tends to remain at rest.

The behavior of the seismograph is determined by the natural frequency of vibration of the inertia member relative to the frequency of the incoming ground vibration. When the two frequencies are identical, then resonance occurs, and this condition is controlled in the seismograph by damping the inertia member either mechanically or electrically. The force necessary to bring the inertia member to rest after one oscillation is known as "critical damping" and an instrument that functions in this way measures ground displacement. However, an instrument critically damped would only permit precise measurement at its characteristic frequency, so that to permit measurements over a range of incoming frequencies, some modification is required. About 0.6 critical damping has been found to provide an instrument in which the resonance factor approximates to unity over a useful range, representing the incoming frequencies likely to be encountered in seismic observations of ground vibrations due to blasting.

If the frequency of the incoming ground vibration is less than the natural frequency of the inertia member in the seismograph then the inertia member acquires a motion of its own, relative to fixed axes in space. This motion is such that, when the natural frequency of vibration of the seismograph is more than twice the frequency of the ground vibrations, then the motion of the seismograph is proportional to the acceleration of the ground. The instrument is then termed an accelerometer. Another form of ground accelerometer may be constructed by using a piezo-electric crystal as the inertia member. The electrical output of the crystal under the motion imposed by an incoming ground wave may then be calibrated in terms of acceleration.

The amount of energy that is transmitted by an oscillating ground wave is closely dependent on the ground wavelength and the amplitude of particle displacement in the wave path. The ground wavelength is a function of wave velocity and frequency, and the particle velocity may be measured directly using an electromagnetic vibration gauge. One such gauge consists of a coil attached to a frame. The coil is placed between the poles of a magnet which is, in turn, suspended from the frame by leaf springs. The magnet acts as the inertia component of a seismograph, while the coil moves under the influence of the incoming ground vibrations. The relative movement between the coil and the magnet produces an electromotive force which is proportional to the particle velocity in the ground wave. Electrical damping is imposed through a suitable resistor in the circuit.

Seismographs may thus be designed to measure ground displacement, acceleration, or particle velocity. The calibration of such instruments may be performed directly, using shaking tables or vibrators, the frequency and motion of which are capable of precise control. The readings of an instrument calibrated in any one of the three modes may be converted to read directly in the other modes by incorporating suitable electronic differentiation or integrating networks into the control circuit.

Using any of these instruments, the wave velocity may be measured by observing the time interval between the instant of impact and the arrival of the first appropriate peak on the seismic trace. The instant of impact is imposed on the observed oscilloscope or oscillograph trace by including a mechanical "trigger" on the impact hammer or breaking a wire at the instant of explosive detonation.

Body wave velocities are usually greater in igneous and crystalline rocks than in sedimentary rocks. In sediments it has been observed that the wave velocity tends to increase with increase in the depth of cover and increase in geological age. It is also common for stratified rocks to display considerable anisotropy in seismic velocity. The wave velocity in a direction parallel to the stratification may be $10 - 15\%$ greater than that normal to the stratification planes. Typical longitudinal wave velocities for rocks are listed by Farmer in Table 13.1.

TABLE 13.1. *Longitudinal Wave Velocities in Rocks (Farmer)*

Rock	V_p m/sec	ρ g/cm^3
Granite	3000–5000	2.65
Basalt	4500–6500	2.85
Dolerite	4500–6500	3.0
Gabbro	4500–6500	3.05
Sandstone	1400–4000	2.55
Shale	1400–3000	2.3
Limestone	2500–6000	2.5
Marble	3500–6000	2.65
Quartzite	5000–6500	2.65
Slate	3500–5500	2.65

Various investigators estimate V_p/V_s in rock materials to be around 1.6.

Attewell and Farmer quote the dynamic properties of two shales as listed in Table 13.2.

TABLE 13.2. *Dynamic Properties of Two Sediments (Attewell and Farmer)*

	Sandy shale	Shale
Longitudinal wave velocity (V_p)	2195 m/sec	1829 m/sec
Shear wave velocity (V_s)	1402 m/sec	1189 m/sec
Surface (Rayleigh) wave velocity(V_r)	762 m/sec	671 m/sec
Velocity ratio V_p/V_s = (r)	1.56	1.54
Poisson's ratio = $\dfrac{r^2/2 - 1}{r - 1}$	0.15	0.13

Determination of Dynamic Elastic Moduli

Two laboratory experimental methods are commonly used to determine the dynamic elastic moduli of a rock material.

Pulse method

A cylindrical or a prismatic specimen is prepared and two barium titanate piezo-electric crystals are placed, one at each end of the rock specimen, with good mechanical contact between them. A pulse generator, connected to the crystal, and to a recorder or an oscilloscope, is used to apply a high voltage, of the order of 1000 volts over 1 to 10 μsec, to the crystal at one end of the specimen. This causes the crystal to generate a compressional or a shear pulse, depending upon which of the crystallographic axes the barium titanate crystal is cut for presentation to the rock specimen. The pulse then travels through the specimen, to be picked up by the crystal at the receiving end, converted back to an electrical signal, and recorded (see Fig. 13.4).

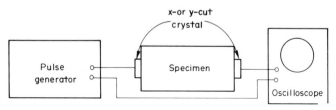

Fig. 13.4. Pulse method of determing dynamic elastic moduli on rock specimens (Obert and Duvall).

The time intervals required for a compression pulse and a shear pulse to pass through the specimen, and the length of the specimen, are measured, to give the longitudinal and shear elastic wave velocities, respectively. Similar measurements over various lengths of specimen, of known density, enable the elastic constants to be calculated from:

$$E = \frac{V_s^2 \rho}{g} \left[\frac{3(V_p/V_s)^2 - 4}{(V_p/V_s)^2 - 1} \right].$$

$$\nu = \frac{\frac{1}{2}(V_p/V_s)^2 - 1}{(V_p/V_s)^2 - 1},$$

$$K = \frac{\rho}{g} V_s^2 \left[\left(\frac{V_p}{V_s} \right)^2 - \frac{4}{3} \right],$$

$$G = \frac{E}{2(1+\nu)}.$$

Resonance method

The USBM method, devised by Obert *et al.*, measures the longitudinal and torsional frequencies of vibration of rock specimens in the form of drill core samples. For such a specimen, the longitudinal elastic wave velocity is related to the fundamental longitudinal frequency of vibration by

$$V_b = 2f_b L$$

where V_b is the longitudinal (bar) velocity,
$\quad\; f_b$ is the fundamental longitudinal frequency,
$\quad\; L$ is the specimen length.

The modulus of elasticity

$$E = V_b^2 \rho/g = 4f_b^2 L^2 \rho/g.$$

When the specimen is set into torsional vibration

$$V_s = 2f_s L$$

where V_S is the shear velocity,

f_S is the fundamental torsional frequency of vibration.

also

$$G = V_S^2 \rho/g = 4f_S L^2 \rho/g$$

while

$$v = \frac{E}{2G} - 1 = \frac{V_b^2}{2V_s^2} - 1.$$

The general arrangement of the USBM apparatus for resonant vibration measurements is shown in Fig. 13.5(a) with details of the torsional and longitudinal vibrators in Fig. 13.5(b).

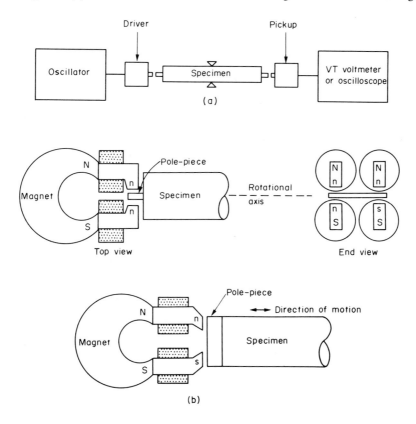

Fig. 13.5.(a and b). Resonance method of determining dynamic elastic moduli on rock specimens (Obert and Duvall).

Wave velocities in soils

The velocity characteristics of wave propagation in soils may be determined by a resonant-column test. This is conducted by supporting a cylindrical sample of the soil in a vertical position on a vibrating drive mechanism, so that the column can be set into longitudinal or torsional vibrations. Dry and saturated samples of non-cohesive soils are contained in thin-walled plastic tubes, while cohesive and partially saturated samples may be unsupported after being compacted by their own weight in a mold. The resonant frequency and the height of soil column are measured, to yield the longitudinal and shear velocities, as in the resonance vibration test on rocks.

Field determination of dynamic elastic moduli

To determine the dynamic moduli from seismic velocity measurements, knowledge of three independent quantities – density, longitudinal wave velocity, and shear velocity – is required. Laboratory observations, by the pulse and resonance techniques, are only representative of relatively small samples of a rock material. The sonic velocities in a rock mass may be expected to vary very considerably from those determined by laboratory test, because they depend on the ambient stress level and other environmental factors, such as porosity and pore fluid content. The rock mass properties should therefore be investigated *in situ*, and this may present problems in that it is sometimes difficult to recognize the arrival of the shear wave unless sophisticated and expensive equipment is available for this purpose. An alternative is to assume a value for Poisson's ratio and measure only the longitudinal wave velocity and the density. Since there may be considerable doubt as to the appropriate value for v this may only yield, at best, an approximation.

TABLE 13.3. *Dynamic Tensile Strength of Rock (Rinehart)*

Rock	Static	Dynamic	Ratio dynamic/static
		kg/cm^2	
Granite	70	390	5.7
Taconite	70	900	13.0
Limestone	40	280	6.5
Marble	60	480	7.8

Carroll suggests that reliable estimates of the dynamic moduli of many rocks can probably be made from a field measurement of compressional velocity, to embrace a specific range of

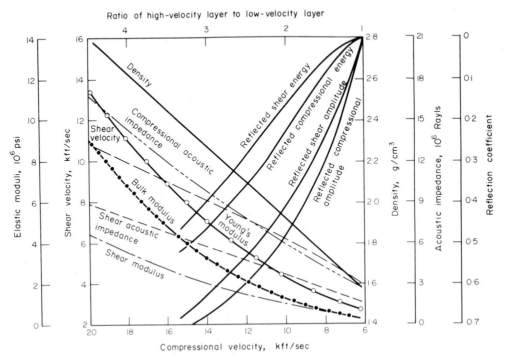

Fig. 13.6. Graph for determining acoustic parameters of rocks
from compressional wave velocities (Carroll).

rock types. Data obtained on a number of volcanic rocks showed approximate linear relationships between longitudinal wave velocity and the dynamic Young's shear, and bulk moduli, shear velocity, shear and compressional characteristic impedance, as well as amplitude and energy reflection coefficients. Carroll's results, subject to certain limitations, are summarized in Fig. 13.6.

In situ seismic velocity measurements may be applied as a means of sub-surface exploration in boreholes drilled to penetrate to considerable depths from the surface. In one system, a logging "sonde" or probe is moved along the length of the borehole to transmit pulses of from 20 to 30 kHz frequency. These pulses are checked by piezo-electric receivers at fixed distances from the transmitter, usually two receivers at 0.3, 0.6 or 1 m apart. The difference in arrival time between the two receivers is recorded directly as the reciprocal of velocity. In full-wave acoustic logging only one receiver is used and two traverses along the borehole are made, each time using a different spacing between receiver and transmitter. Optical recording on photosensitive paper produces records in which successive wave forms are "stacked". From such records the first arrival of the compression wave can be most easily identified, but the separate arrival of the shear wave and the subsequent wave types are less obvious, particularly if the receivers are closely spaced.

The Strength of Rocks Under Dynamic Loading

The strength of rocks is much higher when subjected to dynamic loads than it is under static loading. Comparative values of tensile strength are quoted by Rinehart showing an increase of from about 6.0 to 13 times, the higher ratios being apparent in what are, under static loading, the weaker rock materials.

Ito records comparative static and dynamic rock properties as detailed in Table 13.4.

TABLE 13.4. *Dynamic and Static Rock Properties (Ito)*

Dynamic tests	Marble	Sandstone A	Sandstone B	Granite
Stress rate, kg/cm^2/sec	17×10^6	1.4×10^6	15×10^6	15×10^6
Strain rate, microsec	3.6	3.7	3.3	5.5
Failure stress, kg/cm^2	215	220	190	170
Failure strain, microstrains	490	610	460	630
E. kg/cm^2	51×10^4	64×10^4	40×10^4	30×10^4
Static tests				
Stress rate, kg/cm^2/sec	1.1	1.8	0.5	2.2
Failure stress, kg/cm^2	53	80	29	53
Failure strain, microstrains	145	410	370	510
E. kg/cm^2	47×10^4	19×10^4	10×10^4	12×10^4

Ricketts and Goldsmith, on the basis of a study of dynamic strength properties of a number of rocks, ceramics, plastic mixtures, and aluminum, using the Hopkinson bar technique, quote the results of Table 13.5. The table includes comparative static properties of the materials tested.

Characteristic impedance

The characteristic impedance of a medium through which a stress wave may be transmitted is defined as density × stress wave velocity. The transfer of energy across any junction between

Table 13.5. Summary of Static and Dynamic Properties of the Natural Rocks, Ceramic and Plastic Mixtures and Aluminium (Ricketts and Goldsmith)

Material	Density (g/cm³)	Rod-wave velocity (m/sec)	Specific acoustic impedance (g-sec/dm³)	Compressive Young's modulus (MN/m²)			Static Strength of Specimen (MN/m²)			
				Dynamic	Static (1)	Static (2)	Tensile (1)	Tensile (2)	Compressive (1)	Compressive (2)
Spessartite	2.89	6198	7116	110,320	61,365	19,306-(3); 46,196-(5); 48,265-(12)	17.24	13.79-(3); 4.10-(15); 13.57-(12)	372.3	385.43-(3); 399.20-(5); 399.20-(2)
Basalt	2.72	4928	5318	66,192	44,817	32,061-(3); 24,546-(1)	9.65	8.20-(3)	255.1	272.35-(3); 195.12-(1)
Diorite	2.83	3658	4107	37,922	35,164	22,064-(2)	5.51	3.17-(3)	219.3	145.48-(2)
Leucogranite	2.59	3429	3527	28,959	30,682	28,959-(3)	3.10	2.62-(14)	206.8	186.16-(3)
Limestone	2.23	35.92	3190	28,890	17,927	22,753-(14)	3.58	1.59-(14)	28.1	39.72-(14)
Sandstone	2.11	20.27	1704	8,688	—	7,585-(14)	—	—	—	49.78-(14)
Scoria	1.49	3327	1973	16,548	7,585	4,550-(13)	4.31	2.48-(13)	28.89	19.31-(13)
Pumice	0.428	2530	431	2,758	1,655	1,517-(11)	0.59	0.45-(11)	4.24	3.83-(11)
Diorite concrete	2.30	3391	3102	26,545	24,132	23,443-(1)	1.96	3.52-(1)	30.41	32.20-(1)
Scoria concrete	1.94	2870	2216	15,996	11,721	10,342-(1)	1.14	1.90-(11)	13.93	16.82-(1)
Concrete 7-day	2.17	3645	3140	28,821	19,995	21,374-(1,2)	2.0	—	25.51	28.27-(1,2)
28-day	2.17	3612	3109	28,269	18,616	20,685-(1,2)	2.27	—	32.41	32.41-(1,2)
Lightweight 7-day aggregate	1.34	3132	1654	13,100	10,687	10,756-(1,2)	2.69	1.10-(1,2)	31.99	42.47-(1,2)
concrete 28-day	1.34	3132	1654	13,100	10,687	11,859-(1)	2.69	0.45-(1)	46.89	48.61-(1)
Epoxy aggregate concrete	2.04	2756	2222	16,065	14,065	16,616-(2)	—	1.17-(2)	69.78	71.98-(2)
Prestressed concrete	2.35	3835	3558	34,199	38,956	29,648-(1)	—	—	57.09	51.92-(1)
Epoxy aluminium powder	1.67	2263	1486	8,481	7,033	—	—	—	—	—
Cement paste 7-day	1.97	3416	2672	23,098	21,374	18,616-(1,2)	3.65	—	79.27	72.40-(1,2)
28-day	1.99	3663	2890	26,614	22,064	19,995-(1,2)	2.41	—	87.57	72.40-(1,2)
Epoxy	1.33	1933	1011	4,827	3,516	3,516-(1)	7.17	—	69.98	66.19-(1)
Aluminium 2024T-4	2.72	5182	5587	73,087	73,087	—	227.15	—	—	—

one medium and another is affected, not only by the contact, or coupling, conditions at the junction, but also by any difference in the characteristic impedance on each side of the junction. Leet quotes figures for characteristic impedance of five rock types in Table 13.6.

TABLE 13.6. *Compressional Wave Speeds and Characteristic Impedance for Five Rocks (Leet)*

Rock	Compressional wave velocity (m/s)	Characteristic impedance $(g\text{-}sec/dm^3)$
Granite	5547	5825
Marlstone	3505	2912
Sandstone	3231	2709
Chalk	2774	2798
Shale	1951	1618

Dispersion and Attenuation

The energy transmitted into and through a rock mass by an impacting source is partitioned amongst the various types of wave motion. At any point in the wave path the maximum particle motion is likely to be produced by one or other of the waves rather than by the total wave energy. As a result of this distribution of energy among the different wave types the particle velocity decreases rapidly at first, as the distance from the impact source increases and the waves separate. The separate waves then diverge and the particle motion associated with them decreases at a progressively less rapid rate, until each individual wave ultimately dies out from loss of energy. This attenuation, or rate of energy decay, is determined by the nature of the materials through which the waves are transmitted, attenuation being more rapid in unconsolidated soil and earth than in solid rock.

Attenuation may not be the same for all wavelengths. Auberger and Rinehart, measuring attenuation in rocks over the 100 kHz to 2 MHz range, found that attenuation peaks occurred at high frequencies which could be related to the predominant grain size in the rocks. It appears that when the wavelength of the energy pulse approaches the grain size of the rock the constituent crystals are set into resonance, which causes local increase in the attenuation/frequency relation.

Krishnamurthy and Balakrishna determined attenuation characteristics for several rocks. Attenuation was seen to increase with frequency in fine-grained rocks, but with larger grain sizes in dolerite, limestone, and marble, attenuation was independent of frequency. Attewell and Brentnall have observed the attenuation exponent to vary linearly with frequency over a range of approximately 5 — 100 kHz, and they adduced that the attenuation is due to scattering of energy at grain and pore boundaries (see Figs. 13.7 — 13.8).

From observation of explosion-generated strain pulses in rock, Duvall and Petkof quote a decay in strain according to a law

$$\epsilon = (K/R)e^{-dR}$$

where ϵ = strain,

K = a linear function of the impact pressure and the impedance,

d = a constant depending on the rock type,

$R = \dfrac{distance}{\sqrt{weight\ of\ explosive\ charge}}$ (scaled distance).

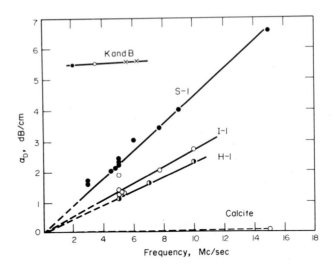

Fig. 13.7. Dilatational wave absorption against frequency for several limestones
(Attewell after Peselnick and Zietz).

d was reported to have values of

0.03	Granite
0.048	Sandstone
0.026	Shale and Chalk
0.08	Marl

In Ricketts and Goldsmith's Hopkinson-bar experiments the attenuation characteristics were determined of strain pulses generated by impact velocities up to 10,000 in./sec. The exponential attenuation coefficient for this level of impulse was small in limestone, but there was much greater dispersion in sandstone.

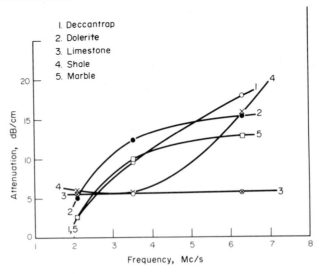

Fig. 13.8. Attenuation against frequency for several rocks (Attewell, after Krishamurthy and Balakrishna).

Dispersion and attenuation change with successive impacts on the same rock specimen, so that repeated shocks produce an increase in the pulse level and a decrease in the longitudinal wave velocity, which are indicative of weakening of the material. These phenomena are of great significance in relation to blasting technology and the operation of rock-breaking tools. They help to explain the cooperative action of the stress waves from an explosive detonation, which weaken the rock, followed by the expanding gas pressure from the explosive charge, which then breaks the rock. They also help to explain the effectiveness of the multiple incident and reflected stress waves produced by rapid-sequence delay blasting, and the effectiveness of percussive tools for hard-rock penetration.

Fatigue Behavior of Rock

While it is apparent that rock materials under dynamic loading generally display strength characteristics much greater than those of the same material under static load, this may only be taken to be so for a limited number of stress reversals. Several investigators have observed that the failure stress on rocks and on concretes diminishes during continuous cyclic compression tests. The earth materials apparently display "fatigue". If the stress reversals are continued until the rock fractures, the strength may be reduced by 30 — 80%, depending upon the load magnitude and frequency of reversal.

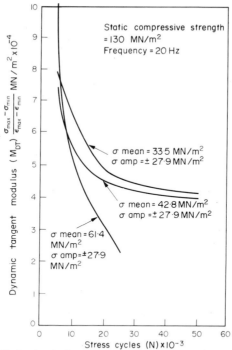

Fig. 13.9. Reduction in dynamic modulus of rocks with increasing load and increase in number of loading cycles (Attewell and Farmer).

Attewell and Farmer describe the results of fatigue tests on dolomite and propose a hypo-thesis to explain the resultant deformation and failure, in terms of crack propagation dependent on strain energy. The basic argument is that, above a stress level at which cracks are initiated, deformation from successive sub-failure load cycles will be cumulative, and failure will occur when the strain energy stored in the specimen exceeds a critical energy level which is approximately equivalent to failure under non-cyclic loading.

The hypothesis is interesting in that one of the techniques used to control rockbursts and gas outbursts in mines, and which has also been proposed as a possible control technique in earthquake-prone fault zones, is to deliberately shatter the rock by blasting in the region of a detected *in situ* stress build-up, so as to dissipate the elastic strain energy stored in the rock before it builds up to release proportions.

The strain energy is thus transferred to surface energy in the fractures and interstices of the shattered rock.

The results obtained by Attewell and Farmer on dolomite are summarized in Figs. 13.9 – 11.

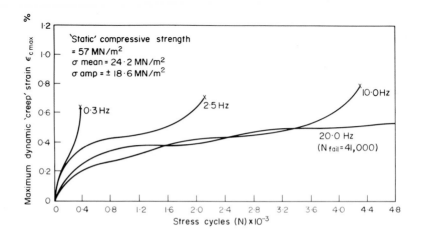

Fig.13.10. Influence of stress frequency upon failure for weaker rocks (Attewell and Farmer).

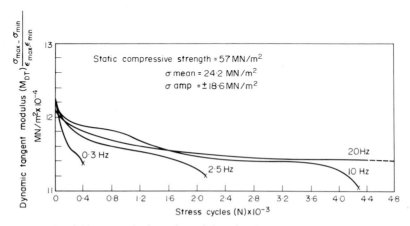

Fig. 13.11. Change in dynamic modulus of rocks with stress frequency
and loading history (Attewell and Farmer).

There is some difference of opinion amongst experts as to whether a low-frequency vibration loading in the range of 10 to 20 Hz should be considered "dynamic", rather than "repetitive".

A precise criterion for dynamic loading is sometimes related to the natural frequency of the sample under test. On this basis, if the load is to be regarded as dynamic

$$f/f_n \geqslant 1/10$$

where f is the frequency of loading,
$\quad$ f_n is the natural frequency of the specimen.

Behavior of Soils Under Dynamic Load

Impact loading tests on soils may be conducted using the drop-hammer technique or by using a combination of falling weight and lever. The test is, in effect, a triaxial test, even though the cylindrical specimen may be unconfined. An effective confining pressure is imposed by the "lateral inertia" of the sample. If lateral expansion of the soil sample is restrained during the test considerable changes in internal pore pressure may be generated during the failure process, which is not regular. Nash and Dixon describe dynamic triaxial tests on saturated sand which produced results similar to those depicted in Fig. 13.12. The pore pressure diminishes with increasing strain during axial load increase, up to the maximum resistance of the specimen. However, during this general pattern there are intermittent partial collapses of the material, in each of which the pore pressure increases temporarily. At each of these periods of intermittent collapse the solid particles are shedding load on to the pore fluids. During the process of failure there is a change in the structure of the material which enables the specimen to build up further resistance to the axial load until a maximum strength value is attained.

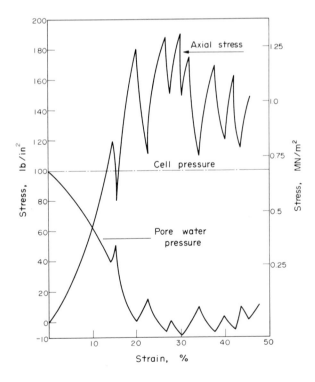

Fig. 13.12. Dynamic triaxial test results on saturated sand (Nash and Dixon).

In a discussion of the evidence presented by various investigators on dynamic triaxial testing of soils, Richart *et al.* conclude that, for soils, the dynamic modulus of deformation and the ultimate strength under single impact dynamic loading may range over values up to three times their equivalent under static conditions. However, while the single-impact or drop-hammer test is useful as a means of exploring the behavior of soils under dynamic load, and provides information on the interaction of solid particles and pore fluids, the strength values so obtained must not be taken as indicative of the resistance of soils to the vibrating loads imposed by machines and machine foundations, or by the vibrating stress pulses that are generated as a result of blasting and earthquakes. In this connection it is also necessary to explore the mode of behavior of the material under cyclic loading, over the range of frequencies likely to be encountered.

Loose, saturated, sands and silty soils with little cohesion may liquefy and become fluid as a result of the pore-pressure build-up during the failure process. In the field this may take place progressively in the loaded foundation sub-stratum. Under the influence of a stress pulse the layer of soil immediately below the loaded contact may liquefy, load being then transferred to the fluid, in which the solid particles readjust to take up the load again as the foundation settles. Succeeding stress pulses repeat the process, so that the zone of liquefaction is propagated downwards. The solid particles in the foundation material are compacted, to a degree and at a rate which depends upon their grain size and permeability, the process being more rapid in coarse-grained soils.

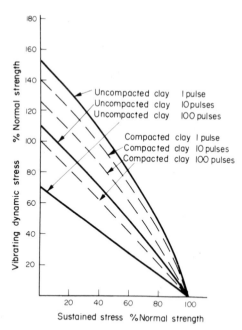

Fig. 13.13. Fatigue stresses in two clay soils (Seed).

Total collapse of the foundation material may occur, after repeated cycles of loading, at a stress level much lower than the single-impact dynamic strength and lower than the strength under static loading. In assessing the resistance of a soil to vibrating loads it is therefore possible to establish a "fatigue failure" criterion, similar to that used for metals, in which the maximum permissible alternating axial stress which, when added to a sustained stress, will produce failure after a given number of load cycles, may be specified. Figure 13.13 shows such information for two soils, as reported by Seed.

The phenomena of liquefaction and compaction of saturated and low-cohesion materials under dynamic loads are matters of grave consequence in earthquake-prone regions. They have great bearing on the stability of foundations and earth slopes. The foundations of buildings and structures that are secure and stable under normal static load may suffer sudden and disastrous settlement in the event of earthquake, and slopes that are normally stable may liquefy and flow.

On the other hand, the phenomena may be put to good use in engineering tools, such as vibrating pile-drivers and earth-penetration devices, where the addition of vibration to static thrust facilitates the penetration process, particularly when the earth materials concerned have a high moisture, or other pore fluid, content. Also in compaction devices, such as vibrating rollers, to increase the density and bearing capacity of weak ground and earth-fill.

Structural Changes in Rocks Under Dynamic Load

The underground detonation of a conventional explosive charge can generate pressures of several hundred kilobars almost instantaneously. With nuclear explosives the shock pressures enter the megabar range and the energy release is completed in microseconds, as compared with milliseconds for a high-yield chemical explosive charge. The total energy from a 100-kiloton nuclear explosive would be released within a microsecond, to produce a peak pressure of around 151×10^6 MN/m^2 at a temperature around 13×10^6 K. In one-fifth of a second the seismic wave would traverse ¾ mile and by that time the explosion cavity would be surrounded by many thousands of tons of molten and plastic rock at temperatures extending to several thousand degrees.

Shock waves of high energy can produce structural changes in the mineral constituents of the rocks through which they are propagated. Such changes, known as "polymorphic transitions". have been observed in quartz, marble, and some ferro-magnesian silicates, in controlled laboratory experiments using explosion shock-waves, so that we can expect them to occur also in the rocks surrounding nuclear blast craters. There is considerable interest in polymorphs of quartz since this is a universally common mineral constituent. Until comparatively recent years, quartz was considered to be so brittle that it could not display ductile behavior in any degree. But observation of the mineralogy and crystallography of the material around large meteoric craters, where the material must have suffered extremely high impact loading, has produced evidence of quartz polymorphs similar to those which may be synthesized at high temperatures and pressures. A feature of these polymorphs is that they often display evidence of plastic deformation. This may take the form of slip along the crystallographic planes, to produce deformation lamellae and twinning, particularly seen in calcitic rocks such as marble. Plastic deformation in quartz is sometimes evidenced as "undulatory extinction" or "strain shadows". This condition may be defined as some angular variation in the position of the optic crystallographic axis over different parts of a particular mineral grain. The crystals are bent, twisted, and tilted.

It has been found that the incidence of undulatory extinction in the quartz of a rock, such as shock-deformed sandstone, shows a pronounced increase over that observed in the same non-shock-loaded rock. Ramez and Attewell have shown that this increase is directly related to the varying intensity of a particular stress field. The crystals in the rock are bent by the passage of a spherical shock front expanding from the impact origin. Halite has also been shown to flow plastically and produce strain birefringence when shock loaded. The cubic salt crystals experience a slight lattice expansion and display a tendency to become amorphous. The clay mineral, montmorillonite, may be destroyed by a shock pulse, with possible recrystallization of its lattice into a feldspar structure. Both montmorillonite and chlorite are unstable under intense compressional loads, to the extent that it may be that in a rock which includes a predominantly clay mineral matrix, a large proportion of the energy in a transient shock wave may be absorbed by the matrix in the process of structural readjustment.

Selected References for Further Reading

ATTEWELL, P. B. Recording and interpretation of shock effects in rocks. *Mining and Minerals Engineering,* London, pp. 21–28 (Sept. 1964).

ATTEWELL, P. B. and FARMER, I. W. Ground vibrations from blasting, their generation, form and detection. *Quarry Managers Journal,* London, pp. 191–198 (May 1964).

ATTEWELL, P. B. and BRENTNALL, D. Internal friction. Some considerations of the frequency response of rocks and other metallic and non-metallic materials. *Int. J. Rock Mech. Min. Sci.,* vol. 1, pp. 231–254 (1964).

ATTEWELL, P. B. and FARMER, I. W. Fatigue behavior of rock. *Int. J. Rock Mech. Min. Sci.,* vol. 10, pp. 1–9 (1973).

AUGBERGER, M. and RINEHART, J. S. Ultrasonic velocity and attenuation of longitudinal waves in rocks. *J. Geophy. Res.,* vol. 66, pp. 191–199 (1961).

CARROL, R. D. The determination of the acoustic parameters of volcanic rocks from compression velocity measurements. *Int. J. Rock Mech. Min. Sci.,* vol. 6, pp. 557–580 (1969).

DUVALL, W. I. and PETKOF, I. *Spherical Propagation of Explosion-generated Strain Pulses in Rock,* U.S. Bur. Min's Rept., Invest. No. 5483 (1959).

FARMER, I. W. *Engineering Properties of Rocks* (Chapter 6, "Dynamic properties"), pp. 70–83, Spon, London, 1968.

FRYER, C. C. Shock deformation of quartz sand. *Int. J. Rock Mech. Min. Sci.,* vol. 3, pp. 81–88 (1966).

GREGORY, A. R. Shear wave velocity measurements of sedimentary rock samples under compression. *Proc. 5th Symposium on Rock Mechs., Univ. of Minnesota,* pp. 439–471 (1963).

HARDY, H. R. and CHUGH, C. S. Failure of geologic materials under low cycle fatigue. *Proc. 6th Canadian Symposium on Rock Mechanics,* McGill Univ., Montreal, 1969.

ITO, I. *et al.* Rock behavior for tension under impulsive load. *Trans. Min. Met. Alumni Ass., Kyoto Univ.,* vol. 15, p. 61 (1963).

KING, M. S. Wave velocities in rocks as a function of changes in overburden pressure and pore fluid saturants. *Geophysics,* vol. 31, pp. 50–73 (1966).

KRISHNAMURTHY, M. and BALAKRISHNA, S. Attenuation of sound in rocks. *Geophysics,* vol. 22, pp. 268–274 (1957).

LEET, L. D. *Vibrations from Blasting Rock,* Harvard Univ. Press, 1960.

LOMBARD, D. B. *The Hugoniot Equation of State for Rocks,* Rept. No. UCRL–0311, U.S. Atomic Energy Commission, Contract No. W–7405–eng 48 (1961).

NASH, K. L. and DIXON, R. K. The measurement of pore pressure in sand under rapid triaxial test. *Proc. Conf. on Pore Pressure and Suction in Soils,* Butterworth's, London, 1960.

OBERT, L. and DUVALL, W. I. *Rock Mechanics and the Design of Structures in Rocks* (pp. 344–359, "Dynamic elastic constants"), Wiley, New York, 1967.

OBERT, L., WINDES, L. S. and DUVALL, W. I. *Standardized Tests for Determining the Physical Properties of Mine Rock,* U.S. Bur. Mines Rept., Invest. No. 3891 (1946).

RAMEZ, M. R. H. and ATTEWELL, P. B. Shock deformation of rocks. *Geophysics,* vol. 28, pp. 1020–1036 (1963).

RICHART, F. E., HALL, J. R. and WOODS, R. D. *Vibration of Soils and Foundations,* Prentice-Hall, New Jersey, 1970.

RICKETTS, T. E. and GOLDSMITH, W. Dynamic properties of rocks and composite structural materials. *Int. J. Rock Mech. Min. Sci.,* vol. 7, pp. 315–335 (1970).

RICKETTS, T. E. and GOLDSMITH, W. Wave propagation in an anisotropic half-space. *Int. J. Rock. Mech. Min. Sci.,* vol. 9, pp. 493–512 (1972).

RINEHART, J. S. Effect of transient stress waves in rock. *Int. Symposium on Mining Research, Rolla, Missouri* (Ed. G. Clark), Pergamon Press, Oxford, 1962.

SEED, H. B. Soil strength during earthquakes. *Proc. 2nd World Conference on Earthquake Engineering,* vol. 1, pp. 103–104 (1960).

CHAPTER 14

The Application of Models to Geotechnology

Experimental Geology

Experimental investigations are an essential part of research in any branch of applied science. Geotechnology is no exception to this rule. Geologists may supplement their field observations by experiments simulating some of the processes of physical geology, such as folding, faulting, jointing, and cleavage, by which the model materials are structurally changed, deformed, and fractured by various systems of loading. The processes of plastic flow that are evidenced in basaltic lava, glacial ice, rock debris, and soil may also be studied by means of material models.

Tests on Natural Rock Materials

Kvapil describes tectonic experiments on natural rocks and mineral materials. The study of these substances in the elastic and plastic states is conducted by methods long established for the purpose of materials testing. Such methods, when applied to rock materials in a brittle condition, can shed important light on the origin of faults, fissures, and fractured strata. When the experimental arrangement is such as to produce plastic deformation in the materials the inter-relationships of time, temperature, and pressure on various rock flow structures may be investigated. As an example of an experiment on rock in the elastic field, Kvapil quotes the reproduction of slip zones in a stratified material with thin laminations. For rock materials in the plastic state, the following relationships apply:

1. The lower the strength of the rock in the elastic field the easier it will attain the plastic state. For the same stress conditions a low-strength rock reaches the plastic state more quickly than does a high-strength rock.
2. Under the same stress conditions a rock may reach the plastic state although other, adjacent, rocks may still be in the elastic field.
3. A rock may exert increasing resistance to compression, although it is plastic. This can explain how, in geotectonics, a hard rock may sometimes appear to be crushed in the contact zone with a rock that displays evidence of plastic deformation. It also explains how the salt strata in a dome structure may appear to have crushed a hard sandstone, or a limestone.
4. A rock with a higher degree of plasticity will crush other, neighboring, rocks of lower plasticity since it will display greater resistance to compression.
5. If two rocks have similar plasticity characteristics, they can, at a high plasticity level, be welded together at their contacts.

Figure 14.1 shows the results of an experiment which reproduced a dome structure in what was originally layered salt strata.

In such experiments the inter-relationship of time, temperature, and pressure may be used, for example to speed up the result, by subjecting the model to increased temperatures or pressures. In nature the time factor would be much greater and the stress levels, and possibly also the temperatures, correspondingly smaller. The experiments therefore may demand specialized laboratory equipment, such as high-pressure triaxial cells with provision for wide temperature variation. As yet there are few laboratories so equipped.

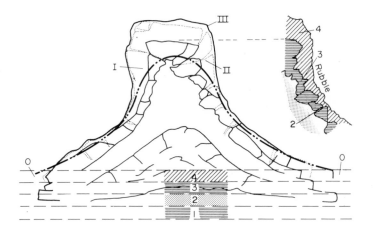

Fig. 14.1. Stages of test on layered salt strata to model the development of a dome structure **(Kvapil)**.

Equivalent Material Models

The exact reproduction of full-scale conditions, or absolute similitude, is not achieved on a small-scale model. For this to be possible the scales of length, time, and pressure (or mass) must be changed, without altering the theoretical equations that describe the behavior of the materials concerned. On such a model the deformations would be observed to the same scale as that of length, and the ratio of stresses in the model, and its prototype, would be equal to the same geometrical scale factor. [The situation is one such as that described by Thakur, in which if the comparative parameters on model and prototype cubes, are of side lengths l and L respectively, and if the ratio l/L is less than unity then the stress in the prototype exceeds the stress in the model.] From this it follows that if the prototype were loaded beyond its ultimate strength then it would fail, but the corresponding stresses generated in the model would not cause that to fail. A simple geometrical scale model therefore cannot reproduce in detail the behavior of a prototype rock or soil structure under load. (See Table 14.1).

In a discussion of the conditions of similitude on scale models in geotechnology, Mandel points out that if both the deformations and the stresses are to be reproduced to scale, then the body forces in the prototype and the model must be in inverse ratio to their linear dimension. When the model is loaded by the forces of gravity, generated by its own mass, it is simpler to reproduce the gravitational body forces in prototype and model. In this case the materials from which the model is constructed must be much softer and less dense than those of the prototype structure. If they are chosen in such a way that the mechanical properties of elastic, plastic, and viscous deformation in the prototype may be deduced from scaled measurements of stresses, times, and deformations on the model, then the model is said to be made of equivalent materials.

When the model is restricted to the reproduction of linear elastic deformations, and the

288

deformations are small, and when the materials are homogeneous and isotropic, it is possible to reduce the required conditions for material equivalence to an equality of Poisson's coefficients for model material and prototype. If the prototype conditions are not homogeneous, for example if they consist of layered strata, then the ratios between the relative moduli of elasticity of these layers must also be preserved. The reproduction of non-linear elasticity and plasticity is more difficult. Nevertheless even here, by restricting the study to a limited range of conditions, and by choosing appropriate scales of time, very useful results may ensue.

TABLE 14.1. *Some Physical Parameters, their Dimensions and the Model Scale Factors (Hobbs)*

Physical parameter	Dimension	Scale factor (a)	Scale factor (b)*
Length	$[L]$	λ	1/50
Mass	$[M]$	μ	$11/25 \times 10^{-5}$
Time	$[T]$	τ	$1/\sqrt{50}$
Density	$[ML^{-3}]$	$\mu\lambda^{-3}$	$\cong 11/20$
Strength	$[ML^{-1}T^{-2}]$	$\mu\lambda^{-1}\tau^{-2}$	1/90
Stress	$[ML^{-1}T^{-2}]$	$\mu\lambda^{-1}\tau^{-2}$	1/90
Young's modulus	$[ML^{-1}T^{-2}]$	$\mu\lambda^{-1}\tau^{-2}$	1/90
Acceleration	$[LT^{-2}]$	$\lambda\tau^{-2}$	$\cong 1$
Poisson's ratio	o	1	1

* Column (b) gives the scale factors employed in the present model work.

Classification of Materials for Structural Modelling

Stimpson has documented a qualitative classification and summary of modeling materials in geotechnology, including an exhaustive bibliography on the subject. In essence, the materials are classed as being either granular or non-granular, with subdivisions as shown in Fig. 14.2.

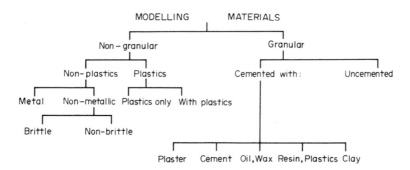

Fig. 14.2. Simple classification of modeling materials in geotechnology (Stimpson).

The range of materials in each subdivision and the purposes for which they were used are described, and some of them are critically assessed in relation to the problem of simulating rock.

Scale Models of Underground Strata Movement

Scale models are used extensively to study the behavior of the strata around mining excavations. Everling describes the methods devised at the Mining Research Station, Essen,

West Germany. Two test rigs were constructed; the first, shown in Fig. 14.3, is intended for models measuring 2 × 2 × 0.4 m. The model is loaded in the rig, consisting of a strong horizontal and vertical steel girder framework, by hydraulic jacks placed four on each side of the biaxial, horizontal, frame. A top cover, to provide "lateral" support in the plane of the model, is also actuated by hydraulic jacks. Control and measurement of the individual jack-pressures, piston strokes, and hydraulic pump capacity, are effected from a central control station. Adjustment of hydraulic pressures and piston travel provides a means of reproducing uniform and non-uniform loading patterns and deformations in any desired sequence. The linear scale of the models is 1/10. Coal, shale, and sandstone are modeled on a strength scale of 1/10 by variations in a mix of Portland cement, fused alumina cement, and quartz powder. Steel support elements are reproduced by extruded sections of aluminum, having a yield point at one-tenth that of steel. Timber support elements are reproduced by balsa wood which has a strength about one-tenth that of pine and spruce.

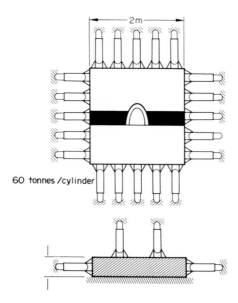

Fig. 14.3. Testing rig for mine roadway-support models (Everling).

The second test rig designed at Essen is intended to accommodate stratified layer models 2 × 10 m, representing a vertical planar section across a horizontal coal face. It is intended for the study of methods of coal-face support and to reproduce the behavior of the roof strata while the working face advances over a distance of 100 m (Fig. 14.4). A point of interest on Everling's test rig is the method of reproducing a variable support resistance, capable of providing quantitative data. The aluminum roadway support is replaced by a flexible bag, having known and constant internal air pressure. When this pressure becomes effective in resisting deformation of the roadway cross-section, the air pushes a corresponding volume of water from an overflow into a measuring vessel. The reduction of cross-section can thus be observed in detail.

A later apparatus, used at the research station of the National Coal Board, London, and designed by Hobbs, tests models 0.6 m × 0.6 m × 127 mm thick, representing a vertical cross-section of horizontal mine roadways, on a geometrical scale of 1/50. Thus, a 3.6 m-wide prototype roadway is represented by a centrally placed roadway of 74 mm on the model, separated from the boundaries of the model by about 3.5 roadway widths. Hobbs gives an account of the application of dimensional analysis to such a model, and on this basis the scale factors listed in Table 14.1 were chosen. The properties of Hobbs' prototype materials and those of the required model material are detailed in Table 14.2. The requirements were met by sand-plaster-water mixtures, and soft lead H-section arch support members.

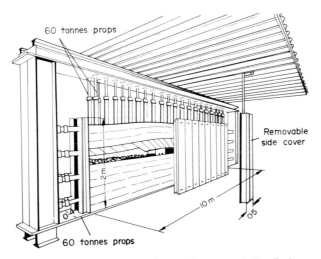

Fig. 14.4. Model testing rig for coal-face support (Everling).

TABLE 14.2. *Compressive Strength, Tensile Strength and Young's Modulus of Rock and Model Material Required*

Rock type	Rock			Required model material		
	Compressive strength	Tensile strength	Young's modulus (MN/m^2)	Compressive strength	Tensile strength	Young's modulus
Pennant sandstone	165.48	65.02	47,575	1.84	0.72	531
Wooley siltstone	131.0	44.82	15,169	1.45	0.50	166
Bilsthorpe silty mud-stone	39.3	–	8274	0.43	–	90
Barnsley hard coal	48.26	–	2896	0.54	–	32
Deep Duffryn coal	6.89	–	1724	0.08	–	19

Typical results on Everling's and Hobbs' rigs are shown in Fig.14.5(a) (Everling) and (b) (Hobbs).

Ideal Continuous and Discontinuous Models

Sometimes the problem being studied is dominated by the influence of one major factor of critical importance, and the physical model may then be applied in such a way as to bring out the effect of this critical parameter. The commonest example of this is when dealing with elastic or pseudo-elastic rock masses, which may be considered to be homogeneous. In this case a photoelastic model may be used. To a limited extent also, photoelastic models can reproduce the conditions which exist in laminated strata in which each layer may be considered to be homogeneous.

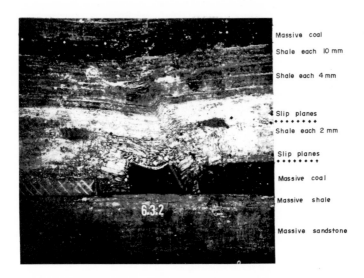

Fig. 14.5(a) Model results. Mine roadway with frame timbering (Everling).

Fig. 14.5(b) Model of unsupported roadway in laminated strata (Hobbs).

TABLE 14.3. *Block Geomechanical Models, Constructed and Tested by Istituto Sperimentale Modelli e Strulture, Bergamo (Fumagalli)*

Model	λ	ξ	ρ	Mountain	Discontinuities	Young's modulus of prototype (kg/cm^2) Homogeneous rock	Overall of rock mass	Hydrostatic load
Vajont(Italy)	85	85	1	3200 precast prismatic bricks, 16 × 14 × 9 cm	Faults and bedding planes	600,000	200,000	With liquids
Pertusillo (Italy)	150	100	0.66	Blocks cast on iron	Fractured bedding planes	350,000	25,000–50,000 (anisotropic rock) 10,000 grés)	With liquids and jacks
Kurobe (Japan)	100	50	0.5	Continuous casting on site	13 faults and weak zones		12,700–72,500	With liquids
Ca'Selva (Italy)	80	128	1.6	8000 precast oblique prismatic bricks of two types for left bank 10 × 10 × 10 cm	Bedding planes and 15 faults	320,000	40,000–80,000	With liquids and jacks
Grancarevo (Yugoslavia)	120	80	0.66	12,000 precast prismatic bricks 5 × 10 cm, h = 3–6 cm	Bedding planes and 14 faults	480,000	Anisotropic rock 70,000–250,000	With liquids
Emosson (Switzerland)	100	150	1.5	Oblique blocks cast on site, 10×10×40 cm	2 faults	400,000	300,000	With liquids
Rapel (Chile)	100	100	1	20,000 precast oblique prismatic bricks of 3 types, 8 × 8 × 8 cm	7 faults	10,000 for the outer zones	70,000–160,000	With liquids

TABLE 14.4. *Interlayering Materials for Various Angles of Friction (Fumagalli)*

Material	Angle of friction (°)
Coating with alcohol varnish and grease	7– 0
Coating with alcohol varnish, talc and grease in various proportions	9–23
Coating with alcohol varnish and talc	24–26
Coating with alcohol varnish	32–34
Limestone powder	35–37
Bare contact surfaces	38–40
Sand of various grain sizes	40–46

Block Models

When the strata are also discontinuous within each layer, or in massive strata intersected by two or more planes of jointing and cleavage, the continuous photoelastic model becomes unrealistic. In view of the critical importance of joint-strength on the stability of large rock cuts, and on the foundations of structures such as large dams, models constructed of blocks, reproducing the joint planes in direction, frequency, and frictional characteristics, may be constructed. Fumagalli describes such models, as applied to the study of foundation problems in the construction of large dams, in the laboratories of ISMES, Bergamo, Italy. Some details of the models, constructed and tested by ISMES, are listed in Table 14.3. The friction and deformability characteristics of the joint contacts were reproduced by Fumagalli, using various materials placed between the blocks (Table 14.4). In this way the peak stress at the yield-point, and the residual strength of the joints, as observed on a field test, may be reproduced on a shear stress-displacement curve obtained on a shear test of the model material in the laboratory.

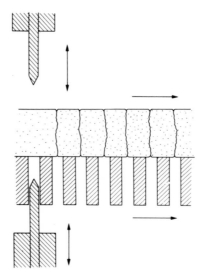

Fig. 14.6. Tension-splitting mechanism to produce blocks for model-joint studies (Barton).

In a review of existing techniques of rock-joint modeling, Barton has concluded that the methods generally used do not provide realistic models of the manner in which rock-joints are mated together, in nature, at their contacts. Fumagalli's methods may be considered to give an oversimplified picture of what really occurs within the interlocking roughnesses between the blocks. By using a tension-splitting mechanism, the principle of which is illustrated in Fig. 14.6, Barton produced rough fractures in modeling material formed from an oven-cured combination of red lead, sand-ballotini, plaster, and water. Choosing a geometric scale factor of 1/500 with this material, the resulting stress scale for a gravitationally loaded model, assuming a prototype density of 2563 kg/m^3, was deduced to be approximately 1 to 666 (Table 14.5).

Using this material Barton compared and evaluated the relationships between shear strength and normal displacement, and observed the pre-peak and the post-peak shear-strength behavior of continuous, cross-jointed, and step-jointed materials (Figs. 14.7 and 14.8).

The shear-strength envelopes are plotted to common displacements, so that Fig. 14.8(a) shows that when there is no joint displacement then the joint has no strength. The joint strength builds up as displacement occurs, to reach a maximum value at the peak strength.

TABLE 14.5. *Model-prototype Properties of Model Joint-study Material (Barton)*

Property	Units	Model	Prototype
Unconfined compressive strength	MN/m^2	0.14	92.39
Tensile strength (Brazilian test)	MN/m^2	0.02	11.03
Young's modulus	MN/m^2	73.78	49161
Axial strain at failure	%	0.31	0.31
Density	kg/m^3	1939	2563

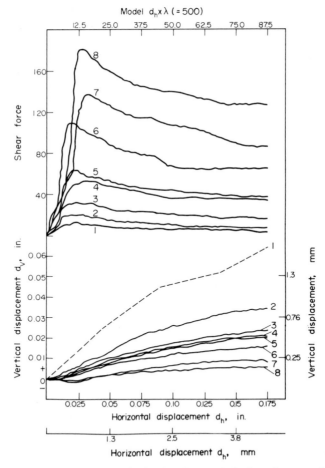

Fig. 14.7. Shear face-displacement and dilation characteristics for primary tension joints in model joint material (Barton).

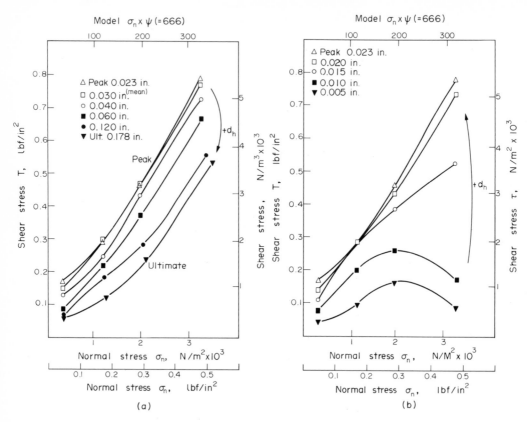

Fig. 14.8(a) Post-peak shear strength behavior of
primary tension joints in model material (Barton).

Fig. 14.8(b) Prepeak shear strength behavior of
primary tension joints in model material (Barton)

Photoelastic Models

Photoelastic model studies, employing the principles outlined in Chapter 7, are widely used in general stress analysis. Their application in geotechnology has mainly been aimed (a) to observe the stress distribution around excavations of various shapes in relation to specific directions of applied load, (b) to observe the interaction of neighboring excavations and their combined effect on the stress distribution in the rock mass as a whole, (c) to determine the stress distribution in the component parts of support members, as an aid to support design, and (d) to study the characteristics of fracture development in rock materials

The primary objectives to be served by studies of type (a) are to determine the location and magnitude of the stress concentrations around the periphery of the excavation and to decide upon an optimum cross-sectional shape and orientation for maximum strength and stability. Typical examples of this type of investigation are described by Hiramatsu and Oka.

Numerous investigators have conducted studies of type (b) since D. W. Phillips first introduced the English-speaking world to photoelastic analysis in geotechnology in the early 1930's. A typical example of the technique is illustrated in Fig. 14.10 (Janelid and Kvapil), which shows the stress distribution around a staggered arrangement of extraction drifts in underground mining by the sub-level caving method. The highest stress concentrations occur at the corners of the drifts, running obliquely from one horizon of drifts to those on the next horizon. The effect of this oblique alignment on the mode of failure when the rock mass is loaded to destruction is shown in Fig. 14.11. At the same critical load, the alignment modeled in Fig. 14.12, with the drifts vertically below one another, remains stable.

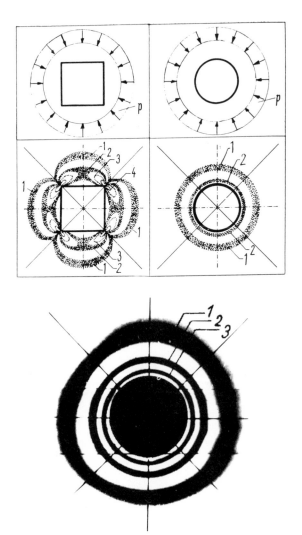

Fig. 14.9. Isochromatics showing stress distribution around a vertical shaft of square and circular cross-section. (Numbers indicate fringe order) (Kvapil).

Photoelastic stress analysis may be applied to underground support structures both with the object of better understanding the principles of operation of the support system and for effective design of the support units. As examples of the former, Fig. 14.13(a) shows a model of a rectangular section tunnel with the high stress concentrations at the corners that are typical of such an excavation. These corners can, however, be "relieved" of stress by cutting slots into the strata or otherwise weakening the ground here so that it can yield and so deflect the high stress concentrations into the interior of the rock mass. In contrast with this, Fig. 14.9 shows a circular tunnel excavation under load, around which there are no localized high-stress concentrations but an approximately uniform stress distribution in concentric

297

Fig. 14.10. Photoelastic isochromatics showing stress distribution around a staggered arrangement of extraction drifts in underground sub-level caving (Janelid and Kvapil).

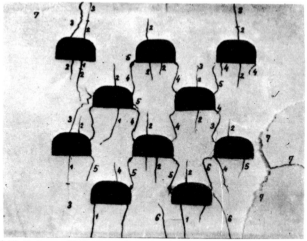

Fig. 14.11. Pattern of failure on plaster model, loaded to destruction, representing staggered drift arrangement (Janelid and Kvapil).

circles, with the fringe orders decreasing with increasing distance into the peripheral wall. The stability of such an excavation depends on the stress gradient in the rock walls, relative to the strength of the wall material. Any excessive yield or creep of the walls could be restrained by inserting rock bolts, screw-tensioned between deep-set anchors and face plates, to provide the shear stress distribution shown in Fig. 14.14.

It can be seen that the compressional restraint imposed by the face plates modifies the

Fig. 14.12. Pattern of failure on plaster model representing extraction drifts
vertically aligned (Janelid and Kvapil).

(a)   (b)

Fig. 14.13. Photoelastic model showing stress concentrations around an opening. (a) High stress
concentrations at drift corners. (b) High stresses transferred to interior of rock mass by "slotting"
at the corners.

original stress distribution to produce an intensified shear stress zone behind them, immediately
around the peripheral wall. However, so long as the restraint of the face plates is maintained,
and if they are situated sufficiently close together, or preferably combined with a ring girder, or
other reinforcement, to give continuous support around the periphery, the effect of this re-
traint is to apply confining pressure and so increase the strength of the wall rock. The stability

of the excavation then depends upon the strength of the walls in relation to the stresses imposed upon them from inside the rock mass. The model, Fig. 14.14, shows that the rock bolts not only impose confining pressure on the peripheral wall, but behind this peripheral ring is produced a broad annulus of low shear stress, one fringe order or less, broken only by the localized high values of shear stress around the bolt anchors. The shear stresses generated in the rock mass as a whole, behind the peripheral stress ring, are therefore reduced, compared with those that existed before the rock bolts were inserted and tensioned. Clearly the effectiveness of such a support system will depend upon the magnitude of the high concentrations of stress at the bolt anchors and immediately behind the face plates. In these high-stress zones some creep of the rock is almost inevitable. The bolts will therefore lose tension after they are set, and the support technique should take account of this fact — minimizing creep by suitable design in the anchors and face plates and periodically reimposing the full extent of designed bolt tension.

Fig. 14.14. Photoelastic isochromatics showing stress distribution around a circular tunnel under uniform biaxial load in the plane of the model with rock-bolt reinforcement.

Lu and Wright applied photoelastic stress analysis to models of prestressed concrete mine sets, as shown in Figs. 14.15 and 14.16. The result provided information as to the degree of reinforcement necessary for economic design. It was shown that a unit member had to provide a tensile resistance of less than 50% of its compression resistance. The value of cushion pieces lined with wood or other soft material was modeled, using rubber buffers between the epoxy-resin set members. The high stress concentrations at the contacts of the two-piece member (Fig. 14.15) are eliminated in the five-piece model (Fig. 14.16).

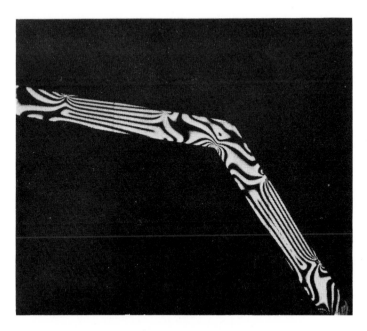

Fig. 14.15. Dark-field isochromatic fringe pattern of a two-piece model of a prestressed concrete mine set (Lu and Wright).

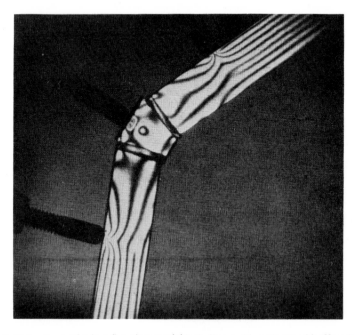

Fig. 14.16. Dark-field isochromatics in a five-piece model including cushion-piece and buffers (Lu and Wright).

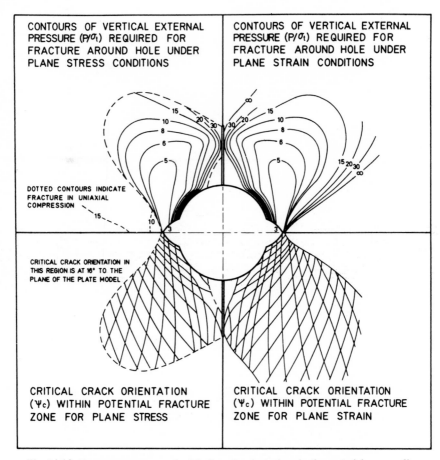

Fig. 14.18. Fracture contours and critical crack orientations in the material surrounding a hole with crown and bottom cracks and side fracture (Hoek).

Nearly all the photoelastic model studies so far mentioned were conducted in the classical manner, using the transmitted-light polariscope. Hoek's studies on the mechanism of brittle fracture also included an examination of fracture propagation in glass, by means of the transmission polariscope. However, by employing photoelastic coatings and the reflection polariscope, Hoek has also investigated in detail the stress distribution around circular holes in rock material, and the fracture mechanism when the rock was loaded to failure. The rock used was a fine-grained siliceous igneous rock, termed chert-dyke, chosen because its elastic properties allowed of an approximate correlation with elastic theory.

Hoek's results, some of which are illustrated in Figs. 14.17 – 14.18, are of considerable interest in that, in addition to the tensile cracks in the crown and at the bottom of the opening (in accordance with the predictions of elastic theory relating to stress concentrations at the periphery of such a hole in elastic material), side fractures are also generated and some of these originate at positions remote from the periphery, and within the interior of the rock walls.

The manner in which the complete failure of a brittle rock material under load may emanate from internal failure at some discontinuity, microcrack, or structural weakness, critically aligned in relation to the ambient stress system, was thus graphically illustrated in these experiments. Hoek further records the critical crack orientations within the potential fracture zones, for plane stress and plane strain conditions, as shown in Fig. 14.18.

Determination of stresses and stress trajectories from photoelastic models

A detailed study of the isoclinic directions allows the principal stress directions over the model to be plotted as a network of "stress trajectories", while from the measured birefringence demonstrated by the isochromatics the distribution of the major and minor principal stresses σ_1 and σ_2, the shear-stress distribution (proportional to $\sigma_1 - \sigma_2$) and the principal stress ratio σ_1/σ_2 may all be deduced. When the model is biaxially symmetrical, as in Hoek's circular hole studies, this information may be plotted graphically as stress contours in the separate quadrants of one diagram (Fig. 14.19).

Measurement of the isochromatic fringe order at a point provides a measure of $(\sigma_1 - \sigma_2)$. The separate identification of σ_1 and σ_2 is then possible by determining the principal stress sum $\sigma_1 + \sigma_2$) at the same point. This principal stress sum may be determined in various ways. One method is to use a conducting paper analog, which is constructed out of a sheet of uniformly graphite-coated paper, cut to the shape of the stressed model.

The conducting paper is then attached to a potentiometer and potential divider. When boundary voltages, determined from the photoelastic model boundary fringe orders and proportional to $(\sigma_1 + \sigma_2)$, are applied to the boundaries of the conducting paper, the voltage recorded by touching the probe to a specific point on the paper is proportional to $(\sigma_1 + \sigma_2)$ at the corresponding point on the photoelastic model. An alternative to the conducting paper method is to use an electrolytic tank of appropriate shape, in which the distribution of electrical potential, in an electrolyte of constant depth, is identical with the distribution of $(\sigma_1 + \sigma_2)$ in the photoelastic model, provided that the boundary stresses in the model correspond to the peripheral voltages applied to the electrolytic tank.

There are also optical methods of determining the separate principal stresses on a photoelastic model and computation of the separate principal stresses is possible, using the "shear difference method" described by Frocht, but to do this manually, on a point-by-point basis over the model, is a long and tedious procedure which the conducting paper and the tank analogs are intended to simplify. However, with the advent of computers, manual point-by-point calculations are no longer necessary, and several investigators have computer programs to bring out the desired information automatically.

Combined Physical Material-photoelastic Models

The photoelastic coating technique, besides being a very useful means of examining the surface strain distribution on heterogeneous materials under load, may also be applied to physical material models, to provide an overall picture of strain in the elastic regime. However, unless the coating material is of negligible rigidity, compared with the model material, it must

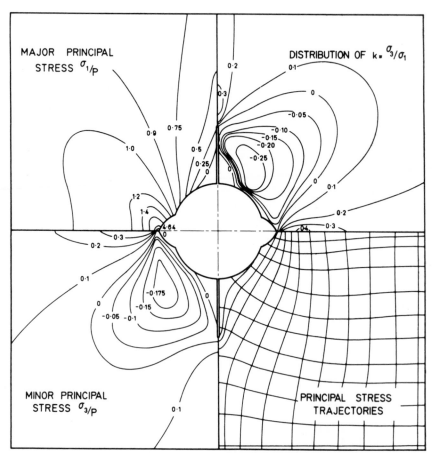

Fig. 14.18. Distribution of principal stresses and principal stress trajectories in the material surrounding a hole with crown and bottom cracks (Hoek).

be removed from the model if appreciable plastic deformation precedes failure, and the approach to failure is to be observed. The technique is particularly useful when a number of similar models are to be compared, with limited time available for analysis. In such circumstances a qualitative assessment of the shear stress distribution may be made visually from the isochromatics, while the principal stress directions at specific points of interest can be made instantaneously apparent by drilling small holes into the coating. The effect of this is that, even at very low fringe orders, two isotropic "strain shadows" appear on the major principal stress axis, one on either side of each hole, while at high fringe orders the stress concentrations around each hole in the coating produce small symmetrical isochromatic patterns which identify both the major and the minor principal stress directions. These local hole-patterns are superimposed upon the general model isochromatic distribution; so that the total pattern presents a combined visual picture of shear stress magnitudes and principal stress directions, without instrumental measurements or computation (Fig. 14.19).

This figure is taken from a report on a model study of rock foundation problems under a concrete gravity dam. The object of the investigation was to study the effect of various alternative structural details in the dam foundations. Several models were therefore constructed each representing a cross-section of the dam and the foundation beds to a scale of 1/200, the dam being represented by a concrete mortar cast 76 mm thick. The foundation beds at the site consisted of laminated limestones and friable brown lignite interlayered with marl which became unconsolidated when wet. A cross-section of the dam and the foundation rocks is shown in Fig. 14.20. Eight possible variants in foundation construction were modelled for study, seven of which are illustrated in Fig. 14.21. The eighth consisted of adding mass by thickening the concrete apron on the downstream side of the dam wall.

Fig. 14.19. Shear stresses and principal stress directions displayed simultaneously on a model under load

In the selection of model materials, similar mechanical strength and rheological characteristics to those demonstrated by the prototype terrain and structure were reproduced, so far as available information allowed. Cement mortar was chosen to represent the dam material and the limestone foundation layers. A gypsum plaster plus a porous filler, in the form of wood sawdust, represented the lignite. An attempt was made to reproduce the lubricating and weakening effects of the wet marl by saturating each model with water for a period of 24 hr after the elastic studies had been completed and before testing the model to destruction.

The elastic studies were performed in the manner already described, after which the coating was carefully removed and the model foundation beds soaked in water. Two possible direct modes of failure were considered: (a) sliding and (b) overturning, under the influence of the reservoir water pressure and resultant horizontal thrust on the dam. In the general case the danger of overturning is resisted by the mass and characteristic triangular cross-section of the structure, possibly reinforced by fastening the upstream side of the dam to the foundation rocks by means of prestressed tendons or cables, represented in model no. 5. Shear forces that result from the hydraulic thrust are resisted by the material properties of the structure and the foundations upon which it rests. There is little difficulty in providing guaranteed safety against sliding in the body of the dam, but there are two possible sources of weakness in a laminated foundation where special precautions may be necessary. One obvious source of weakness is the joint between the dam and its foundation, generally strengthened by some such precautions as lowering the base of the structure several feet into the rock, giving a slope to the contact plane by lowering the upstream side, or by providing reinforced keys to the structure, extending these into the rock below the contact plane. Another, sometimes less obvious, source of weakness is the possible existence of discontinuities such as bedding planes, cleavage, joints, and fissures in the foundation beds. When they are horizontally dispersed, such discontinuities provide possible additional planes of sliding below the contact between the dam and its foundations. Generally speaking the same precautions in design as have already been outlined, must be used, but since the danger is in this case extended in depth the precautionary measures must also be extended and intensified. It is with such a case that the study was concerned.

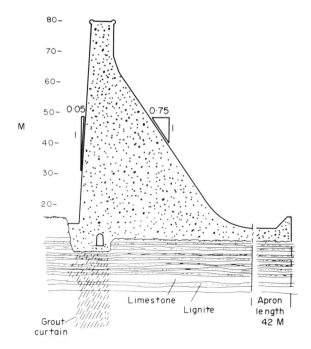

Fig. 14.20. Typical section of dam and foundation beds modeled.

When testing each model to failure, after surplus water had drained off, the model was placed slowly under load, first by gradually increasing both the gravity load and the horizontal thrust to the maximum prototype condition. Then, keeping the gravity load constant the horizontal thrust was increased until the dam began to slide. The onset of this condition was indicated by a sudden instability in the pressure gauge attached to the horizontal hydraulic jack, indicating that the model was ceasing to resist thrust.

A point of interest when comparing the behavior of the various models was that in models 6 and 7, which were reinforced by a system of rock bolts extending through the apron of the

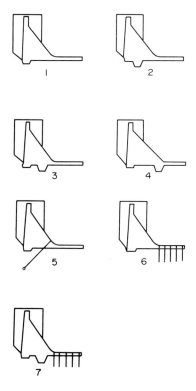

Fig. 14.21. General proportions of the models tested.

dam and the foundation beds below, the rock bolts acted as passive reinforcement over the normal range of **prototype** load. Indeed, when the horizontal thrust was increased from zero to the maximum prototype condition, representing the reservoir full, the tension in the rock bolts decreased slightly due to the compression of the foundation strata by the resultant thrust of the vertical and horizontal loads. The rock bolts only became active strengthening elements when slip between the dam and the foundation beds, or between separate layers of the foundation beds, actually began. Sliding of the rock-bolt reinforced model no. 6 began when the applied horizontal thrust represented 2.5 times the maximum prototype condition. In other words, this construction provided a factor of safety of 2.5. A similar experiment was conducted to investigate the effect of increased confinement of the strata below the apron, such as would be obtained either by increasing the pretension in the rock-bolt system or by increasing the weight of the apron. The results suggested that the rock bolt pretension, or the apron dead weight load, would need to be increased by a factor of 3 to raise the factor of safety against sliding from 2.5 to 3.0. The most effective foundation reinforcement system appeared to be that represented in model no. 5, which included steel tendons anchored deep in the strata below the upstream face of the dam and tensioned against the downstream face of the dam. This gave a factor of safety against sliding of approximately 4.0, as compared with 3.0 for the extension key (model no. 3), 2.5 for the rock-bolt system (model no. 6), and 2.25 for model no. 1.

Laminated Photoelastic Models

Many transparent and stress-birefringent materials are available for use in photoelastic models. Some details of those most commonly used (up to 1969) are listed by Hendry in Table 14.6. While any of the materials may model elastic deformations in a homogeneous isotropic prototype, detailed translation of the results obtained on a homogeneous model to a non-homogeneous prototype is seldom possible. To a limited extent non-homogeneous prototypes may be represented by composite models, made of two or more different stress-birefringent materials, but certain problems then arise.

It is conventional, when modeling a homogeneous elastic prototype in two dimension, to neglect Poisson's ratio, if the model is sufficiently rigid to stand up to the applied loads without supporting restraint on its front and rear faces. That is, the study is made under plane stress conditions, and Poisson's ratio is assumed to be the same, all over the model. However, the effects of Poisson's ratio become more important if the faces of the model are restrained, or if the model is thick, such as one studied by the three-dimensional frozen stress technique. In this case the observed conditions correspond to plane strain. It is possible to apply corrections for the Poisson effect, to the measurements made on a plane stress model, but this is seldom done. It may be argued that, when modeling earth masses, the result is seldom so precise as to render any correction meaningful. In any event, any errors are likely to be on the safe side. That is, the plane stress model is weaker than the plane strain prototype, if the criterion of failure is assumed to be maximum permissible principal stress.

However, if a layered or composite model is made up of different photoelastic materials those materials will have different elastic properties. It may be possible, from the limited range of available materials, to match model and prototype in terms of relative moduli of elasticity, but if this is done it is most unlikely that the comparative layers will also match in terms of Poisson's ratio. Conversely, if Poisson's ratio is used as a matching criterion the relative values of Young's modulus in the different layers will be wrong. The different photoelastic materials will also have different stress-optical properties, which means that the isochromatics seen in a composite photoelastic model present a distorted picture of the shear stress distribution in the model itself, and therefore a still less representative picture of the prototype situation. Furthermore, the surface finish of the materials at points of contact between the different layers must be very precise, to eliminate local stress concentrations.

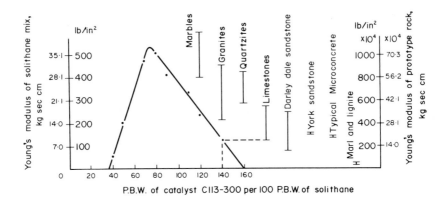

Fig. 14.22. Young's modulus for different mixes of urethane rubber, with comparative
rock types (Rankilor and McNicholas).

TABLE 14.6. Properties of Materials for Photoelastic Models (Hendry)

Type of Material	Commercial names	Tensile strength (MN/m²)	Young's modulus (E) (MN/m²)	Poisson's ratio	Unit fringe value (F) (MN/m²)	Figure of merit (E/F)	Remarks
Epoxy resin	Araldite CT 200	72.4	2895.9	0.35	0.41	7100	At room temperature
			13.45	–	0.0096	1400	At 100°C
Epoxy resin	Araldite MY 753	62.06		0.35	0.31	10,000	At room temperature
			13.10		0.0093	1410	At 80°C
Glyptal resin	Bakelite BT-61-893	117.21	4240.4	0.36	0.63	6700	At room temperature
		0.96	7.58	0.5	0.011	690	At 114°C
Phenol-formaldehyde	Catalin-800	48.27	2068.5	0.42	0.27	7600	At room temperature
					0.018		At 80°C
Polystyrene	Marcon				–	–	At 60°C
Polystyrene	Fosterite				–	–	At room temperature
Polystyrene	Castolite	55.85	4860.9	0.36	1.09	4460	At room temperature
		2.41	27.99	0.5	0.057	490	At 118°C
Cellulose nitrate	Celluloid	48.27	2068.5	0.4	1.55	1350	At room temperature
Plate glass			62,055.0	0.25	21.93	2820	At room temperature
Acrylic resin	Perspex (Plexiglass)	48.27	3102.7	0.38	5.90	525	At room temperature
Allyl diglycol	CR-39		2068.5	–	0.59	3510	At room temperature
Gelatine 13% aqueous solution	–		0.04	–	0.001	–	At room temperature

Rankilor and McNicholas have described a method of reducing some of these difficulties by employing one material — urethane rubber — to represent layered prototype materials. By making various mixes of catalyst and polymer, and testing samples for Young's modulus and Poisson's ratio, a suitable range of rubbers was produced, to simulate rock types, keeping the relative Young's moduli in comparative order (Fig. 14.22).

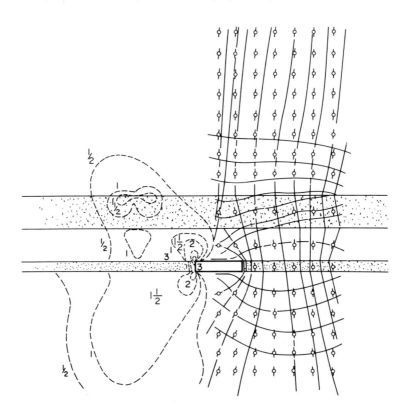

Fig. 14.23. Stress directions and stress concentrations on a layered photoelastic model
(Rankilor and McNicholas).

Rankilor applied his materials to study the stress concentrations on a layered photoelastic model simulating a mine excavation. The drillhole technique to determine principal stress directions was also applied here (Fig. 14.23). A subsequent application of the technique to model the process of earth subsidence over mine workings is also described by Rankilor. The simulated excavations were generated after the model had been assembled, by the manner indicated in Fig. 14.24. A photograph of the model, under load, is shown in Fig. 14.25. This model is of particular interest in that it displays the formation of an inverted wedge of deformed ground of progressively increasing extent above the zone in which the lateral extent of deformation decreases with extension into the roof strata. This is because the width of excavation exceeds the critical width for that depth. Had the excavation been less than the critical width then deformation at the layer interfaces would have decreased to a minimum lateral extent to a point above which all the superincumbent layers in the model deform more or less as a continuous beam (Fig. 14.26).

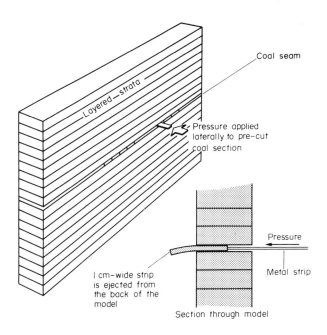

Coal seam

Layered—strata

Pressure applied
laterally to pre-cut
coal section

Pressure

I cm-wide strip
is ejected from
the back of the
model

Metal strip

Section through model

Fig. 14.24. Technique for generating the "mining excavations in a layered rubber model (Rankilor).

Gelatin Models

There are two commonly used methods of reproducing the mass distribution of body forces in a model. One is to spin the model in a centrifuge. If the model is to be of any useful size, however, the equipment and facilities required for this are costly. While centrifugal loading on small specimens is more feasible and sometimes very useful, for example for tensile testing of brittle materials, when the reproduction of gravitational body forces on a scale model are concerned an alternative method is available that will bring out the desired information far more simply and at considerably less cost. That method is to observe the birefringence in a gelatin model.

Gelatin is a most sensitive photoelastic material. Although in itself it is too soft for shaping and forming, when mixed with glycerin its mechanical strength and useful life are increased. The modulus of elasticity of the mix increases with increase in gelatin content, as also does the stress/fringe constant. The relationships are recorded by Farquharson in Fig. 14.27. A method of determining the stress/fringe value is described by Frocht (Vol. 1, p. 346).

Gomah describes the use of gelatin models to study open-pit slope stability. A 15–25–60% by weight mix of gelatin, glycerin, and distilled water was used, molded into sheets 100 mm thick. Samples of each mold were tested for density, and each sheet trimmed and calibrated to determine its material fringe value. The sheets were trimmed to 35.5 X 30.5 cm from which open pits slopes of 30°, 45°, 60°, and vertical, to depths of 127 mm and 178 mm, were modeled.

The objective of the investigation was to determine the probable stress distribution in a rock mass intersected by slopes and cuts of specific geometry, assuming the rock mass to be homogeneous, isotropic, and elastic. It is of considerable interest to compare the fringe patterns observed by Gomah under plane stress conditions, with those observed under plane strain. One

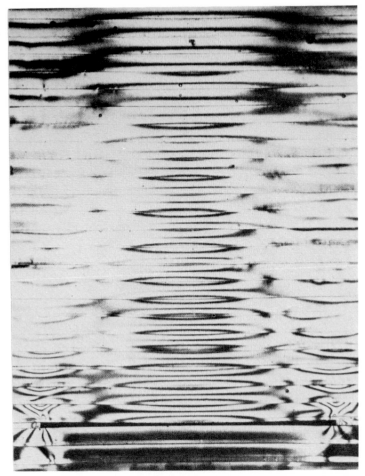

Fig. 14.25. Photoelastic layered model subsidence over an excavation wider than the
critical width for that depth of cover (Rankilor).

of the plane stress models, tested without confinement of its vertical front and rear planes,
displayed a fringe pattern similar to that shown in Fig. 14.28. Another similar model, tested
with its front and rear faces confined as a sandwich between glass plates each 25 mm thick,
to reproduce plane strain conditions in the model, displayed isochromatics similar to those
shown in Fig. 14.29. Two things are immediately apparent. (1) The model is too small. The
stress concentrations at the lower corners of the model, due to boundary restriction, affect
the stress distribution in the area of interest, particularly at the "toe" of the slope. (2) The
isochromatics displayed under plane stress and plane strain are not the same. There are important differences, again at the toe of the slope, where the confined model shows much higher
values of shear stress, and a higher stress concentration, than does the unconfined model. The
unconfined, plane stress, model shows the fringes following the slope contour, indicating a
uniform shear stress with increasing depth along the slope. This is a most unlikely situation
in practice. Of the two fringe diagrams the information offered by the plane strain, confined,
model more nearly approaches reality. The results of Gomah's study thus may be summarized

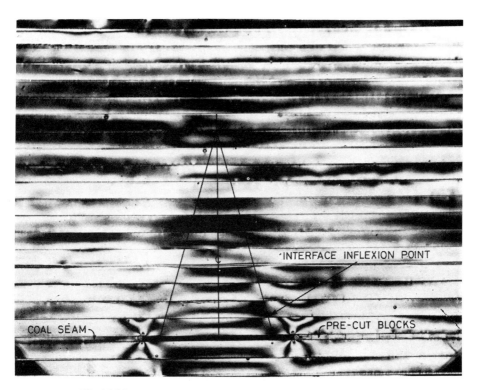

Fig. 14.26. Photoelastic layered model of subsidence over an excavation
less than the critical width (Rankilor).

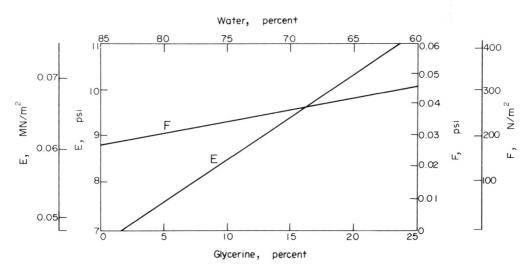

Fig. 14.27. Modulus of elasticity and stress-optic constant of
gelatin-glycerine-water mixtures (Farquharson).

by Fig. 14.30(a) and (b) showing the principal stress and shear stress trajectories for the slope.

In a subsequent stress analysis of open-pit slopes, comparing a finite-element mathematical solution with a three-dimensional photoelastic model, Young also compares the plane stress and plane strain solutions, and points out that the lower stress concentration at the toe, developed under the plane stress state, may lead to significant errors in slope design, especially if the slope is excavated in material having a high Poisson's ratio and if the slope failure is determined by the maximum shear stress at the toe. A plane strain condition, therefore, should be assumed in the analysis of slope stability, in such cases.

Fig. 14.28. Isochromatic fringe pattern in a 45° slope gelatin model, 10 cm thick. Cut 18 cm high. Model unconfined. (After Gomah.)

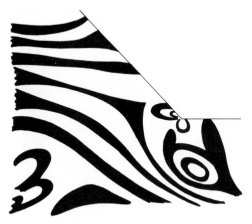

Fig. 14.29. Isochromatic fringe pattern in a 45° slope gelatin model, 10 cm thick. Cut 18 cm high. Model confined on front and rear. (After Gomah.)

Three-dimensional Photoelastic Models

Three-dimensional photoelasticity by the frozen stress technique has been applied only to a very limited extent in geotechnology. Agarwal describes a method of analysing the stress distribution in an elastic material in which two circular openings of equal diameter pass at right angles, one over the other, in a uniaxial stress field. The variable studied in this problem was the distance between the centerlines of the two openings. Young's slope-stability study included a three-dimensional gelatin model, in which the gravitational body stresses were frozen by placing the model, of 0.3 m cube, in a refrigerator to set and leaving it there for a period of 1 week. By experimenting with various mixes a composition was found which remained stable for a long enough period, when removed from the refrigerator, to enable the cube to be sliced for examination by polariscope, in the customary manner.

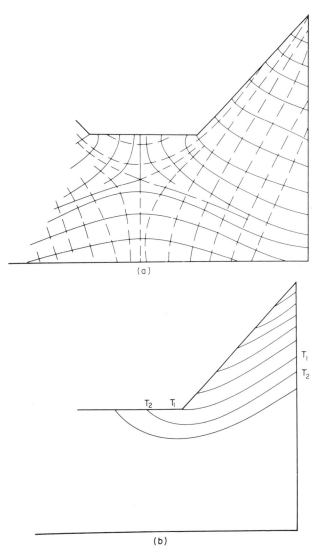

(a)

(b)

Fig. 14.30. Stress trajectories in a 45° slope model (a) Principal stresses. (b) Shear stresses. (After Gomah.)

315

Analog Computers

A photoelastic model is a form of optical analog, and it can depict what may be, in reality, a complex stress distribution, in a form that the observer may assimilate quickly, as a visual image. However, it is only when the strength or failure criterion of the prototype is one of shear stress magnitude, that the isochromatics are very meaningful in themselves. In other circumstances the individual principal stress magnitudes and directions must be determined, and sometimes this can be done very conveniently and precisely by mathematical analysis, aided by computer, without the need for a photoelastic model at all. Photoelastic models are mainly useful as educational aids, as descriptive aids in reports, and as reconnaissance aids to provide a rapid qualitative assessment of the situation before embarking upon a more elaborate and costly geotechnical study. Reflection photoelasticity, using the coating technique, is also invaluable as a method of observing the strain distribution over composite models, and over heterogeneous materials too complicated for mathematical treatment.

The conducting-paper and electrolytic tank methods of determining the principal stress summation, used in combination with photoelastic models, are, in fact, electrical analogs, in which the measurement of voltage can represent the measurement of stress, if appropriate boundary conditions are applied to the analog.

The observed elastic behavior of the hard rocks in South African gold mines justifies the application of elastic theory to design problems in those mines, using the classical concepts of continuum mechanics in a homogeneous isotropic material. By assuming that layered sediments are transversely isotropic, elastic theory may also be applied to predict the surface displacements induced by coal mining. Both the isotropic and transversely isotropic models may be represented by electrical analogs. By feeding the readings of an electrical analog into a digital computer, the design problem can be translated into numerical terms. This combined analog-numerical technique, or hydrid computer, allows of the numerical solution of many problems in geotechnology. It is proving to be particularly useful in connection with the mining of tabular deposits and the design of excavation support systems incorporating strata pillars.

Finite Element Models

The essence of classical elastic theory is that the behavior of materials under load may be described in terms of infinitesimal dimensions and then extended to the material as a whole, by a process of integration, assuming the material to be a homogeneous and isotropic continuum. In geotechnology, the question as to whether or not the assumption of a continuum is justified is primarily a matter of size and scale. The science of geophysics, for example, is built on assumptions that the Earth's crust and mantle, under certain conditions, behaves as an elastic continuum. But to an observer viewing an isolated exposed rock cut it will be obvious that this cannot be so when the exposed rock is seen to be fissured, laminated, or jointed. On a smaller scale the behavior of rock materials, which may be obviously granular or crystalline and therefore discontinuous, is often described in terms of the results obtained by testing specimens in which the internal discontinuities in the material are small, relative to the overall dimensions of the specimen. An elastic analysis may then be made, assuming that the stress-strain relationships at any point and in any direction will be the same throughout.

But a continuum may exist in which isotropy and homogeneity do not occur. The material may include directional factors and its stress-strain relationships may vary, both in time and

316

space. The analysis of such a system may be approached by the finte element method. This is an extremely useful stress analysis tool and one which has been rendered practicable by the advent of computers. It can take into account both linear and non-linear stress-strain relationships and can also include the effects of heterogeneity and anisotropy in the model.

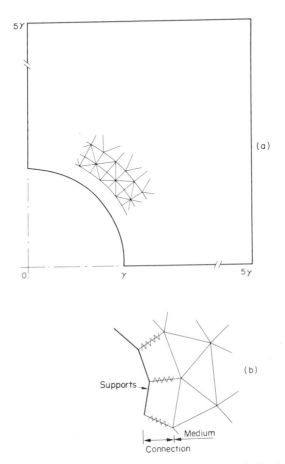

Fig. 14.31. Finite element model of a circular tunnel lining and idealization of interaction between support and rock, on a finite element model (Young).

In this method the continuum is assembled from a network of finite elements. Each element is of a geometrical shape, usually triangular but sometimes quadrilateral, connected to other elements in the continuum. The pattern of displacements and forces at the mesh junctions or "nodes", in relation to the external loads applied to the network and the stiffness of the network, is described in terms of a number of simultaneous equations and analyzed numerically to compute the stresses and strains over the model. Zienkiewicz discusses continuum mechanics as an approach to geotechnical problems in rock masses, and gives examples of the application of the finite element technique to typical dam foundations and tunnel lining projects. Blake also describes the application of the method to boundary-value problems in rock mechanics, and compares the finite element solution with the classical elastic and photoelastic solutions to the problem of stresses round a circular excavation.

317

Geotechnology

Applications of the finite element technique to geotechnology have proliferated greatly during the past 5 years, and several investigators have compiled computer programs to deal with specific problems. A 1972 symposium on the subject, sponsored by the U.S. Army Engineer Waterways Experiment Station, included a total of thirty-five authoritative papers concerning dams, excavations, and slope stability, foundations and pavements, seepage, consolidation, creep, earthquake analysis, and earth-structure interaction. While some investigators may appear to have confused the merit of computer programs to generate finite element solutions with the merit of the solution itself, there can be no doubt that the method is a very powerful tool with which to handle complex problems of analysis and design. The success of the method, however, in practical terms, is entirely dependent on the validity of the input data.

Limitations on the capability of computers to deal with complex problems also exist. While future developments will undoubtedly improve the situation, the present generation of computers does not possess the storage capacity that is needed to represent a real geotechnical situation in all its detail.

Discontinuous Mathematical Models

Mathematical models describing the behavior of earth materials may be classified as being linear (in which cause and effect are directly proportional to one another) or non-linear. A non-linear relationship between cause and effect, within certain limits, may be considered to be comprised of a number of linear increments, each one over correspondingly smaller limits, the total effect due to a number of causes then being equal to the sum of the individual effects. In such a manner rheological models may be assembled, to describe not only elasticity but also visco-elastic and plastic behavior, in earth materials.

Another classification may be made on the basis of the structure of the earth material, which may be considered to be either continuous or discontinuous. The separate elements or constituents of a continuous material are considered to be in smooth contact at all points on all contiguous surfaces. During the processes of deformation this uniform contact is maintained. That is, there is no internal separation within the medium, and no new points of contact are formed, nor are any removed. We have observed that the concept of a continuum may be legitimately applied to earth materials that are obviously not continuous, provided that the proportions of size and scale are appropriate. We have observed also, that by dividing the continuum into a number of separate finite elements the mathematical treatment may, to a limited extent, take account of changes in the physical properties of the material, over the continuum.

A discontinuous earth material is comprised of separate, discrete, particles or grains. The conditions of contact between the grains changes during the deformation process, and contact is not contiguous. There may be many places where open separation exists between the constituent grains. Discontinuous mechanics applies mathematical concepts to such materials, to describe both linear and non-linear behavior. There are various methods of approach. Since we are dealing with assemblages of discrete fragments statistical methods are obviously appropriate. These may be applied in conjunction with probability concepts, for example to estimate the strength of a material, or to forecast the deformational behavior of an earth mass. Litwinisyn analyzes the formation of subsidence troughs in the ground overlying mine workings, making use of such an approach.

Discontinuous mechanics has proven to be a valuable tool for the examination of problems of soil mechanics, and it is being extended to rock masses. Trollope describes the procedure,

318

with typical applications, in relation to the determination of static and dynamic stresses, the establishment of failure criteria for rock materials and rock masses, and problems of fluid flow through porous media.

Assessment of the Results from Model Studies

It must never be forgotten that the results obtained by means of a model study, be it mathematical, physical material, or photoelastic, describe only what has been calculated for, or observed on, the model. The application of the results to reality, on the prototype, depends upon the veracity of the results and the validity of their transfer. It is seldom, if ever, that absolute similitude between model and prototype exists. The model results, in themselves, have no value. It is what happens on the prototype, in the field, that is the engineer's concern. Therefore any calculated result, or observed behavior on a model, should be regarded as suspect unless and until it has been checked against field observations. Only when theory and practice have been shown to agree is it possible to use the results of theory to forecast what is likely to happen in practice, that is, to use mathematical or other model studies as a secure basis for design. For example, the subsidence profile determined by a probabilistic interpretation of data to which statistical methods have been applied, when compared with the surveyed and measured profile, will enable a correction factor to be determined for that site. This factor may then be applied in calculated estimates of subsidence for similar conditions at other sites.

Again, the information gained from a numerical analysis of a pillar support system represented by an electrical analog can be compared with underground measurements or experience, then adjusted, or appropriately modified, and fed back into the computer. In this way, by a process of successive approximation, we may hope to arrive at a computer solution reasonably close to reality and this may provide a basis for projection into future designs.

The dilemma of the geotechnical engineer is that, all too often, he is expected to provide quick solutions to problems for which there are no easy answers, and very often not one answer, but a number of possible alternatives. Without adequate information upon which to draw a firm conclusion, even the expert cannot say which of the alternatives is likely to provide the optimum result. The practical engineer may well feel, in such a situation, that he is just as able, and perhaps better qualified intuitively, to guess the answer, than would be a more academic or research-minded consultant.

Hence there has grown, in geotechnology, and particularly in engineering geology and rock mechanics, a division of interest between the engineer and the geologist, and between practice and theory. In rock mechanics, the gap is widening, with theoretical aspects advancing, at an accelerating pace, away from practical applications. Engineering geology is being pursued, on the one hand by engineers who have virtually no background in geology other than that gained by experience in the field, and on the other by geologists whose training has omitted all the engineering sciences.

It has been the object of the author, in this text, to attempt to bridge this gap at an elementary level and to prepare the groundwork for further reading, dealing with field problems in geotechnology and applied rock mechanics. If, in so doing, he has interested the reader sufficiently to stimulate him into going further into the subject, and possibly acquiring the necessary expertise in geology and engineering to apply his knowledge towards the control and improvement of the human environment, and to the practice of engineering in earth materials, his object will have been served.

Selected References for Further Reading

AGARWAL, R. Three-dimensional photoelasticity involving mining problems. *4th Int. Conference on Strata Control, Columbia Univ.,* New York, 1964.

BARTON, N. R. A model study of rock-joint deformation. *Int. J. Rock Mech. Min. Sci.,* vol. 9, pp. 579–602 (1972).

BERRY, S. The ground considered as a transversely isotropic material. *Int. J. Rock Mech. Min. Sci.,* vol. 1, pp. 159–167 (1964).

BLAKE, W. Application of the finite element method of analysis in solving boundary value problems in rock mechanics. *Int. J. Rock Mech. Min. Sci.,* vol. 3, pp. 169–179 (1966).

CROUCH, S. L. Two dimensional analysis of near-surface single-seam extraction. *Int. J. Rock. Mech. Min. Sci.,* vol. 10, pp. 85–96 (1973).

CUNIE, J. B. Experimental structural geology. *Earth Sci. Rev.,* vol. 1, pp. 51–67 (1966).

DESAI, C. S. (Ed.) *Application of the Finite Element Method in Geotechnical Engineering,* Soil Mechs. Information Analysis Center, U.S. Army Engineer Waterways Exptl. Station, Vicksburg, Miss., 1972.

DHAR, B. B. and COATES, D. F. A three-dimensional method of predicting pillar stresses. *Int. J. Rock Mech. Min. Sci.,* vol. 9, pp. 789–802 (1972).

DUVALL, W. I. *Stress Analysis Applied to Underground Mining Problems,* U. S. Bur. Mines Rept., Invest. No. 4192 (1948).

EVERLING, G. Model tests concerning the interaction of ground and roof support in gate roads. *Int. J. Rock Mech. Min. Sci.,* vol. 1, pp. 319–326 (1964).

FARQUHARSON, F. B. and HENNES, R. G. Gelatin models for photoelastic analysis of earth masses. *Civil Engineering,* N. Y., vol. 10, pp. 211–214 (1940).

FROCHT, M. M. *Photoelasticity,* vol. 1, Wiley, N. Y., 1941.

FUMAGALLI, E. Model simulation of rock mechanics problems. In *Rock Mechanics in Engineering Practice,* Stagg and Zienkiewicz (Eds), pp. 353–384, Wiley, London, 1968.

GOMAH, A. H. Gelatin models for photoelastic study of slope stability in open-pit mines. Ph. D. Thesis, Univ. of Utah (1965).

GYENGE, M. *A Computer Program for Calculating Principal Stresses in Photoelasticity,* Dept. of Energy, Mines & Resources, TB 88, Canada, 1967.

HIRAMATSU, Y. and OKA, Y. Stress on the wall surface of levels with cross-sections of various shapes. *Int. J. Rock Mech. Min. Sci.,* vol. 1, pp. 199–216 (1964).

HOBBS, D. W. Scale model studies of strata movement around mine roadways. *Int. J. Rock Mech. Min. Sci.,* vol. 3, pp. 101–127 (1966).

HOEK, E. *The Application of Photoelastic Models as Means of Solving Problems of Stress and Fracture around Excavations at Depth,* CSSR South Africa Publication, RN 117 (1960).

HOEK, E. Rock fracture around mining excavations. *4th Int. Symposium on Strata Control,* Columbia Univ., N. Y., 1964.

HOEK, E. *Rock Fracture Under Static Stress Conditions,* CSIR South Africa, Report ME 9383 (1965).

JANELID, I. and KVAPIL, R. Sublevel caving. *Int. J. Rock Mech. Min. Sci.,* vol. 3, pp. 129–153 (1966).

JOHNSON, A. M. *Physical Processes in Geology.* Freeman, Cooper & Co., San Francisco, 1970.

KVAPIL, R. Photoelastic research in rock mechanics. *SNTL Technical Digest,* Prague, Czechoslovakia, vol. 2 (1960).

KVAPIL, R. Tectonic experiments in natural rocks. *Int. J. Rock Mech. Min. Sci.,* vol. 1, pp. 17–30 (1963).

KVAPIL, R. Gravity flow of granular materials in hoppers and bins in mines. *Int. J. Rock Mech. Min. Sci.,* vol. 2, pp. 277–304 (1965).

LITWINISZYN, J. On certain linear and non-linear models. *4th Int. Conf. on Strata Control,* Columbia Univ., New York, 1964.

LU, P. H. and WRIGHT, F. D. Stress analysis and design criteria for prestressed concrete mine sets. *Int. J. Rock. Mech. Min. Sci.,* vol. 4, pp. 345–351 (1967).

MANDEL, J. Tests on reduced scale models in soil and rock mechanics, a study of the conditions of similitude. *Int. J. Rock Mech. Min. Sci.,* vol. 1, pp. 31–42 (1963).

PHILLIPS, D. W. and HUDSPETH, J. M. The forces induced by the extraction of coal, and their effects on coal-measure strata. *Trans. Inst. Min. Engrs, London,* vol. 85, pp. 37–56 (19   ).

POTTS, E. L. J. Stress distribution, rock pressure, and support. *J. Leeds Univ. Mining Soc.,* No. 30, pp. 107–121 (1954).

RANKILOR, P. R. The construction of a photoelastic model simulating mining subsidence phenomena. *Int. J. Rock Mech. Min. Sci.,* vol. 8, pp. 433–444 (1971).

RANKILOR, P. R. and McNICHOLAS, J. B. The preparation and use of a stress-sensitive material in multi-layer photoelastic models. *Int. J. Rock Mech. Min. Sci.,* vol. 5, pp. 465–474 (1968).

ROBERTS, A. A model study of rock foundation problems underneath a concrete gravity dam. *Engng. Geol.,* vol. 1, pp. 349–372 (1966).

SOLOMON, M. D. G. Two-dimensional treatment of problems arising from mining tabular deposits in isotropic or transversely isotropic ground. *Int. J. Rock Mech. Min. Sci.,* vol. 5, pp. 159–185 (1968).

SOLOMON, M. D. G. Stability, instability and design of pillar workings. *Ibid.*, vol. 7, pp. 613–631 (1970).

STARFIELD, A. M. and FAIRHURST, C. How high speed computers advance design of practical mine pillar systems. *Engng. Min. J.*, vol. 169, no. 5, pp. 78–84 (May 1968).

STIMPSON, B. Modelling materials for engineering rock mechanics. *Int. J. Rock Mech. Min. Sci.*, vol. 7, pp. 77–121 (1970).

THAKUR, D. N. Some aspects of rock pressure modelling on equivalent materials. *Int. J. Rock Mech. Min. Sci.*, vol. 5, pp. 355–369 (1968).

TROLLOPE, D. H. The mechanics of discontinua or elastic mechanics in rock problems. In *Rock Mechanics in Engineering Practice,* Stagg and Zienkiewicz (Eds.), pp. 275–319, Wiley 1968.

WANG, C., BOSHCOV, S. H. and WANE, M. T. The application of barodynamic photostress technique to the study of behavior of rock beams loaded by their own weight. *9th Symposium on Rock Mechs., Colorado, 1967,* AIME, New York, 1968.

WANG, F. D., SUN, M. C. and WOLGAMETT. A computer program for generating finite element models of mine structures. U. S. Bur. Mines Inf. Circular 8510 (1971).

YOUNG, D. S. Stress analysis in open-pit mines, Ph. D. Thesis, Univ. of Utah, 1971.

ZIENKIEWICZ, O. C. Continuum mechanics as an approach to rock mass problems. In *Rock Mechanics in Engineering Practice,* Stagg and Zienkiewicz (Eds.), pp. 237–273, 1968.

CHAPTER 15

Gathering and Recording Data on Geology, Rock Structure, and Rock Classification

Recording Discontinuities

The geotechnologist involved in site investigation has at his command all the techniques of geology and geophysics, allied to soil and rock mechanics. While standard procedures for soils are generally applied no similar concensus of agreement yet exists for rocks. Since many engineering problems of strength and stability are dominated by the discontinuities in the rock mass rather than by the material structure of the rock itself, it is important that joint, cleavage, and fault fracture patterns be observed in detail, and recorded.

Discontinuities in rocks may be studied over a size range from thin sections observed under the microscope to regional complexes involving whole mountain ranges, observed by satellite. Measurements at the microscopic level are made on oriented samples using a petrological microscope fitted with a universal stage. The techniques of petrofabric analysis, thus applied, are described by Emmons, Fairbarn, Turner and Weiss, and by Ramsay. Measurements on the site may be made by photogrammatic methods, or directly observed point-by-point by means of a compass and clinometer. Whichever method is applied, the required information includes: (i) Strike of joint or fracture traces. (ii) The length of the fracture traces. (iii) Joint or fracture density. (iv) Spacing between fractures. (v) Dip of the fracture planes. Many hundreds of joint attitudes may have to be observed over even a modest site area. The next step is to display these measurements in a manner suitable for subsequent analysis, so that an interpretation of the joint system can be made.

Optical Data Processing

The "preferred directions" of jointing, cleavage, and fracture patterns may be deduced from a statistical analysis of counted fractures, as described by Muller. In recent years optical data processing has been allied to photogrammetry for this purpose. Virtually instantaneous read-out of analysis is possible from what at first appears, on visual inspection, to be the most complex geological fabric and structure.

Methods of Recording Joint Data

Various graphical techniques are applied, to record the observed data. The most elementary system is to draw, on a map, symbols to represent the dip and strike of the observed discontinuities. The method has obvious limitations, as to the extent to which detail may be recorded. The deduction of a valid analysis is only possible if sufficient information is available, and the almost inevitable result of trying to plot all the necessary detail directly to scale is to render the map virtually indecipherable.

To help resolve these difficulties various statistical plots may be used. The simplest of these is a histogram on which the joint frequency is plotted along one axis and the strike direction on another axis. Directional information, obtained by subsequent treatment of this data, may be represented by a vector system in which the arrow shows the direction of maximum joint frequency, or "preferred direction" while the length of the arrow denotes the relative magnitude or frequency of the preferred joint system.

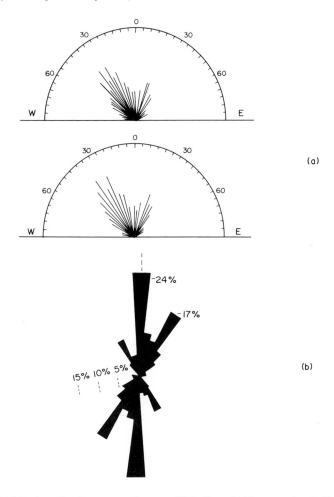

Fig. 15.1(a) Fault strike. Frequency diagrams. (b) Strikes of 410 range-edges in Nevada, as measured from USGS topographical sheets (Freidman).

A graphical representation to show the angular relationships may also consist of a "joint rose" or "star" diagram. Here the strikes and joint frequencies are represented by the directions and lengths of vectors plotted either on a half-circle, or a full-circle. Directions are usually plotted to represent samples contained in arcs of from 1° to 5°, while magnitudes are plotted to scale, the radius of the circle being made equivalent to a specific number of joints.

Spherical Projections

A two-dimensional projection to represent a sphere in space may be constructed to contain

323

the data which in nature is observed in three dimensions. The sphere is thus represented by a circle and the location of points on the surface of the sphere is projected on to the circle and plotted by reference to a network of coordinates. Depending upon whether the sphere is viewed along its polar axis, or at right angles to that axis, the circle is divided into either a polar or an equatorial network (Fig. 15.2).

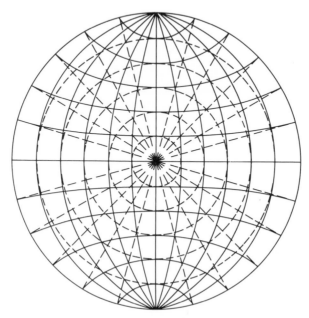

Fig. 15.2. Comparison between normal (dashed lines) and equatorial (full lines) nets in equal-interval projections (Vistelius).

Data may be projected from the sphere on to the circle in any one of three ways: (a) So that angles measured between specific curves on the sphere are represented by the same angles between the corresponding curves on the projection. This is the equiangular or stereographic projection. (b) Corresponding areas on the sphere and on the projection are equal. This is the equal-area or equal-size projection. The equality of corresponding areas means that the scale changes in different directions and between different locations on the projection. (c) Equal-interval projections, in which the scale in one of the principal directions has a constant value.

Both the stereographic and the equal-area projections are used in geotechnology, in connection with petrofabric analysis, the study of structural geology, and in slope stability analysis. In both projections planar discontinuities are pictured as passing through the center of the reference sphere, so that they cut the spherical surface along great circles which plot as circular arcs.

Stereographic projection

It is conventional to project a stereographic plot on to a horizontal plane either up or down, so that the plot represents either the upper or the lower hemisphere. The projection is made on to an equatorial stereonet or Wulf net (Fig. 15.4). On such a net the perimeter circle

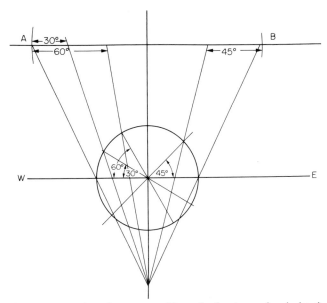

Fig. 15.3. Construction of a stereographic projection (upper hemisphere).

represents the horizontal plane, and diameters of this circle represent either vertical planes or the strike directions of inclined planes. For example, Fig. 15.5 shows three inclined planes dipping at angles 60° and 30° due east and 45° due west, respectively. To project these in the upper hemisphere, the stereonet perimeter represents a horizontal circle of diameter AB. On the net, AB is aligned in the W–E direction, which is the line of maximum slope for each of the inclined planes. The great circle arc representing each of these respective planes is then transferred on to a sheet of tracing paper laid over the net. The traces of planes dipping in other directions are traced in a similar manner, by first marking the appropriate strike direction on the tracing paper and then rotating this about the stereonet center until the full dip direction coincides with one of the major axes of the net. The arc is then traced through from the net below. For example, on Fig. 15.6 the arc AB represents an inclined plane dipping 50° in a direction 170° azimuth. The diameter AB is the strike direction of this plane, and a line normal to AB in the plane is along the line of full dip. The point P is the intersection of the normal to this plane and the surface of the sphere. P is termed the "pole" of this plane, and the point P on the projection uniquely describes the plane concerned. Other planar discontinuities can be represented by points on the projection, in a similar manner.

Equal-area projection

Of the three spherical representations only the stereographic projection can be produced directly by geometric plot. The others are plotted by mathematical scales.

The advantage of the equal-area projection is that the plotted points may be statistically analyzed and contoured. A projection showing the location of poles, representing planar discontinuities, is termed a scatter diagram (Fig. 15.7), which may be produced by computer print-out. Manual counting is aided by grid methods, such as the Schmidt net (Fig. 15.8) or the use of contoured point diagrams. In the latter technique if circles 2 cm diameter are drawn about each

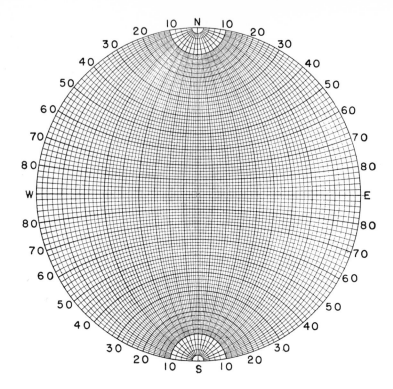

Fig. 15.4. Two-degree meridian stereonet.

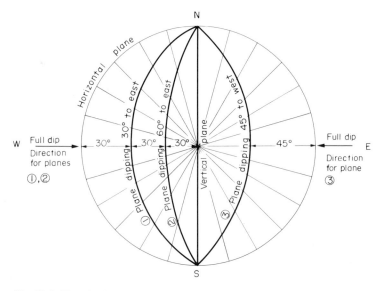

Fig. 15.5. Three incline planes striking N–S on a stereonet. (Compare Fig. 15.3.)

point to represent 1% concentrations, then the 1% contour is the limit of the circles and the 3% contour is located by the overlap of three or more circles. Mechanical aids to counting include peripheral and grid counters. Having determined the number of poles within the area of the counter, at specific intervals over the grid, each number is expressed as a percentage of the total, and contours are drawn through points of equal percentage concentration. The diagrams may then be applied to locate the preferred directions of planar discontinuities in relation to the engineering problem concerned, such as the orientation and support of excavations, the layout of blasting and mechanical boring patterns, and rock slope stability.

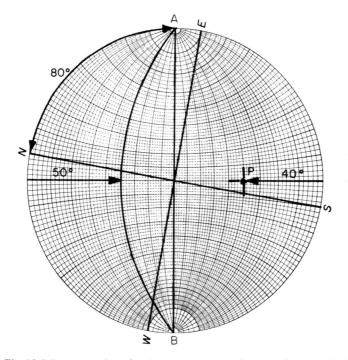

Fig. 15.6. Representation of a plane on a stereonet, by arc and pole (Calder).

Examples of fracture orientation analysis on an engineering site are reported by Mahtab *et al.,* who present a new procedure for the purpose. This procedure, coded in a computer program, identifies clusters or groupings among the fracture orientations and calculates the mean orientation of the fractures within each cluster and the dispersion or scatter among the fracture orientations. The method is important because, as a result of the increasing availability of large computers, coupled with advances in analytical methods such as the finite-element technique, boundary-value problems in jointed rock are becoming increasingly tractable. In mathematical models of geotechnical problems the orientation, spacing, and character of discontinuous surfaces, as well as their mechanical properties, are important input parameters.

Logging Rock Boreholes for Engineering Purposes

Considerable expenditure of time and money is involved in a detailed site investigation to develop a mineral property or to plan an important engineering project. Much of this expenditure goes into the drilling of exploratory boreholes, and it is therefore important that the

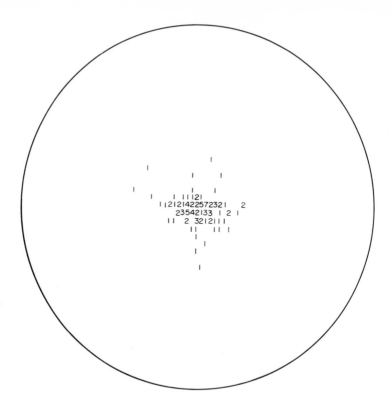

Fig. 15.7. Scatter diagram.

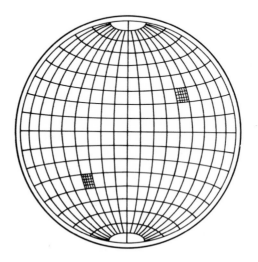

Fig. 15.8. Schmidt net.

maximum use be made of these, to bring out as much information as possible about the geological character, hydrology, and engineering properties of the strata penetrated.

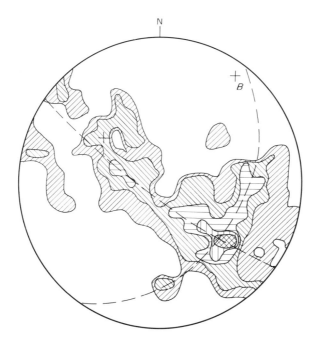

Fig. 15.9. Schmidt plot.

Geophysical Logging

Geophysical borehole logging has evolved over the past 50 years, beginning with early experiments on the measurement of magnetism, and also the electrical resistivity of rocks in the field. Subsequently, measurements of spontaneous potential, radioactivity, thermal conductivity, gravitational, and seismic properties were added.

Resistivity log

The rocks of the Earth's crust contain mineralized water to an extent depending upon their porosity. This water acts as an electrolyte, to conduct electric current between electrodes inserted into the rock. Solid minerals such as silicon are perfect insulators when dry but their electrical resistivity decreases if water is present. Measurements of electrical resistivity are generally observed in terms of a unit, the ohm-metre, which is the electrical resistance of a cube of rock of side length 1 metre. The measurements are commonly designated as "resistivity" (ohm m^2/m). Observed figures may range over typical values from 0.15 ohm m^2/m in loose water-bearing sands, to over 10,000 ohm m^2/m in massive limestones.

A resistivity log consists of a graphical plot of the values of resistivity measured at various depths along the borehole, and the information from a number of such curves over the site enables accurate sub-surface contours to be drawn. Anomalies on these contours identify the possible location of discontinuities, which may be faults, unconformities, or major fissures.

Geotechnology

The resistivity of the rock may be expressed as

$$R_0 = F \times R_w$$

where R_0 is the true resistivity of the rock,

R_w is the resistivity of the water contained in the rock,

F is termed the "formation resistivity factor".

Theoretically F varies from 1 to infinity, for when $F = 1$, $R_0 = R_w$, and when $F = $ infinity no current flows. In practice F ranges from about 4.0 for loose, porous, sand, to several hundreds in compact massive sediments.

F may be expressed in terms of a "cementation factor" m, such that

$$F = 0.62/P^m$$

where P is the effective porosity, that is, the amount of fluid contained in the rock. (For porous sediments $m = 2.15$.)

These observed relationships provide a means of exploring the hydrological properties of the rocks penetrated in boreholes. When exploring in oil sands the pores contain oil and gas, which are insulators, as well as water. An oil sand therefore has a true resistivity R_t, higher than R_0 (which it would have if its pores were filled with water only). R_t and R_0 can be computed from the resistivity log, and the amount of water in an oil sand is related to R_t and R_0 by the expression

$$S^a = R_0/R_t$$

where S is the percentage saturation (% of effective porosity) and a is an exponent (commonly 2.0).

The amount of oil in the sand is then $1 - S$.

Spontaneous potential log

Observations of rock resistivity in boreholes also present evidence of naturally occurring electrical current flow. Measurements of this natural current consist of recording the difference of potential between a moving electrode (which is attached to an insulated cable and lowered down the borehole), and a stationary electrode at constant potential, anchored at the surface. The SP log, thus obtained, usually consists of a straight-line plot in impervious rock, taken to be zero potential, with deflections from this base peaking when permeable strata are encountered. The record can be used to show the relative permeability of the strata penetrated, to correlate with sections obtained in neighboring holes, and to compute the resistivity of the strata encountered.

Microlog and lateralog

The ability of standard electrical logging to identify resistivity changes occurring in thin strata is limited by the spacing between observations and by the influence of saline mud. When the borehole being surveyed is filled with mud of low resistivity the induced current is diverted through it and through the borehole, rather than through the rock. Apparent resistivities are then recorded which are very much smaller than the true values. The techniques of "micrologging" and "lateralogging" were devised to overcome these difficulties. In micrologging the resistivity of a very small volume of rock is measured, close to the wall of the borehole. This is done with the electrodes set at very small spacing, the points of measurement being distributed over the rock in such a way that the observation is as free as possible from the influence

330

Gathering and Recording Data

of the mud. In lateralogging a multiple system of electrodes and potential controls is used, so as to focus a "sheet" current into the strata opposite the electrode system. This gives better definition of the strata and it functions very favorably in saline mud.

Radioactive logging

The radioactivity of rocks may be expressed in terms of the radiation equivalent of radium, per gram of rock. Many rocks are sufficiently radioactive throughout geological time to render this property useful as a means of differentiating between them. For example, the natural radioactivity associated with sedimentary rocks is as seen in Table 15.1.

TABLE 15.1.

Rock	Residual Radioactivity (g rad equiv./g)
Coal, rock salt, anhydrite	1×10^{-12} to 2×10^{-12}
Dolomote, limestone, sandstone	2×10^{-12} to 10×10^{-12}
Organic marine shales, potash	60×10^{-12} to 100×10^{-12}

Various methods of radioactive logging are widely employed. In the gamma-ray log a Geiger-Muller detector is lowered down the borehole. This is a device which makes use of the property of gamma-rays to render certain gases electrically conductive, when normally they would be infinitely resistive. The gamma radiators thus produce electrical impulses, which are impressed on the surface of a condenser. The potential across the condenser is then proportional to the number of pulses and is measured. Sometimes the Geiger-Muller detector is replaced by a scintillation detector, which has a higher efficiency of resolution because of its smaller length and increased speed of recording. This detector makes use of crystals such as sodium iodide, which become luminescent when bombarded by gamma-rays. The generated luminescence is observed by a photomultiplier, and transformed into electrical impulses which are amplified electronically and then recorded.

In the neutron-gamma log, artificial radioactivity is applied. High-speed neutrons, electrically neutral and having about the same mass as the hydrogen atom, are emitted from a mixture of radium and beryllium, in the borehole probe. These radiations collide with the atoms of various substances present in the strata encountered, so that their energy is absorbed. At the lower energy levels the neutrons can be captured by many common substances and especially by hydrogen. The capturing or absorbing atom then emits a "gamma-ray of capture" which has high energy. These rays penetrate a detector, to be counted in a similar manner as are natural radiations. However, the detector is sensitive only to the energetic gamma-ray of capture and not to natural radiation or to the gamma-rays emitted from the neutron source (see Figs. 15.10 and 15.11).

Temperature log

The thermal conductivity of rocks varies widely in rocks of different character, so that this physical property may be used as a means of differentiation in a borehole probe. The method is particularly useful in carbonaceous sediments, in which typical ranges of thermal conductivity are as shown in Table 15.2.

Observation of Dip

In exploratory boreholes it is important to determine the amount and direction of strata

331

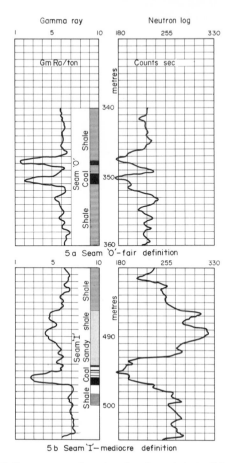

Fig. 15.10. Gamma-ray and neutron logs in coal strata (Schwaetzer).

TABLE 15.2.

Rocks	Thermal conductivity (kcal/m hr °C)
Sandstones	0.5 – 2.0
Clays, shales and mudstones	0.6 – 0.8
Coal	0.12–0.15

inclination. Various instruments are available for this purpose. One, which is widely used, is based on three simultaneous records of curves registering resistivity or spontaneous potential. The magnitude and direction of strata dip are computed from the relative displacement of the recorded curves. In all probability the borehole itself will not be truly vertical, and its orientation and inclination must also be determined. This can be done by photographing the position of a pendulum and compass system, relative to a glass graticule incorporated into a borehole camera or photoclinometer.

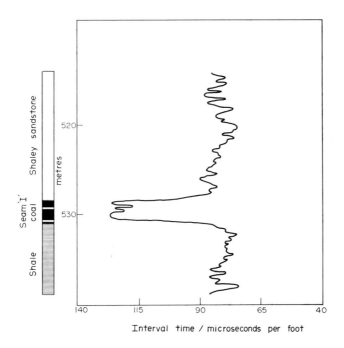

Fig. 15.11. Sonic velocity log in coal strata (Schwaetzer).

Optical Scanning in Boreholes

The number and orientation of discontinuities in the rock walls of a borehole may be observed and recorded on film, or directly monitored on closed-circuit television, using a borehole camera coupled to a surface monitor and videotape recorder.

Observation of Mechanical Properties

Direct observation of the mechanical strength of the strata encountered in a borehole may be performed by using a section gauge, which measures changes in borehole diameter. These changes are the result of erosion caused by the circulating drilling fluids, or by solution in the case of saline evaporites, and hence are related to changes in lithology. Caving of the sidewalls in friable strata may also be an aid to correlation between adjacent boreholes. In cased holes drilled to form wells for water or oil the section gauge log will be useful when selecting stations for packer insertion and pumping tests. Also, if the holes are to be used for cementation and strata reinforcement it may be necessary to compute hole and cavity volumes. While it is possible, to a limited extent, to take samples from the side of a borehole at depth, in general a detailed study of engineering properties is only possible if the hole is cored throughout, to provide samples for mechanical testing and for the determination of comparative indices such as RQD, C-Factor, the Point Load Index, and the Stability Index. The major advantage of geophysical logging is that it provides methods of exploring the rock characteristics in boreholes that are not cored, and hence which can be drilled very quickly and relatively cheaply.

333

Drilling and Sampling Techniques

The validity of the RQD and the Swedish C-Factor as indices of rock quality depends upon the assumption that consistent drilling and sampling techniques are maintained. The methods also assume that the friability of core samples is representative of the physical bond strength of the same material in the mass, under static load. This may not always be so. Some rock materials are so affected by the impact and abrasive action imposed by the rotating drill pipe, or are so weakened by the circulating drill fluid, that intact core samples are virtually impossible to obtain, although the rock mass *in situ* may have substantial mechanical strength. Dual-wall pipe and core barrels have, for many years, played an important part in minerals exploration, as an aid to obtaining intact core samples from friable materials. The technique of dual-wall pipe drilling is illustrated in Fig. 15.12. The inner pipe forms the core barrel and, together with the core breaker, remains stationary while the outer pipe forms the rotating drill-stem. The core may be broken into sections about 5 in. long, to be delivered at the surface along with the rock cuttings. Drilling fluid or mud is pumped down the outer pipe, to return up the inner pipe carrying the cuttings and core. If cylindrical core is not required the bit may take the form of a standard tri-cone roller bit and air may replace water and mud as the drill fluid.

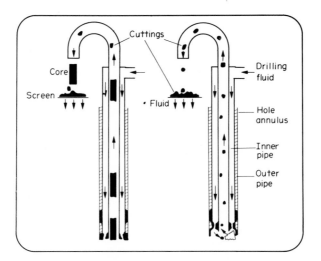

Fig. 15.12. Dual wall pipe drilling (Wilderman).

Chippings and core are examined to extract geological information, using a careful system of sample collection. The physical condition of the sample is used as a means of identifying the properties of the material penetrated by the drill. Size analysis is assisted by the use of separating cyclones, belt-type samplers, and vibrating filters. The drilling equipment may also be instrumented to control and monitor the various drilling parameters. These include the thrust, rotary speed, torque, and penetration rate. In essence, all the measured data that are normally associated with a controlled drilling experiment in the laboratory can be obtained in the field, during normal drilling operations, by making use of continuous monitors attached to the drilling machine. The data so obtained are then applied to bring out the relative geological, hydrological, and engineering characteristics in the strata at various depths in the borehole (see Fig. 15.13).

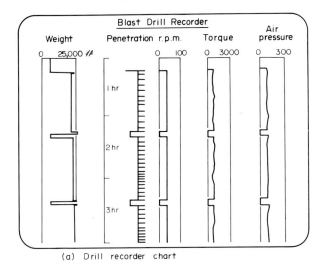

(a) Drill recorder chart

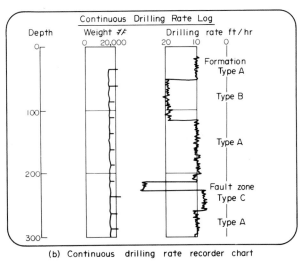

(b) Continuous drilling rate recorder chart

Fig. 15.13. Drill recorder charts (Wilderman).

Geophysical Exploration in Advance of Geotechnical Projects

Seismic Exploration

The application of seismic techniques to explore changes in the character of rock walls around excavations, and to explore for the possible location of stress concentration zones, has been described elsewhere in this text. The method is of great importance in exploration geophysics applied to engineering geotechnology. In addition to the indirect examination of the strength of rock foundations by determining the dynamic deformation moduli, the method is applied to locate pressure zones and discontinuities within the rock mass. In soil mechanics

335

it is applied to identify the sub-surface conditions, to locate the depth to bedrock, and to provide a rapid preliminary assessment of the "rippability" of bedrock materials, in relation to mechanical excavation techniques (see Fig. 15.14). The technique is also applied for basic route surveys in highway engineering, since it is possible to survey many miles per day and obtain information on the sub-surface character of the planned route.

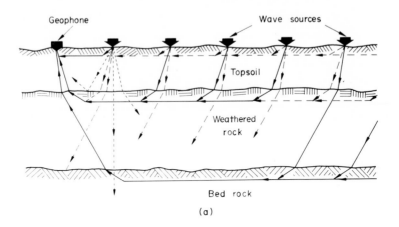

(a)

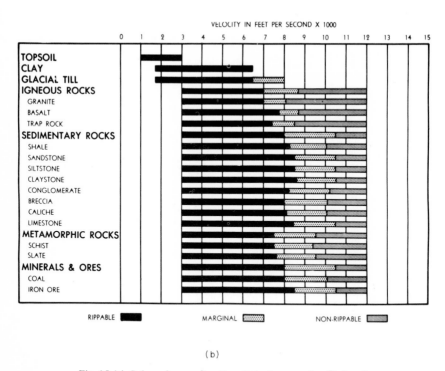

(b)

Fig. 15.14. Sub-surface exploration. Seismic surveying (Soiltest).

The relationship of geophysical measurements to engineering and construction parameters on geotechnical projects is described by Scott *et al.,* with reference to the Straight Creek Pilot

Tunnel. This 4 m tunnel under the Continental Divide in Colorado formed the pilot bore for the Straight Creek Highway Tunnels. These are twin bores, each 12 m diameter and about 2500 m long. Seismic refraction and electrical resistivity measurements made along the walls of the pilot bore indicated that both a low-velocity and a high-resistivity layer existed in the disturbed rock surrounding the excavation.

The seismic measurements were analyzed, to obtain the thickness and seismic velocity of rock in the low velocity layer, the velocity in the rock behind this layer, and the amplitude of seismic energy received at the detectors. The electrical resistivity measurements were analyzed to obtain the thickness and resistivity of similar layers. The electrical resistivity and the seismic velocity of rock at depth, the thickness of rock in the low-velocity layer, and the relative amplitude of seismic energy were then correlated against height of the tension arch in the strata above the tunnel, the stable vertical rock load, RQD, rate of construction and cost per foot, percentage of lagging and blocking in the tunnel lining, support type, spacing, and the amount of steel support required.

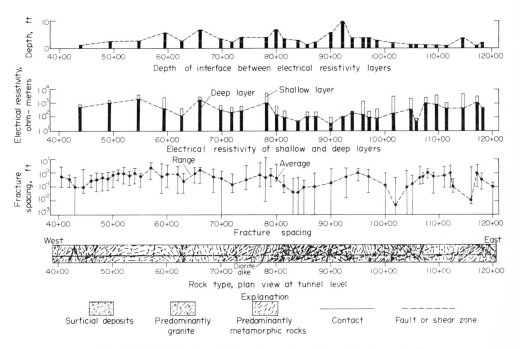

Fig. 15.15. Electrical resistivity and rock type information (Scott *et al.*).

Typical records of seismic, resistivity, and rock-type information are shown in Figs. 15.15 and 15.16. The correlations were observed to be statistically meaningful and they suggested that the prediction of engineering and performance parameters might be possible on the basis of information collected by geophysical measurements in feeler holes drilled ahead of a working face.

Gentry *et al.* followed up this line of investigation by a model study, in which the advance rate associated with the Straight Creek Pilot Tunnel was compared, using a step-wise regression analysis, with the engineering and geological variables determined during construction. As a prediction of the mean advance rate the model was found to be satisfactory within 10% of the range for which it was designed. Accurately predicting the advance rate associated with the

337

Geotechnology

driving of a tunnel can obviously help the contractor in the maintenance of his schedules, equipment, and services, as well as in the overall planning of his project. The project can be completely upset, and its cost multiplied several-fold, by inaccurate predictions, and in the absence of information gained by exploration geology and geophysics predictions can only be, at best, intuitive guesswork.

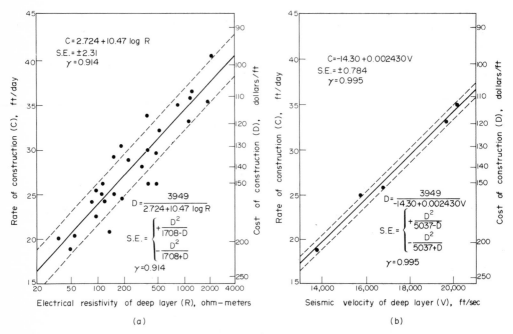

Fig. 15.16. Rate and cost of construction plotted against geophysical data.
Straight Creek tunnel (Scott *et al.*).

Geophysical Studies of Strata Continuity

Advance detection of the presence and the location of strata discontinuities is critical to the economic viability of underground excavation techniques employing high capital-cost equipment, such as tunnel borers, continuous miners, and automated support systems, which cannot easily be adapted to change, once they have been put into commission.

Offset shooting

In oil and natural gas reservoir engineering the technique of "offset shooting" is used to detect the presence of faults, to identify their direction, and estimate their throw. The method consists of lowering a geophone down a central borehole and then detonating explosive charges at the surface, at circumferential points on a circle of average radius several hundred meters. The differences between the seismic wave velocities over the radial distance are measured, for the various directions.

Gravimetric measurements

Gravimetric observations, which measure density, may also be used to detect the presence of gravity anomalies beneath the ground surface, or within the walls of an excavation. Localized changes in density, which are shown as anomalies on gravity contours of the areas surveyed, may indicate the position of geological discontinuities, such as fault zones and unconformities. However, in common with magnetic and electrical resistivity methods, gravimetric methods are self-integrating, and they do not permit of detailed and precise separation of individual layers. In certain circumstances an exception might be made of the radioactive gamma-gamma log in boreholes, which also may be regarded as a measurement of formation density. In exploring carboniferous strata this technique can establish the boundary between coal (around 1.2) and the adjacent sediments (about 2.6) on a decimetric scale, as shown in Fig. 15.17.

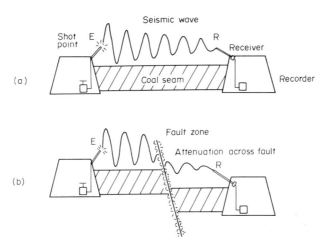

Fig. 15.17. Undisturbed and disturbed seismic waves (Saul and Higson).

Saul and Higson describe an assessment of the Borges seismic transmission method of locating dislocations in carboniferous sediments. This system measures seismic velocities in the stratum immediately above or below that being mined (Figs. 15.18 and 15.19). Seismic waves are initiated by firing charges inset into boreholes in the rock stratum. The charges are all fired in succession from a roadway on one side of the area to be explored, and the shock waves are received by geophones mounted on rock bolts, secured into the stratum on the other side of the exploration area, and coupled to recording equipment. By firing several shots from different locations a large number of wave paths can be examined. An intensity/distance relation can be established which is characteristic of undisturbed rays. Disturbed rays are then identified by their deviation from this characteristic, and an interpretation of continuity is deduced from the relative geometry of the undisturbed and disturbed wave paths.

Recording the Log of Cored Boreholes

The conventional geological log, giving descriptive details of the strata encountered, usually incorporating a symbolic diagram indicating variations in the strata, is quite inadequate as an engineering record. A more complete log would be recorded in separate columns containing rock description, core-loss details, casing details, core size, elevation, and depth, fracture analysis, lithology, water level, strata permeability, and drilling parameters. However, it is very seldom that drilling records are so comprehensive. It is important that the core be examined

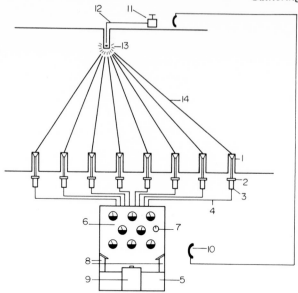

(1) Expanding shell-roof bolt
(2) Geophone
(3) Amplifier
(4) Cable to monitoring station
(5) Box enclosing recording equipment
(6) Ammeters mounted on panel
(7) Clock

(8) Lamps
(9) Movie camera
(10) Telephone
(11) Detonator
(12) Cable to shot
(13) Shot
(14) Seismic rays crossing the coal panel

Fig. 15.18(a) Layout of the seismic equipment.

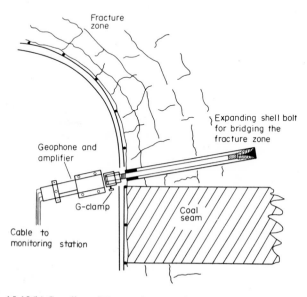

Fig. 15.18(b) Coupling of the geophone to the strata. (Saul and Higson.)

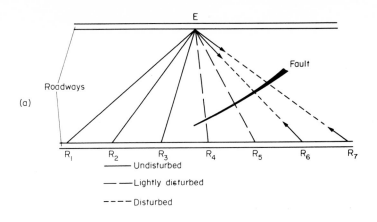

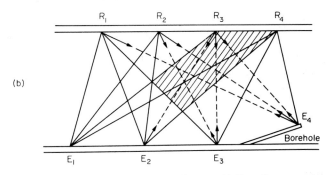

Fig. 15.19. Seismic waves crossing faulted strata. (a) Fan of rays crossing a fault. (b) Location of fault within a coal panel. (The fault must lie within the shaded area as it must not cross an undisturbed ray.) (Saul and Higson.)

as soon as possible after its recovery, preferably on the site so that it is not damaged by vibration and shock during transport. Subsequently it may be removed to a convenient place for detailed examination, study and storage.

A recommended format for the log is:

1. Identification. Hole number, purpose, drilling authority.
2. Location. Project name. Site details, grid reference. Elevation. Orientation. Diameter.
3. Drilling technique. System of drilling. Type of bit. Flushing medium. Drill performance details. (A machine recorder should supplement the driller's daily record sheet or diary.) The diary should list.

    (i) Depth of hole at start and end of work shift.
    (ii) Depth at start and finish of each core run.
    (iii) Depth and size of casing at start and end of each core run.
    (iv) Core diameter and changes in core size.
    (v) Condition of bit.
    (vi) Time to drill each core run.
    (vii) Character and volume of flushing fluid returned to surface.
    (viii) Standing water level in hole at start and end of each working period.

(ix) Description of strata encountered.

(x) Core recovery with note as to possible location of core losses.

(xi) Details of rock fragments sampled.

(xii) Details of delays and breakdowns.

(xiii) Details of *in situ* tests and instrumentation.

(xiv) Details of any backfilling or grouting used.

Selected References for Further Reading

ATTEWELL, P. B. and TAYLOR, R. K. Jointing in Robin Hood's Bay, north Yorkshire coast, England. *Int. J. Rock Mech. Min. Sci.,* vol. 8, pp. 477—482 (1971).

EMMONS, R. C. The universal stage. *Geol. Soc. Am.* Mem. 8 (1943).

FAIRBARN, H. W. *Structural Geology of Deformed Rocks,* Addison Wesley Press Inc., Cambridge, Mass., 1949.

FREIDMAN, M. Petrofabric techniques for the determination of principal stress directions in rocks. *Int. Conf. on State of Stress in the Earth's Crust. Santa Monica, Calif.,* pp. 451—551, Rand Corpn., 1964.

GENTRY, D. W., KENDORSKI, F. S. and ABEL, J. R. Tunnel advance rate prediction based on geologic and engineering observations. *Int. J. Rock Mech. Min. Sci.,* vol. 8, pp. 451—476 (1971).

GOODMAN, R. E. The resolution of stresses in rock using stereographic projection. *Int. J. Rock Mech. Min. Sci.,* vol. 1, pp. 93—103 (1964).

MAHTAB, M. A., BOLSTAD, D. D., ALLDREDGE, J. R. and SHANLEY, R. J. *Analysis of Fracture Orientations for Input to Structural Models of Discontinuous Rock,* U. S. Bur. Mines Rept., Invest. No. 7669 (1972).

MAHTAB, M. A., BOLSTAD, D. D. and KENDORSKI, F. S. *Analysis of the Geometry of Fractures in San Manuel Copper Mine, Arizona,* U. S. Bur. Mines Rept., Invest. No. 7715 (1973).

MULLER, L. Stability studies of steep rock slopes in Europe. *Trans. Soc. Min. Engrs. AIME,* vol. 226, pp. 326—332 (1963).

PINCUS, H. Sensitivity of optical data processing to changes in rock fabric. *Int. J. Rock Mech. Min. Sci.,* vol. 6, pp. 259—276 (1969).

RAMSAY, J. G. *Folding and Fracturing of Rock*, McGraw Hill Book Co. Inc., San Francisco, 1967.

SAUL, T., and HIGSON, G. R. The detection of faults in coal panels by a seismic transmission method. *Int. J. Rock Mech. Min. Sci.,* vol. 8, pp. 483—499 (1971).

SCHWAETZER, T. Geophysical studies of the continuity of coal seams. *Int. J. Rock Mech. Min. Sci.,* vol. 2, pp. 167—196 (1965).

SCOTT, J. H., LEE, F. T., CARROLL, R. D. and ROBINSON, C. S. The relationship of geophysical measurements to engineering and construction parameters in the Straight Creek Tunnel Pilot Boring, Colorado. *Int. J. Rock Mech. Min. Sci.,* vol. 5, pp. 1—30 (1968).

TURNER, F. J. and WEISS, L. E. *Structural Analysis of Metamorphic Tectonics,* McGraw Hill Book Co. Inc., San Francisco, 1963.

VISTELIUS, A. B. *Structural Diagrams,* Pergamon Press, Oxford, 1966.

WILDERMAN, G. H. Exploration drilling techniques used as a field determinator and data gatherer in mill and plant design. Annual Gen. Meeting CIM Vancouver, April 1973. *CIM Bull.,* vol. 65, pp. 110—116 (1974).

Index